Limestones and caves
of Wales

Limestones and caves
of Wales

EDITED BY
TREVOR D. FORD
Department of Geology, University of Leicester

for the British Cave Research Association

*The right of the
University of Cambridge
to print and sell
all manner of books
was granted by
Henry VIII in 1534.
The University has printed
and published continuously
since 1584.*

CAMBRIDGE UNIVERSITY PRESS

CAMBRIDGE

NEW YORK · PORT CHESTER · MELBOURNE · SYDNEY

Published by the Press Syndicate of the University of Cambridge
The Pitt Building, Trumpington Street, Cambridge CB2 1RP
40 West 20th Street, New York NY 10011, USA
10 Stamford Road, Oakleigh, Melbourne, 3166, Australia

First published 1989

Printed in Great Britain at the University Press, Cambridge

British Library cataloguing in publication data

Limestones and caves of Wales.
1. Wales. Limestone regions. Geological features
I. Ford, Trevor D. (Trevor David), *1925–*
554.29

Library of Congress cataloguing in publication data

Limestones and caves of Wales.

Includes bibliographies.
1. Caves – Wales.   I. Ford, Trevor David.
GB608.465.L56   1989   551.4′47′09429   88–28513

ISBN 0 521 32438 6

Contents

Contributors

Peter Appleton. Metallurgist, Brymbo Steel, North Wales.

Peter A. Bull, B.Sc., Ph.D. Lecturer, School of Geography, University of Oxford.

Noel S. J. Christopher, B.Sc., Ph.D. Water treatment chemist, Housman (Burnham) Ltd, Manchester.

Alan C. Coase. Former Warden, Thornbridge Hall Educational Centre, near Bakewell. (deceased)

John Crowther, B.A., M.A., Ph.D. Lecturer in Geography, St Davids University College, Lampeter, Dyfed.

Mel Davies, B.Sc. Scientist, Nature Conservancy Council, Oxwich Nature Reserve, Gower.

Dai P. Ede, B.Sc., Ph.D. Consulting hydrologist, Messrs Dames & Moore, the Netherlands.

John V. Elliott, Builder, Bramble Cottage, Camomile Road, Upper Lydbrook, Gloucestershire.

D. G. Everett. 131 South Rd, Taunton TA1 3ED.

Martyn Farr. Schoolteacher, South Wales.

Trevor D. Ford, B.Sc., Ph.D., F.G.S. Senior Lecturer in Geology, University of Leicester.

Clive G. Gardener. Film editor, Cardiff.

W. Gascoine, B.Sc. Lecturer in Chemistry, Crosskeys Tertiary College, Gwent.

H. Stephen Green, Ph.D., F.S.A. Acting Keeper of Archaeology and Numismatics, National Museum of Wales, Cardiff.

G. T. Jefferson. Former Reader in Zoology, University College, Cardiff. (deceased)

O. C. Lloyd. Former Reader in Pathology, Bristol University. (deceased)

David J. Lowe, B.Sc. Geologist, British Geological Survey, Keyworth, Nottingham.

Sam Moore. Senior Technologist, British Coal's liquefaction project, Point of Ayr, North Wales.

Peter L. Smart, B.Sc., Ph.D. Lecturer in Geography, University of Bristol.

Mark E. Tringham. Chief Geologist, Petrofina Exploration, Melbourne, Australia.

Tony Waltham, B.Sc., Ph.D. Lecturer in Geology, Trent Polytechnic, Nottingham.

Clive D. Westlake, B.A. Schoolteacher, Tamworth, Staffordshire.

Preface

This book is the fourth in the series *Limestones and Caves of...* covering the major caving regions of Britain. These arose out of the growth of knowledge of Britain's caves as summarized in Chapter 5 of *British Caving – An Introduction to Speleology* edited by Cecil Cullingford in 1953 (2nd edn, 1962). The aim of the series of regional books has been to summarize the state of knowledge of the origin and development of the limestone archaeology as stimuli to future research and some 20 authors have co-operated in the chapters which follow. Bringing it all together has been a daunting task but one which has given me great satisfaction in the long run. The authors and editor have all given their services without charge so that the proceeds can go to the British Cave Research Association in order to further speleological research.

Four co-ordinating editors have each in turn accepted the task of bringing this book to completion, but each has found the pressures of job changes, promotion, illness, family affairs or other personal problems to be too much to allow the time necessary. These four editors must be thanked most sincerely for their efforts: Dai Ede, Gillian Groom, Doug Bassett and Peter Bull. More recently Mark Noel and Tony Waltham have helped considerably with editing certain chapters and in chasing recalcitrant authors.

During the 15 years of gestation many changes have occurred, each necessitating revision of parts of the text. The nomenclature, classification and correlation of the Carboniferous Limestone has changed completely demanding the revision of several chapters. The recognition of palaeosols (ancient soil horizons) within the limestone sequence has called attention to previously unrecognized lithological controls on cave passage formation.

In later geological times, the glaciations of the Pleistocene Ice Age and their attendant interglacials have been the focus of intensive study, though the detailed sequence of events and cross-country correlations are still in a state of flux, and only tentative suggestions can be made even now.

These and other problems have delayed completion. The problems have been compounded by the deaths of three contributors, Alan Coase, Oliver Lloyd and Geoff Jefferson. All are sadly missed but their work will serve as a memorial. Other contributors have found their efforts overtaken by events: the major cave discoveries at Llangattwg have meant that the chapter on the vast cave systems of Agen Allwedd, Craig y Ffynnon and Daren Cilau has had to be totally rewritten. Discoveries are still being made so that some parts of this book may well be out of date by the time it is published.

The level of treatment varies somewhat between chapters but this is at least in part due to different amounts of study and resultant knowledge in different caves and/or valleys. The authors are duly credited in the chapter headings, but, behind the scenes, many others, whose names do not appear as authors, have helped in this compendium: they include George Bray, Bob Charity, Peter Cousins, W. F. Grimes, Peter Hart, David Leitch, Norman Martell, Bill Maxwell, Roy Musgrove, Ian Penney, Paddy O'Reilly, Roger Smith, Ian Standing, Peter Standing and John Weaver.

Some of Britain's foremost cave photographers have helped with illustrative material, notably Paul Deakin, Chris Howes, Jem Rowland and Jerry Wooldridge (all F.R.P.S.), as well as T. Bennett, Ian Davinson, B. A. Speyer, Tony Waltham and Clive Westlake. Photos not specifically credited are by chapter authors. Others, too numerous to mention, have assisted with surveying and drafting, as well as in stimulating ideas and discussion.

TREVOR D. FORD
Geology Department
University of Leicester   March 1988

1

Introduction

TREVOR D. FORD

Although Wales is best known for its mountains of volcanic and slaty rocks it also contains some striking limestone landscapes along the fringes of the South Wales Coalfield and on the Clwyd moors of northeast Wales, as well as fine limestone cliffs along both north and south coasts. More important, since 1945, the Welsh limestones have yielded up some of the secrets of long-suspected cave systems and some of the caves now rank among the world's leaders in length and complexity. Ogof Ffynnon Ddu now totals some 43 km of passages and when the inevitable link-up of the Llangattwg caves is made they may exceed that figure. South Wales can now boast two of the longest cave systems in Britain, the largest passage and also Britain's deepest cave.

This book is a survey of the extent, distribution and genesis of both the surface karstic features and the cave systems. While most chapters also outline the history of discovery and exploration of the caves, the primary objective is to consider the form of the cave features and to discuss their origin and development in the light of the general principles of geomorphology of limestone regions.

The Welsh caves owe their nature and layout at least in part to the lithology of the enclosing limestones, so that a geological basis for the arguments that follow has been set out in Chapter 2. Erosional processes have acted on these Carboniferous limestones at several periods since their deposition: palaeokarsts were developed during Carboniferous sedimentation, again in Permo–Triassic times, and in the Tertiary. Relics of these have affected limestone landscape development in Quaternary times, and Chapter 3 on the karst geomorphology surveys these factors in some detail. In particular, the solution-induced collapse features, where Millstone Grit rock has been let down into the limestone, appear to be on a uniquely large scale in the Welsh karst. The effects of glaciation or of several glaciations are still being unravelled, but the caves clearly owe much of their evolution to successive high run-off stages in the waxing and waning phases of glacial episodes.

Caves owe their existence partly to the lithology of the enclosing limestones, partly to topography and partly to the movement of water. The main outcrop of Carboniferous Limestone north of the coalfield is composed of upland generally between 400 and 500 m above sea level. The valleys with the resurgences are mostly at around 200 m giving a possible 300 m of relief and potential depth to the cave systems. Rainfall reaches a maximum annual average of about 2750 mm in the central part of the outcrop, with less than half that figure in coastal areas. It is spread throughout the year although effective precipitation is greatest in winter. Occasional storms can give rise to as much as 50 mm in one day, but much of the precipitation is slow. However, peat-covered moorlands soak up much of this and yield rather acidic run-off through swallets and percolation into the cave conduits. Underground drainage routes mostly more or less parallel the major valleys, but much tracing has been necessary to establish catchments, as described in Chapter 4. The pattern so established should provide a basis for future research into solution and hydrogeochemical processes. While some important local studies have been made, as at Ogof Ffynnon Ddu, few have been published and there is no comprehensive survey available.

Scientific studies of caves include several other specialties, but in particular biology and archaeology are significant here. Being dark, cold and wet, the cave environment is often thought of as being unfavourable to life, but the late Geoff Jefferson's Chapter 5 on cave biology reveals that a large number of species, mainly crustaceans and insects, can and do complete their life cycle underground. Some 50 Welsh caves have yielded records of the occupation by ice-age mammals and man, and Chapters 6–8 provide the first comprehensive archaeological survey of Welsh caves.

Details of the methods of cave exploration are out of place in this volume, although several chapters bring out the excitement of discovery and of setting foot where no man has ever been before. In the world of caves, it is still possible for a Saturday afternoon trip to give such a sense of achievement – expeditions to the Poles or to Himalayan peaks are not necessary for all discoveries! Cave diving, discussed in Chapter 9, provides the ultimate challenge and has borne fruit in Wales with the discovery of much previously unknown passage and the first 'through trip' of a British mountain in the penetration of Mynydd Llangattwg from north to south.

The discussions of specific caving areas (Chapters 10–20) each bring out certain aspects of the evolution of their particular bit of the Welsh karst, and speak for themselves in showing the current state of knowledge of the caves, their drainage patterns and developmental history. The authors vary in the depth of their treatment, but this is partly because the different areas contrast in the extent of geomorphological development, partly because the current stage of exploration varies and partly due to the intensity of systematic geomorphological study. The different areas described herein demonstrate a general sequence of events. The Hepste River caves are 'young' as they only just breach the cover of Millstone Grit; the Little Neath River Cave is rather more developed and accessible, while Porth yr Ogof is 'old' in the sense that it has been truncated by collapse and gorge formation. Ogof Ffynnon Ddu is predominantly due to vadose development along the strike while the neighbouring Dan yr Ogof has strong joint control in its layout. Parts of the complex cave systems of Mynydd Llangattwg have, to some degree, been left high and dry above present water levels.

One particular theme that comes through from the area studies is the contrast in cave development between North and South Wales, and between these and other karst regions, both British and overseas. It is certainly not wise to assume that all British karst areas are alike. Most of the major South Wales caves are in the flanks of valleys draining southwards from the older rock areas of central Wales towards the Bristol Channel. These valleys have been deepened by glaciation and at least some of the caves have been left above present water tables; others remain active conduits partly in the phreatic zone. The limestone beds mostly dip southwards and vadose cave passages tend to follow the dip, so that allogenic water enters the cave systems low in the stratigraphic sequence close to the Devonian Old Red Sandstone and resurges with or without phreatic lifts near the base of the Millstone Grit cover. Whilst jointing more or less parallel to the dip is common, bedding-plane passages predominate. High rates of run-off during deglaciation doubtless had their effects in cave development, particularly of vadose passages.

By contrast many of the North Wales caves are phreatic systems trending oblique to the dip and only drained by the activities of lead miners in recent centuries. Many kilometres of drainage tunnels have both intersected caves and drastically lowered the water table. South Wales does not have such mineral veins and the need for drainage tunnels does not exist there. The presence of primary vein cavities no doubt gave an advantage for deep phreatic circulation resulting in today's predominance of passages near the water table. The gloom of the bare rock in some recently drained North Wales caves contrasts with the widespread stalactites of South Wales, although some older high-level caves in North Wales are superbly decorated.

Both North and South Wales caves contrast with the Pennine and Mendip caves in having very extensive development in limestones concealed beneath the cover of Millstone Grit rocks: the Welsh caves thus provide the main British example of covered karstic features. As a result they are far more extensive than the narrow outcrops of limestone on the geological map would suggest.

The angle of dip of the limestones in both South and North Wales is generally around 10–15° in marked contrast to the near-horizontal disposition of most of the Pennine limestones. As underground water movement could follow bedding planes or bedding-joint intersections in Wales there has been little opportunity for the development of numerous vertical potholes as in the Ingleborough region of the Pennines, where the horizontal bedding-plane developments are linked vertically. The lack of steep dips in the Welsh limestones contrasts with the tightly folded Mendip limestones so that deep phreatic loops are few in Wales. There are steeply dipping limestones on the South Crop of the South Wales coalfield and in Gower and south Pembrokeshire, but most of the caves there are of limited extent. Reef limestones, which influence some of the passage morphology in the Peak District, are almost absent in Wales.

This book does not set out to be a guide to the caves of Wales. Instead the authors and editor hope that in summarizing the present state of knowledge, readers' eyes are opened to what remains unknown as well as what is already established. After all, at the end of World War II in 1945, who would have forecast the presence or extent of such wonders as Ogof Ffynnon Ddu, Dan yr Ogof and Ogof Agen Allwedd? Who would have predicted the large and superbly decorated Otter Hole beneath Chepstow racecourse? And, after years of the North Wales lead miners trying to understand their perennial water problems, who would have thought that the persistence of a few dedicated cavers would provide at least some of the answers? Who would have forecast that Pontnewydd, an obscure North Wales cave, once used as a Home Guard ammunition store, would prove to be a most important site for proving the presence of early man in Wales? Who would have thought that some of the little-known invertebrate animals in some Welsh caves might turn out to be relics of the pre-glacial British fauna? The above shows what has been accomplished so far, but there is a long way to go, and this book aims to point out the areas where future research is likely to pay dividends.

The geology of the Carboniferous Limestone of South Wales

DAVID J. LOWE

For the purpose of this account South Wales will be considered as the area as far north as the northernmost outcrops of Lower Carboniferous (Dinantian) Limestone (Fig. 2.1). The region has a long and varied geological history with all systems from Pre-Cambrian to Jurassic, with the exception of the Permian, being represented (Table 2.1) by a wide range of rock types. Although the rocks of Dinantian age are of fundamental importance in the present context, brief details of the overall stratigraphy and structure of the area will be considered before the major karst areas are described in greater detail.

Brief geological history

The oldest rocks recognized in South Wales are found only in Pembrokeshire and are of Pre-Cambrian volcanic and intrusive igneous origin (Fig. 2.1). Generally acidic lavas and

tuffs of the Pebidian 'Series' are believed to pre-date acidic fine- and coarse-grained intrusive rocks of Dimetian age. Small, isolated inliers of Pre-Cambrian sediments are present near Carmarthen and Radnor. Prior to Lower Palaeozoic deposition the Pre-Cambrian rocks were folded, faulted, uplifted and heavily eroded. The similarity of the Pebidian and Dimetian rocks to the Pre-Cambrian rocks of Shropshire suggests that the whole of the South Wales area might be underlain by complexly deformed mainly igneous basement.

The entire region subsided at the beginning of Cambrian times and a major depositional basin or geosyncline was established, effectively persisting until the end of Silurian times despite pulsatory uplift and resubsidence during this time. Great thicknesses of geosynclinal sediments including shales and greywackes were laid down, together with shal-

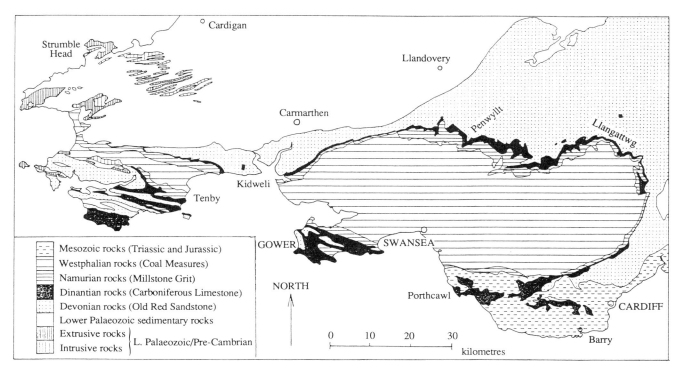

Fig. 2.1. Generalized map of the solid geology of South Wales.

Table 2.1 *The Major divisions of geological time*

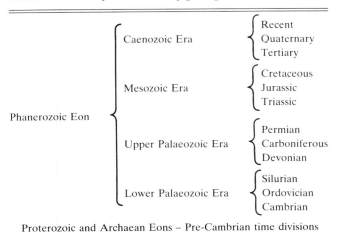

Phanerozoic Eon	Caenozoic Era	Recent Quaternary Tertiary
	Mesozoic Era	Cretaceous Jurassic Triassic
	Upper Palaeozoic Era	Permian Carboniferous Devonian
	Lower Palaeozoic Era	Silurian Ordovician Cambrian

Proterozoic and Archaean Eons – Pre-Cambrian time divisions

Table 2.2 *Chronostratigraphical nomenclature of the Carboniferous*

System	Sub-system	Series	Stage
Carboniferous	Silesian	Stephanian	Stephanian C[a] Stephanian B[a] Stephanian A[a] Cantabrian [b]
		Westphalian	Westphalian D Westphalian C Westphalian B Westphalian A
		Namurian	Yeadonian Marsdenian Kinderscoutian Alportian Chokierian Arnsbergian Pendleian
	Dinantian	'Viséan'	Brigantian Asbian Holkerian Arundian Chadian
		'Tournaisian'	Courceyan

[a] Stages believed to be absent in South Wales.
[b] Beds of basal Cantabrian age are possibly present in South Wales.

lower water deposits, as the basin was periodically infilled, uplifted, eroded and re-established. Worldwide, sporadic volcanic activity occurred throughout the Lower Palaeozoic, reaching its peak in South Wales during the Llanvirn Stage of the Ordovician. Volcanic rocks of sub-aerial and submarine origin are preserved in Pembroke, while sediments of Arenig to Caradoc age are intruded by dolerite sills and laccoliths which are probably of late Ordovician age.

It has been estimated (George, 1970) that at least 12 000 m of sedimentary rock was deposited in the geosyncline during the Lower Palaeozoic, but generally only a fraction of this is preserved in any one area. Towards the end of Silurian times the area was affected by earth movements at the climax of the Caledonian Orogeny and the geosynclinal seas retreated southwards as a new landmass, St George's Land, was raised in what is now central and southwest Wales, possibly extending westwards into Ireland. On the southern flanks of this landmass the Devonian 'Old Red Sandstone' was laid down as a series of river and floodplain deposits which locally overlie the folded Lower Palaeozoic rocks with marked discordance. Renewed uplift occurred in mid-Devonian times and deposition ceased until late Devonian subsidence led to the 'Upper Old Red Sandstone' being laid down unconformably on the 'Lower Old Red Sandstone', with the 'Middle Old Red Sandstone' absent. Whereas the early Devonian rocks in the area were entirely freshwater in origin, a gradual marine encroachment from the south took place in late Devonian times and intercalations of freshwater and marine sediment mark the upward transition into wholly marine conditions in the Carboniferous. Essentially marine deposition, mainly of relatively pure limestone, continued throughout the Lower Carboniferous (Dinantian), which is discussed in more detail below, but a gradual oscillatory reversion to terrestrial conditions with sandstone and shale deposition through the Namurian (Millstone Grit) Series was more or less complete by the start of Westphalian times. Coal Measures strata deposited during the Westphalian were the products of fresh or brackish water swamps and deltas, although deposition was cyclic and many cyclothems include a thin horizon of marine shale, indicating periodic incursion by the sea. These 'Marine Bands' often

have a great lateral extent and distinctive fossil content, providing a useful and reliable means of correlation, but they become less frequent higher in the Westphalian sequence.

Sometime after the deposition of the highest preserved Carboniferous strata, which may be slightly younger than Westphalian (Table 2.2), the area underwent further uplift and intense folding during the Hercynian Orogeny. It is uncertain whether rocks of very late Carboniferous and Permian age were deposited and subsequently removed by erosion or whether rocks of this age were not laid down in the area. When the Hercynian earth movements died down a new phase of erosion, submergence and sedimentation began, with first Triassic and then Jurassic rocks being deposited against the southern flanks of the greatly denuded Hercynian mountains. The Triassic rocks are essentially red, fluviatile and lacustrine deposits and lie unconformably on the older rocks, being derived from the sub-aerial wastage of the Devonian and Carboniferous strata. A gradual transition to marine conditions took place towards the end of the Triassic and much of the uppermost stage, the Rhaetian, is marine.

Jurassic marine sediments follow conformably upon the Triassic rocks and these shales and impure limestones are the youngest consolidated strata preserved in the area. It is likely that no Upper Jurassic or Lower Cretaceous beds were deposited in South Wales, but younger Cretaceous deposits (Chalk) might have been laid down and subsequently removed by Tertiary erosion. Little is known of geological events in the area since the deposition of the early Jurassic rocks. Uplift and non-deposition might have commenced at once and persisted until a possible late Cretaceous submergence, as mentioned above, and uplift and folding certainly took place in mid-Tertiary (Miocene) times. Small clay-rich pocket deposits accumulated during the Tertiary, but essentially erosive conditions were dominant through into the Quaternary.

Over much of South Wales the solid rocks described above are masked by unconsolidated deposits of Quaternary age, especially glacial drift laid down in late Pleistocene times. Much of this glacial material, which includes till, outwash sand and gravel and moraine, is believed to have derived from stagnation of the ice of the last (late Devensian) cold episode, but deposits from early Devensian or the earlier Wolstonian cold periods might be present locally. The Devensian ice finally stagnated about 10 000 years ago, an event accompanied by a rise in sea level. Since that time highly variable unconsolidated deposits including blown sand, peat and marine, estuarine or fluvial alluvium have been laid down in suitable environments, often in highly complex sequences due to minor sea-level variations. At present most of the area is undergoing erosion, although various alluvial deposits, blown sand and scree are being laid down locally.

General geological structure

From the brief details above it will be apparent that the tectonic history of South Wales is long and complex. Little is known of the earliest, pre-Palaeozoic, orogenic episodes, but the area has undergone considerable uplift and deformation on two occasions since the commencement of the Phanerozoic. The earth movements associated with the Caledonian Orogeny were more long-lived than elsewhere in the British Isles and the structures produced are in many cases cumulative and multiple. Major unconformities within the Lower Palaeozoic sequence suggest that the principal Caledonian events began during the late stages of the Cambrian and continued spasmodically through into the Devonian, a total time span of more than 100 million years. As in other Caledonide areas the major structural grain is northeast to southwest, and folds of this trend are well preserved in mid-Wales and probably preserved (although modified by later tectonic overprinting) in the basement rocks of the area to the south.

The second major Phanerozoic tectonic episode began in late Carboniferous times and involved the fringe effects of the European Hercynian Orogeny. Stress from the south produced a series of predominantly east–west structures affecting the Upper Palaeozoic rocks, acute in the south but becoming less so northwards where the orogenic pressures dissipated against the bulk of St George's Land. The major Hercynian structure in the area is the South Wales Coalfield Syncline (Fig. 2.2) which can be traced from St Bride's Bay in the west, through the old counties of Carmarthenshire and Glamorgan into Monmouthshire, where the fold is terminated by the SSW–NNE-trending Usk Anticline. In detail the structure is a complex synclinorium which includes several en-echelon periclinal flexures (Fig. 2.2) which are individually impersistent when traced along strike. Dinantian limestone sequences are preserved and exposed around the outer rim of the major structure (Fig. 2.1) and are also found in sub-parallel periclinal folds to the south, in Pembroke and Gower.

Faults are numerous and were often formed contemporaneously with the Hercynian folding. Strike faults of normal or reverse type, locally of low enough angle to be termed thrusts, may take the place of fold limbs. In the more

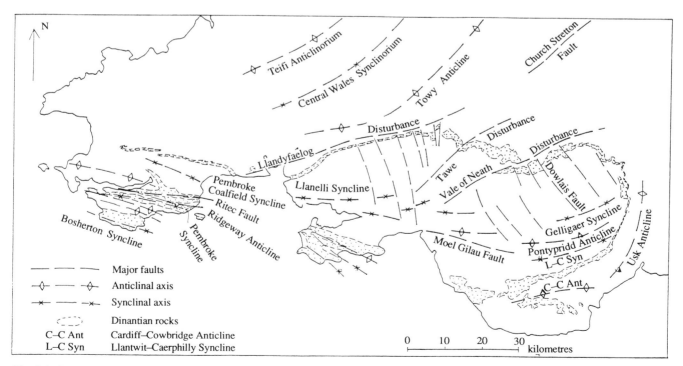

Fig. 2.2. Generalized map of the structural elements of South Wales (based on George, 1975).

extreme cases, such as where the Ritec Fault (or Thrust) has carried the Pembroke Syncline onto the Sageston Anticline, fault movement in excess of 1 km is involved. Contemporaneous Hercynian faulting is also indicated by numerous fractures running obliquely to the regional east–west grain, usually on a northwest–southeast line. These fractures often displace fold axes and juxtapose anomalous structural elements along their length, in addition to exhibiting apparent variation of throw direction and magnitude – all features indicative of tear faults with substantial horizontal strike-slip movement. Their contemporaneity is confirmed where some of the faults end abruptly against the plane of a Hercynian thrust or high-angle reverse fault.

In addition to those Hercynian structures mentioned above, a number of major zones of disturbance trending northeast–southwest cross the area. The contrast of stratigraphy across these zones (Owen, 1954) shows that they were active as early as Dinantian times. The two largest, the Neath and Tawe (Swansea Valley) disturbances, appear to combine sinistral strike-slip with compression towards the northwest, and it is possible that the Hercynian movement in these zones merely continued to develop pre-existing, deep-seated, Caledonide structures. Similar disturbance zones, such as the Careg Cennen Disturbance, occur to the west (Fig. 2.2) where again they are sub-parallel to the Caledonian grain, and probably of deep-seated origin. Minor tectonic features associated with both these major orogenic episodes are common, including small-scale folding in rocks of all ages, cleavage in the Lower Palaeozoic rocks and tectonic joints in the more recent deposits. Joints in the Dinantian limestones are of particular importance and will be discussed in the individual area descriptions below.

The final recognizable orogenic episode occurred in mid-Tertiary times, the South Wales area this time being on the fringe of the Alpine Orogen. Mesozoic rocks laid down after cessation of the Hercynian movements were faulted and deformed into relatively broad, gentle folds. In contrast to the earlier structures these are fairly minor, the folds having amplitudes of only tens of metres and the faults rarely exceeding 100 m in throw. The effect of this orogenic phase on the Palaeozoic rocks is negligible and generally unrecognizable against the background of Caledonian and Hercynian structures.

While there have been minor oscillations of sea level since Tertiary times, which might have been caused by isostatic uplift or simply by climatic change on a world-wide scale, there is no evidence of any more recent tectonic activity than the Alpine fringe effects in the South Wales area.

Stratigraphical terminology

Broadly, *stratigraphy* is the study of the rock sequences which comprise the geological record and the events that produced them. In order to correlate rocks between different areas a concise and unambiguous nomenclature is required. Inevitably, as stratigraphy evolves and becomes more refined, certain terms become obsolete, others change their meaning or significance; some become totally invalid and new terms may be introduced. A useful and fairly definitive guide to stratigraphical terminology and procedure is provided by Holland (1978) and details specific to the cavernous rocks of South Wales are summarized in George *et al.* (1976). A brief explanation is given below of the derivation, usage and implications of the terminology used in this account to describe the limestone sequences of South Wales.

The Carboniferous Period began about 360 million years ago and ended about 280 million years ago. All the rocks laid down during this time constitute the Carboniferous System, the major chronostratigraphical unit covering this part of the geological record. *Chronostratigraphy* is an attempt to provide a fixed time-scale for expressing the relative ages of rocks by correlation or comparison with standard divisions, each with its base defined at a suitable boundary stratotype in an exposed rock sequence. This correlation may be simple and direct but in some circumstances it can involve several intermediate correlations of different diagnostic features held partly in common by a number of sequences. The divisions comprising the Carboniferous System are shown in Table 2.2. By definition the base of the Carboniferous System is at the same point as the base of the Dinantian Subsystem, which should in turn correspond to the base of the lowest series in the Dinantian, the base of the lowest stage in that series and so on in the smaller divisions. As defined at present, however, the status of the two series which in common usage comprise the Dinantian does not satisfy this criterion. The Tournaisian Series at its type locality includes rocks which are pre-Dinantian in age. As a partial and temporary solution to this problem, the British regional stage, the Courceyan (divided on the Continent into the Hastarian below and Ivorian above), is so defined as to include all those rocks in the Tournaisian which are of Dinantian age. Thus the base of the Courceyan Stage corresponds to the base of the Dinantian Subsystem and the Carboniferous System. Similar objections preclude the use of the Viséan as a series in the British Isles, where the defined boundary of the Viséan and Tournaisian is unrecognizable. Additionally, to be a valid division, the base of the Viséan Series should correspond to the base of its lowest component stage. Since the top of the Tournaisian Series is equivalent to the top of the Courceyan Stage, and since this boundary is defined by the base of the overlying Chadian Stage, it follows that the base of the Viséan Series ought to be the same as the base of the Chadian Stage, which is not the case. Since the Chadian Stage is a regional stage stratotyped in Britain, and the Chadian–Courceyan boundary lies above the base of the Viséan as defined in Belgium, the use of the term Viséan Series should be avoided in Britain. However, since the two divisions Tournaisian and Viséan are in loose and common usage in Britain, and pending the definition of more meaningful divisions of the British Dinantian, the terms are informally retained as Tournaisian 'Series' and Viséan 'Series'. Most standard stratigraphical textbooks and descriptive guides to South Wales will still be found to use this division, and generally to subdivide the 'series' on a purely biostratigraphical basis.

Biostratigraphy is in many ways bound to chronostratigraphy and involves the use of fossils as a means of correlation. The units of biostratigraphy are *biozones*, often referred to loosely as *zones*, which are specified by the fossils that they contain. Higher in the Carboniferous, particularly in the Namurian, it has been recognized that biozones are bounded by virtual time-planes and can therefore be regarded as *chronozones*, the sub-division of the stage. Again, in the Namurian the biostratigraphy is so precise and well controlled by evolving fossils that the individual biozones can be readily divided into *sub-biozones* and hence *sub-chronozones*. In Dinantian sequences, however, although a number of biozonal schemes have been worked out in different areas based upon different types of macro- or micro-fossils (George *et al.*, 1976) it has so far proven impossible to define time-plane-bounded biozones or chronozones. Some biozonal schemes suffer from their fossil content being facies-controlled and therefore of only local significance at any one time. Other biozones are defined by distinctive faunal marker bands which are often of limited lateral extent. Still others were defined in rock successions with unrecognized local non-sequences, which led to confusion when attempting correlation with more complete successions elsewhere. The biozonal scheme most often applied to the rocks of the South Wales Dinantian is a refined form of that originally proposed by Dixon & Vaughan (1912) which comprises the following zones (biozones):

D   *Dibunophyllum* Zone
S_2 *Seminula* Zone
C_2S_1 Upper *Caninia* Zone
C_1 Lower *Caninia* Zone

Z   *Zaphrentis* Zone
K   *Cleistopora* Zone

How this biozonal scheme can be approximately tied to the stratigraphy of the area is shown in Table 2.3.

The final aspect of stratigraphy to be considered in the area is *lithostratigraphy*, which involves the description of rock successions in terms of named and recognizable units. These units are generally of local extent and are described and defined on the basis of their observable characters, such as gross lithology (e.g., limestone or shale), colour, mineralogy, gross palaeontology (e.g., barren or fossiliferous) or geochemical properties. Unit boundaries are placed at levels where a significant change of rock type occurs and a boundary might be a sharp contrast or an arbitrary point in a gradational sequence. Generally lithostratigraphical boundaries do not represent time planes and on the contrary might be strongly diachronous, but the rock units provide a framework for correlation and geological interpretation.

The terminology of lithostratigraphy is particularly chaotic, having been in use since the earliest days of geology and being modified and rationalized over the years. Many old terms remain in the nomenclature, some of which would not be used in a modern context, some of which are easily confused with modern chronostratigraphical usage and others which are simply incorrect and misleading. The modern approach to lithostratigraphy is to divide the rock sequence into hierarchical units on the basis:

Table 2.3. *Correlations of Dinantian nomenclature in South Wales*

Regional Stages	Major cycle (Ramsbottom)	Schematic non-sequences (striped)	Examples of rock units	Regional gross rock units	Zones (Biozones)		'Series'	Regional Stages
BRIGANTIAN	6th group of minor cycles		Oystermouth Beds	UPPER LIMESTONE SHALES	*Dibuno-phyllum* (D)	D_3 to D_2	'Viséan'	BRIGANTIAN
			Penwyllt Lst. Oxwich Head Lst.					
ASBIAN	5th group of minor cycles		Oxwich Head Lst. Penderyn Oolite			D_1		ASBIAN
			Honeycombed Sst.					
			Greenhall Lst.					
HOLKERIAN	4		Dowlais Lst.	MAIN LIMESTONE	(S_2) *Seminula*			HOLKERIAN
			Cil yr Ychan Lst.					
			Hunts Bay Oolite					
ARUNDIAN	3		High Tor Lst.		*Upper Caninia* (C_2S_1)	S_1		ARUNDIAN
			Llanelly Fm.			C_2		
CHADIAN	2		Caswell Bay Oolite					CHADIAN
			Gilwern Oolite		*Lower Caninia* (C_1)			
			Clydach Beds					
COURCEYAN	1		Penmaen Burrows Limestone				'Tournaisian'	COURCEYAN
			Blaen Onneu Oolite		*Zaphrentis* (Z)			
			Pwll y Cwm Oolite					
			Cefn Bryn Shale	LOWER LIMESTONE SHALES	*Cleistopora* (K)			
			Lower Limestone Shales					
			Grey Grits etc.	OLD RED SST.				

supergroup   member
group   bed
formation

The *formation* is the primary local unit and may be divided into units of smaller size, *members* and *beds*, although it need not be subdivided. A formation should, however, be mappable at the surface or underground. A *group* comprises two or more adjacent formations which are naturally related. If adjacent formations show no obvious association they need not be gathered into a group. Similarly a *supergroup* contains two or more adjacent and related groups.

Each lithostratigraphical unit should have been formally defined at a type locality, ideally with a completely exposed type section, and its name should reflect lithology, geographical location and nature of unit, e.g., Penwyllt Limestone Formation. In less formal usage, once defined, this unit might be called the Penwyllt Limestone or the Penwyllt Formation. Although current practice is to follow this laid-down procedure wherever possible, abundant older nomenclature remains. Such terms as Oolite Group, Light Oolite, Calcite–Mudstone Group and Clydach Beds (not to mention Carboniferous Limestone Series) remain in use, but are gradually being replaced by precisely defined formations in the modern context (George *et al.*, 1976).

A relatively recent development in stratigraphy involves the recognition and correlation of sedimentary cycles (Ramsbottom 1973, 1977). The fundamental unit in this system is the *mesothem*, which comprises the rocks deposited during the maximum time of a recognizable cycle in a conformable succession (e.g., in a basinal sequence). Away from the basinal sequence the mesothem may still be recognized, in adjacent shelf areas for instance, but here may be bounded by unconformities. The base of the mesothem is defined as the base of the lowest chronozone it contains, and hence the mesothem is time-plane-bounded and exists in parallel with those chronostratigraphical divisions previously described. Certain mesothems are exactly equivalent to an existing stage, but depending which chronozone is chosen as the base, the mesothem may be bigger or smaller than its nearest equivalent stage. Mesothems are in turn composed of smaller cyclic units called *cyclothems* (a term already in common usage in describing sub-units of, for instance, Coal Measures-type sedimentation). Mesothems may also be grouped together as larger units, which also exhibit cyclicity and are called *synthems*. The major cycles of Dinantian sedimentation described by Ramsbottom (1973) are generally accepted as being approximately coincident with the British regional stages (Table 2.3).

From the above it will be apparent that modern stratigraphy has still to achieve an idealized and logical form, mainly due to a vast legacy of imprecise terminology compounded over the years. In the following account, wherever possible, rock units will be referred to by modern defined lithostratigraphical names (e.g., Dowlais Limestone Formation) within the chronostratigraphical framework of the British regional stages. Where modern nomenclature does not yet exist, older names will be used, with quotation marks indicating particularly dubious terminology, and the biozonal indices such as S_2

or D_1 may be used, with reference to the assumed approximately equivalent regional stage (see Table 2.3).

The Dinantian Limestones of South Wales

As mentioned above, towards the end of Devonian times intercalations of marine sediment began to be laid down in the essentially terrestrial sequence, indicating the first pulsatory encroachment of the sea onto the slowly subsiding southern flank of St George's Land. The main transgression appears to have been gentle, so that in south Pembrokeshire the earliest Carboniferous (Dinantian) beds follow conformably on the Devonian Skrinkle Sandstone (Upper Old Red Sandstone). Elsewhere the transgressive nature of the marine deposits is indicated by the lowest Dinantian beds overlapping northwards onto older strata. Relative change of sea level against St George's Land continued in pulsatory fashion throughout Dinantian times and is evidenced by the cycles of marine transgression–regression recorded in the rock sequence (Ramsbottom, 1973). There are several important implications of this cyclicity and periodic relative uplift. Only in south Pembrokeshire was a more or less complete Dinantian sequence deposited and preserved, as this area sustained effectively marine conditions throughout the major cycles of deposition. Northwards and eastwards there is a general overall thinning of the limestone succession, reflecting only partial deposition or locally non-deposition during some of the stages. Figure 2.3 shows the thickness variation of the total Dinantian sequence and its component stages (or cycles) based upon generalized sections around the present outcrop. The constant variation of water depth is indicated by lateral facies variation such that, for example, at the same time relatively deep-water bioclastic ('standard') limestones might have developed in one area, shallower-water oolitic deposits in another and reef limestones fringing lagoonal areas of calcite–mudstone deposition elsewhere, while in contemporary positive areas no deposition would have occurred. From the preceding points it will be apparent that a number of local unconformities exist where the sediments of the various cycles overstep older rocks either below or within the Dinantian sequence, and that whilst the Dinantian sequence follows conformably on Devonian rocks in Pembroke and southern outcrops in general, there is a tendency for the base of the Carboniferous to become unconformable as the contemporary flanks of St George's Land are approached (George, 1970; George *et al.*, 1976).

As a result of the cyclic variation of depositional environment the lithostratigraphy across the area is also highly variable. On a broad level, however, it is possible to divide the limestone sequence into three:

Upper Limestone Shales
Main Limestone
Lower Limestone Shales (or Cefn Bryn Shale in Gower)
(Skrinkle Sandstone)

The higher part of the Skrinkle Sandstone, which lithologically belongs to the Upper Old Red Sandstone facies, is of Dinantian age and forms the lower part of the Courceyan Stage in south Pembrokeshire, where the Lower Limestone Shales follow it conformably, and reach maximum development.

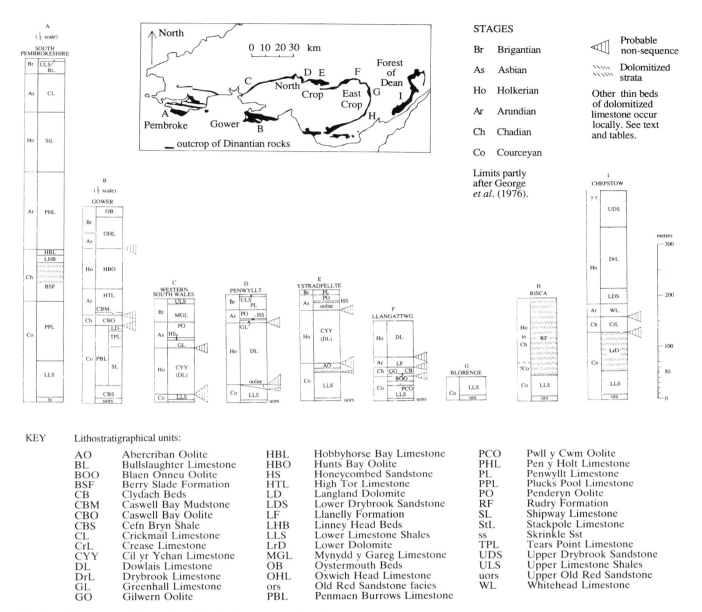

KEY　Lithostratigraphical units:

AO	Abercriban Oolite	HBL	Hobbyhorse Bay Limestone	PCO	Pwll y Cwm Oolite
BL	Bullslaughter Limestone	HBO	Hunts Bay Oolite	PHL	Pen y Holt Limestone
BOO	Blaen Onneu Oolite	HS	Honeycombed Sandstone	PL	Penwyllt Limestone
BSF	Berry Slade Formation	HTL	High Tor Limestone	PPL	Plucks Pool Limestone
CB	Clydach Beds	LD	Langland Dolomite	PO	Penderyn Oolite
CBM	Caswell Bay Mudstone	LDS	Lower Drybrook Sandstone	RF	Rudry Formation
CBO	Caswell Bay Oolite	LF	Llanelly Formation	SL	Shipway Limestone
CBS	Cefn Bryn Shale	LHB	Linney Head Beds	StL	Stackpole Limestone
CL	Crickmail Limestone	LLS	Lower Limestone Shales	ss	Skrinkle Sst
CrL	Crease Limestone	LrD	Lower Dolomite	TPL	Tears Point Limestone
CYY	Cil yr Ychan Limestone	MGL	Mynydd y Gareg Limestone	UDS	Upper Drybrook Sandstone
DL	Dowlais Limestone	OB	Oystermouth Beds	ULS	Upper Limestone Shales
DrL	Drybrook Limestone	OHL	Oxwich Head Limestone	uors	Upper Old Red Sandstone
GL	Greenhall Limestone	ors	Old Red Sandstone facies	WL	Whitehead Limestone
GO	Gilwern Oolite	PBL	Penmaen Burrows Limestone		

Fig. 2.3. Comparative lithostratigraphical and chronostratigraphic nomenclature in selected sections around South Wales (partly after George et al., 1976).

Elsewhere in South Wales a thinner, incomplete Lower Limestone Shales sequence was deposited. In south Glamorgan the lowest Dinantian beds are not exposed, although beds of Lower Limestone Shales lithology almost certainly underlie the exposed sequence. As a generalization the Lower Limestone Shales comprise alternations of calcareous mudstone and thin beds of limestone, some of which is impure. The overall impression is that throughout its deposition abundant terrigenous material was being washed into a fairly shallow-water environment.

In the light of modern stratigraphy the whole of the Lower Limestone Shales can be seen as comprising just the first part of a major depositional cycle, in this case the cycle which approximates to the Courceyan Stage. As the depositional environment changed and the sediment lost its terrigenous component later in the Courceyan, the first of the thick, pure limestones of the Main Limestone was deposited (Figs. 2.4, 2.5). In effect the Main Limestone is an informal collection of predominantly pure limestone units representing parts of the depositional cycles approximating to the Courceyan, Chadian, Arundian, Holkerian, Asbian and Brigantian stages. In different areas different parts of the various cycles were deposited and may be preserved, these generally being described by local names. The grouping together of these local units within the term Main Limestone has previously concealed the presence of a number of local non-sequences and implied a lithological and stratigraphical constancy which is now understood not be present. A number of mudstone and sandstone horizons of limited lateral extent are present in different areas, giving an indication of the incomplete cyclic nature of the sedimen-

Fig. 2.4. Natural arch in vertically dipping limestones of Courceyan age, Skrinkle Haven, southwest Dyfed (photo by A. L. Leach, courtesy of the National Museum of Wales).

Fig. 2.5. Gently dipping Dinantian limestones at The Wash, west of Stack Rocks, southwest Dyfed (photo by A. L. Leach, courtesy of the National Museum of Wales).

tation. In the detailed area descriptions below the limits of the Main Limestone will be compared to those of the regional stages and of the local lithostratigraphical nomenclature.

The Upper Limestone Shales, the highest part of the Dinantian sequence, are not present everywhere around the South Wales outcrop; a strong sub-Namurian non-sequence existing

in some areas. Where present the unit represents closing stages of the final Dinantian cyclic episode, the upper part of the Brigantian Stage, and reflects a return to depositional conditions similar to those pertaining in early Courceyan times, with abundant terrigenous material entering the sedimentary basin. Similar, but more constantly terrigenous, conditions persisted throughout the deposition of the overlying Namurian rocks.

In completing this general account of the local Dinantian stratigraphy it must be pointed out that in some areas and at certain levels in the sequence the limestone beds have undergone alteration or dolomitization since they were laid down. The effect of this alteration has been to destroy or mask the details of original lithologies and to obliterate micro- and macro-fossils within the rocks, thus removing the most readily applicable means of stratigraphic correlation. Such dolomitization has taken place particularly in the east of the South Wales area, where no realistic distinction can be made between rocks assumed to form part of the Courceyan, Chadian, Arundian and early Holkerian stages. Elsewhere the dolomitized parts of the sequence are thinner and can generally be located by reference to the beds above and below. Geological maps of South Wales at the one-inch (or 1:50 000) and six-inch (or 1:10 000) scales are available from the British Geological Survey. Although in some cases these include obsolete nomenclature, the limits of Dinantian carbonate sequences and the structures affecting them are shown in detail.

Gower

The Dinantian sequence in Gower (Fig. 2.3, Table 2.4) commences with beds of Upper Old Red Sandstone facies (Quartz Conglomerate) which pass upwards into the alternating calcareous shales and generally impure limestones of the Cefn Bryn Shales. This unit varies in thickness up to about 150 m and is the local representative of the Lower Limestone Shales of early Courceyan (K Zone) age. The overlying Penmaen Burrows Limestone typically comprises fossiliferous, crinoidal limestones and may be dolomitized towards its top in some areas. In modern terminology the unit, which forms the upper part of the Courceyan sequence, has been referred to as the Penmaen Burrows Limestone Group (George *et al.*, 1976). It is divided into three sub-units, the Shipway Limestone Formation at the base (including that part originally referred to the Z Zone), the Tears Point Limestone Formation (of similar lithology, but with a more typically C_1 Zone fauna) and the Langland Dolomite Formation, which is possibly a secondarily dolomitized bioclastic limestone, originally of C_1 Zone affinities.

A period of non-deposition probably occurred locally in early Chadian times, following which a series of relatively pure oolitic limestones, the Caswell Bay Oolite Formation, was laid down, reaching a thickness of about 40 m. An uneven and eroded surface separates these oolites from the Caswell Bay Mudstone Formation (previously the 'Calcite–Mudstone Group') above, suggesting that rather than being of Chadian age (George *et al.*, 1976) the upper unit is Arundian. In fact the 'Calcite–Mudstone Group' was originally placed in the Upper *Caninia* (C_2S_1) Zone (George, 1954), and this zonation is in agreement with a Cycle 3 (Ramsbottom, 1973) or Arundian

Table 2.4 *Generalized stratigraphy of Gower*

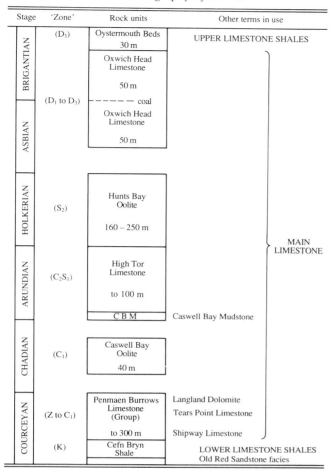

Stage	'Zone'	Rock units	Other terms in use
BRIGANTIAN	(D₃)	Oystermouth Beds 30 m	UPPER LIMESTONE SHALES
BRIGANTIAN	(D₁ to D₃)	Oxwich Head Limestone 50 m — — — — coal Oxwich Head Limestone 50 m	
ASBIAN	(D₁ to D₃)		
HOLKERIAN	(S₂)	Hunts Bay Oolite 160 – 250 m	MAIN LIMESTONE
ARUNDIAN	(C₂S₁)	High Tor Limestone to 100 m C B M	Caswell Bay Mudstone
CHADIAN	(C₁)	Caswell Bay Oolite 40 m	
COURCEYAN	(Z to C₁)	Penmaen Burrows Limestone (Group) to 300 m	Langland Dolomite Tears Point Limestone Shipway Limestone
COURCEYAN	(K)	Cefn Bryn Shale	LOWER LIMESTONE SHALES Old Red Sandstone facies

age. The High Tor Limestone (Formation) which overlies the Caswell Bay Mudstone is undoubtedly of Arundian age and comprises up to 100 m of predominantly 'standard' crinoidal–bioclastic limestones with local developments of oolitic and porcellanous limestone indicating periodic shallowing of the depositional basin.

There is no apparent widespread discontinuity between the High Tor Limestone and the overlying Hunts Bay Oolite (Formation), although there may be a minor non-sequence locally. The Hunts Bay Oolite, formerly the *Seminula* Oolite (S₂ Zone), is assigned to the Holkerian Stage and comprises more than 160 m of dominantly massive, coarse-grained oolites, the product of shallow water, shelf deposition. Towards the top of the Holkerian sequence thin calcite–mudstone beds may be present locally, indicating a further shallowing of depositional conditions and the short-lived establishment of hypersaline lagoons or lakes.

The Hunts Bay Oolite is disconformably overlain by the Oxwich Head Limestone (Formation), a series of generally massive, crinoidal, bioclastic limestones with some pseudo-breccias in the lower part, which extends through the Asbian into the Brigantian Stage. A thin and impersistent coal seam in the middle of the Formation probably marks the cycle boundary between the two stages, with sub-aerial coal swamp conditions marking the final stage of the Asbian mesothem.

Above the Oxwich Head Limestone about 30 m of shales and muddy limestones comprise the Oystermouth Beds (Formation). These beds, once referred to as the 'Black Lias', are the local equivalent of the Upper Limestone Shales and are the youngest Dinantian (Brigantian Stage) rocks preserved in the area.

Whilst the Gower sequence contains several apparently speleogenic limestone units, the Penmaen Burrows Limestone, the High Tor Limestone, the Hunts Bay Oolite and the Oxwich Head Limestone, the equivalents of which show significant cave development elsewhere in South Wales, only the Oxwich Head Limestone seems to contain major cave systems. Tooth Cave (SS 5320 9110) and Llethrid Swallet (SS 5310 9120) both lie in the Oxwich Head Limestone, Llethrid's entrance being close to the top of the unit and that of Tooth Cave being close to the level of the thin coal seam marking the chronostratigraphical break in the middle of the Formation. No detailed study of the relationships between caves, lithology and geological structure has been carried out, but the evidence of sumps at a relatively high topographical level, together with the flood-liable nature of both systems, suggests that phreatic cave development has been lithologically controlled and that passage formation has been confined to specific, favourable horizons. On the somewhat scanty evidence of Tooth Cave and Llethrid Swallet it seems that the initial stages of cavern formation might have been restricted to the beds of the Oxwich Head Limestone above the coal seam previously mentioned. Not only would such a bed and any associated seatearth or mudstone beds present an aquiclude to phreatic water movement, but at this level there is the additional possibility of acids being generated by the spontaneous breakdown of any sulphide or sulphate minerals deposited during the regressive phase at the end of the Asbian Stage. If cave development has been limited lithologically in this way, the extent of open cave passage would be expected to be controlled by minor faults which would locally raise or lower the favourable horizon. If significant systems have formed in stratigraphically lower beds, as they have in the Penwyllt and Llangattwg areas for instance, any evidence of them is yet to be found, although the remote possibility exists that they have been lost to erosion, or that they are simply blocked by recent sedimentary infill.

Western South Wales

Although the most complete Dinantian sequence was deposited and preserved in south Pembrokeshire (Fig. 2.3, Table 2.5), few speleological sites have been located or described in this area. Eastwards into the old county of Carmarthenshire Dinantian deposition began slightly later than in Pembrokeshire. A relatively thin sequence of shaly mudstones and muddy limestones, the Lower Limestone Shales, follows on conformably from Dinantian beds of Upper Old Red Sandstone facies ('Grey Grits') in some areas, whilst elsewhere the Dinantian rocks overstep onto older Devonian rocks. As elsewhere the Lower Limestone Shales are part of the Courceyan Stage (K Zone); no younger Courceyan rocks are known in the Carmarthenshire area and likewise no Chadian Stage or Arundian Stage deposits have been recognized. Westwards beds of Courceyan to Chadian age, the

Table 2.5. *Generalized stratigraphy of western South Wales*

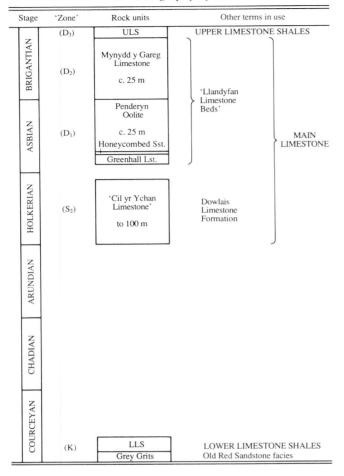

Mynydd y Gareg Limestone Formation, a series of fossi-liferous crinoidal limestones belonging to the Brigantian Stage and equivalent to the Penwyllt Limestone Formation further east. The highest Dinantian rocks preserved, of late Brigantian age, are a thin (≤10 m) development of Upper Limestone Shales.

As in Gower, few major cave systems have so far been explored in the western area, although innumerable minor sites are recorded. Two significant caves, Llygad Llwchwr (SN 6688 1780) and Ogof Foel Fawr (SN 7342 1875), have entrances close to the top of the Cil yr Ychan Limestone (Dowlais Limestone), a rock which has proven to be highly speleogenic along most of the North Crop. No detailed study of the formation of these caves has been undertaken, but the moors around are pock-marked by numerous dolines, which are probably of solutional origin rather than subsidence features. A possibility exists that intense solution of the limestone and cave inception under phreatic conditions at this level are linked to the disconformity between the Holkerian and Asbian stages. More directly the link could be with thin argillaceous bands deposited at this level, which might act as aquicludes to phreatic water movement, and to associated sulphate minerals (such as gypsum) formed during the desiccating conditions common in the closing episodes of a stage or mesothem. Decomposition of such minerals, with or without the presence of oxygen, has been shown to produce small quantities of strong acids which might be capable of removing sufficient limestone to facilitate groundwater movement and establish suitable conditions for cave development to proceed.

The only other major cave system in the area is Ogof Pwll Swnd (SN 7622 1839). This complex cave has its entrance in the Greenhall Limestone Formation below the level of the Honeycombed Sandstone, but probably penetrates into the underlying Cil yr Ychan Limestone. Here again the inception of at least part of the system could have been stratigraphically controlled in the way described above, with upward ramification into the Asbian beds being a later stage of development under phreatic, mixture–corrosion conditions. Pending a serious scientific investigation of these systems, the above hypothesis must be considered as only one of several equally unsubstantiated possibilities. Other factors undoubtedly had an influence on cavern development following inception. Within the lithological constraint of the Cil yr Ychan Limestone, underground drainage seems generally to follow the dip of the beds, this being locally affected by faulting and, as might be expected, a degree of control over passage direction and morphology has been exercised by joints and minor faults.

The Penwyllt area

The relationship of caves to geology is perhaps better known in the Penwyllt area than in any other part of South Wales (Fig. 2.3, Table 2.6). East of the River Tawe much work has been carried out over the years at Ogof Ffynnon Ddu, the results having been synthesized by O'Reilly (1969). West of the Tawe the detailed research of Coase in Dan yr Ogof (Coase & Judson, 1977) is unlikely to be challenged, although as elsewhere in South Wales stratigraphical detail has been refined and processes of cavern initiation may become better

Pendine Oolite and Pendine Conglomerate are preserved. This sequence lies west of the Llandyfaelog (Careg Cennen) Disturbance and it seems likely that this structural feature, possibly a continuation of the Church Stretton fault line, was active in early Dinantian times and exerted control on the size and nature of local depositional basins through until the end of the Arundian Stage.

By the beginning of the Holkerian Stage the area was open to the sea again and although the earliest Holkerian (S₂ Zone) rocks may not be present sedimentation was more or less continuous throughout the Stage, producing the massive oolitic and algal limestones of the Cil yr Ychan Limestone (Formation). These beds are the direct equivalent of the Dowlais Limestone Formation further east and broadly comparable to the Hunts Bay Oolite of Gower. Disconformity probably exists around the boundary of the Holkerian rocks with those of the overlying Asbian Stage. Formerly considered as part of the 'Llandyfan Limestone Beds', the Asbian rocks are now divided into three units (George *et al.*, 1976), the Greenhall Limestone Formation (crinoidal and oolitic limestones) at the base, followed by the primarily calcareous, but often decalcified, Honeycombed Sandstone (Formation) and finally the Penderyn Oolite Formation (formerly the 'Light Oolite'). Above these formations of Asbian age the upper part of the former 'Llandyfan Limestone Beds' is now known as the

Table 2.6 *Generalized stratigraphy of Penwyllt area*

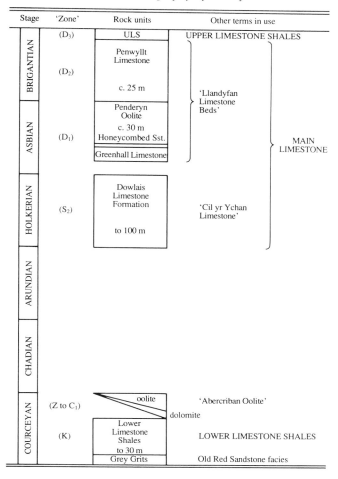

Table 2.7 *Comparison of lithostratigraphy and chronostratigraphy in the Dan yr Ogof and Ogof Ffynnon Ddu area (stages not to scale)*

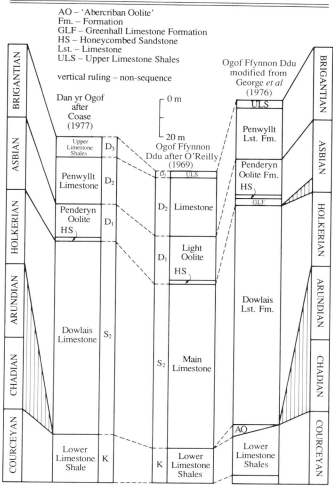

understood. The Dinantian successions in the immediate vicinities of Dan yr Ogof and Ogof Ffynnon Ddu are essentially the same, although they differ slightly in both thickness and the minor detail of lithologies low in the sequence (Table 2.7).

Surface exposure shows the earliest Dinantian rocks (Courceyan Stage) to be of Upper Old Red Sandstone facies ('Grey Grits') followed with no break in sedimentation by the mixed shales and impure limestones of the Lower Limestone Shales, which may reach 30 m in thickness. Over part of the area there was a significant break in sedimentation above the Lower Limestone Shales, most obvious west of the Tawe where at least over part of the outcrop beds of early Holkerian (S_2 Zone) age, the Dowlais Limestone Formation, rest on the Courceyan rocks. To the east the remnants of an oolite sequence are preserved between the Lower Limestone Shales and the Dowlais Limestone. These beds, known locally as the 'Abercriban Oolite' (see also Llangattwg–Clydach section) lie on top of the Lower Limestone Shales with a small thickness of dolomitized limestone sometimes preserved in between. By comparison with sequences elsewhere it is assumed that these beds equate with the lower part of the 'Oolite Group' and may correlate with the Pwll y Cwm Oolite (Formation) or the Blaen Onneu Oolite (Formation) of late Courceyan age. These oolites have not been recognized west of the Tawe and it is

possible that different depositional conditions prevailed on either side of the Tawe (Cribarth) Disturbance, which by comparison with the Careg Cennen Disturbance to the west and the Neath Disturbance to the east, was probably active in early Dinantian times. If this was the case, it was probably for similar reasons that conditions of non-deposition continued through the Chadian and Arundian stages.

Following on from the Lower Limestone Shales (or the oolite described above) is a series of crystalline and oolitic limestones up to 100 m thick (Dowlais Limestone Formation) belonging to the Holkerian Stage (S_2 Zone). Thin shaly partings are exposed in the caves but not always visible in surface exposures. If these could be studied in detail they might give an insight into minor cyclicity in the lower part of the Holkerian sequence or could indicate that previously unrecognized representatives of earlier stages are hidden within the lower part of the Dowlais Limestone.

As elsewhere on the North Crop there was probably a break in sedimentation at the start of the Asbian Stage. The boundary between the S_2 and D_1 zones (which approximates to that between the Holkerian and Asbian stages) was traditionally

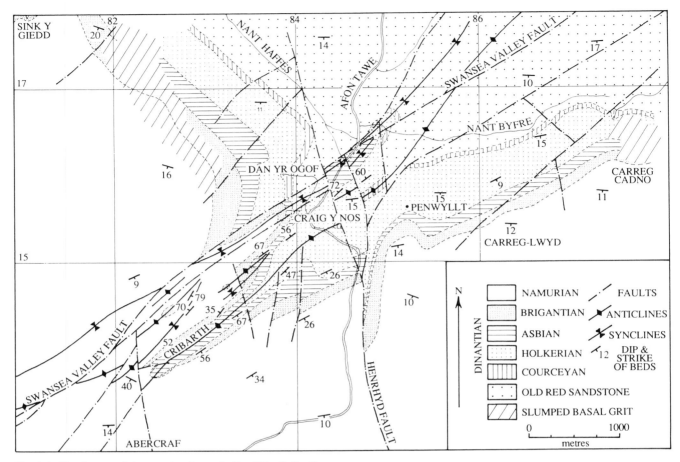

Fig. 2.6. Sketch map of the structural complexity of the Cribarth–Penwyllt area (John Weaver).

taken at the Honeycombed Sandstone, a thin bed of calcareous sandstone, often decalcified to produce an open, cellular structure. By comparison with areas to the west it seems likely that a relatively thin (≤35 m) development of oolitic and crinoidal limestones below the Honeycombed Sandstone, the Greenhall Limestone Formation, is of Asbian age. The Greenhall Limestone, Honeycombed Sandstone and overlying Penderyn Oolite Formation (formerly 'Light Oolite', up to 30 m thick) form the lower part of the 'Llandyfan Limestone Beds', the upper part of which is now referred to as the Penwyllt Limestone Formation (George *et al.*, 1976) and is of Brigantian age. The Penwyllt Limestone consists of beds of highly siliceous limestone with intercalations of chert and quartzite. A thin development of Upper Limestone Shales, which may be locally hidden by the overstep of younger rocks, completes the local Dinantian sequence.

Structurally the Penwyllt area is fairly complex (Fig. 2.6). The line of the Tawe Disturbance crosses from southwest to northeast and includes a number of folds and faults on this trend. In addition there are a number of NNE–SSW and NNW–SSE fractures of limited throw, which probably formed as compensation faults to the long-term wrench movement associated with the main disturbance. Coase (Coase & Judson, 1977) suggested that much of the generally north–south passage development in Dan yr Ogof was influenced by the minor fractures on this trend or by the joints which approximately

parallel them. This north–south trend also directs the passages down-dip towards the axis of the Dan yr Ogof Syncline (Coase & Judson, 1977) and this relatively gentle structure, part of the Tawe Disturbance belt, appears to have concentrated and directed drainage and passage formation between Mazeways II and the Show Cave.

In Ogof Ffynnon Ddu the effect of faults is less marked, although a strong joint influence on passage direction is recognized, one joint-set being parallel to the local minor fault trend. The degree of correlation of joint–fault trend to passage direction has been the subject of some controversy (Weaver, 1974; Coase & Judson, 1977) and the extent to which the fracture controlled passage formation or merely exerted an influence is still uncertain. Elsewhere in the Penwyllt area similar relationship of passage direction to joint and fault trends can be recognized in most cave systems and is particularly striking in Cathedral (or Tunnel) Cave, while Pant Mawr Pot parallels a significant fault line throughout its length.

The other element of geological influence on cavern formation in the Penwyllt area is the lithostratigraphy. Both major cave systems, despite having developed appreciable depth due to the effects of dip, are largely confined to the Holkerian (S_2 Zone) Dowlais Limestone Formation. The only significant cave development in younger beds is at Sink y Giedd where a joint-parallel shaft penetrates the Brigantian Penwyllt Limestone Formation, probably to the level of the Brigantian–

Asbian boundary. Coase (Coase & Judson, 1977) pointed out the apparent importance of thin 'shales' between the upper and lower parts of the S_2 Zone to passage development in the Dan yr Ogof area. Modern interpretation shows that these beds lie at the Asbian–Holkerian (D_1–S_2) boundary. Similar lithological control is exhibited at Ogof Ffynnon Ddu, and on both sides of the Tawe at smaller systems such as Pwll Dwfn, with its entrance on the outcrop of the stage boundary. The Top Entrance of Ogof Ffynnon Ddu and the lower entrance to the same system, although a considerable vertical distance apart, are at this same geological horizon. As in the other major cave areas described on the North Crop, the significance of this cave development below the Asbian–Holkerian disconformity has yet to be fully investigated, but the possibility of inception being accelerated by acids due to weathering processes cannot be discounted. Without this, or some similar mechanism, it is difficult to account for the abundance of cave passage in the speleogenic limestone beneath this stratigraphic boundary and the paucity of development in speleogenic limestones elsewhere in the sequence. Nor is it possible to reconcile the controlling influences suggested by various earlier authors (water tables, surface erosion levels, highly aggressive but localized peat bog drainage, and so on) with the almost exclusive local cave development within the Holkerian rocks. (However, see also the description of the Llangattwg–Clydach area.) In both Dan yr Ogof and Ogof Ffynnon Ddu a number of 'avens' penetrate upwards through the disconformity and into the Asbian beds. It is assumed here that rather than being primary inception features these were formed at a later stage and were enlarged upwards by mixture–corrosion in an already well-developed phreatic drainage system.

The Ystradfellte area

Beds of Upper Old Red Sandstone facies are the earliest Dinantian (Courceyan Stage) deposits in the Ystradfellte area, and these pass transitionally upwards into a typical mixed shale and limestone sequence of the Lower Limestone Shales (K Zone) up to 50 m thick (Fig. 2.3, Table 2.8).

As at the eastern side of the Penwyllt area the Lower Limestone Shales are followed by a relatively thin oolite succession, the 'Abercriban Oolite', which is generally dolomitized in its lower part. By comparison with Penwyllt on the one hand and Llangattwg on the other, these oolites are equivalent to one or other of the oolite formations forming the lower part of the Abercriban Oolite Group ('Oolite Group'), either the Pwll y Cwm Oolite or Blaen Onneu Oolite. There may be a non-sequence between the Lower Limestone Shales and the oolite, as suggested by the dolomitization at their junction, but the Oolite is probably of late Courceyan age.

Above the oolite there is a definite break in sedimentation, as elsewhere on the North Crop. There are no beds of Chadian or Arundian age (C_1–C_2S_1 zones) recognized. The 'Abercriban Oolite' is followed unconformably by a thick sequence of sparry and oolitic limestones, the Dowlais Limestone Formation ('Cil yr Ychan Limestone') of Holkerian (S_2 Zone) age. Whether there was deposition right through the Holkerian is uncertain, but there are no typical regressive sediments at the top of the sequence. A thin discontinuous bed

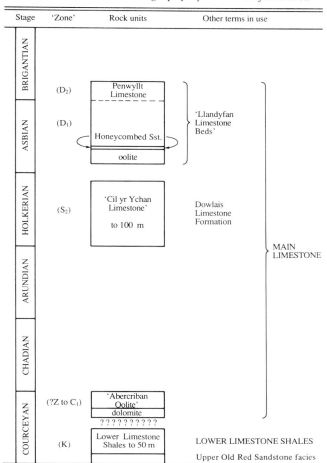

Table 2.8 *Generalized stratigraphy of the Ystradfellte area*

Stage	'Zone'	Rock units	Other terms in use
BRIGANTIAN	(D_2)	Penwyllt Limestone	'Llandyfan Limestone Beds'
ASBIAN	(D_1)	Honeycombed Sst. / oolite	
HOLKERIAN	(S_2)	'Cil yr Ychan Limestone' to 100 m	Dowlais Limestone Formation
ARUNDIAN			
CHADIAN			MAIN LIMESTONE
COURCEYAN	(?Z to C_1)	'Abercriban Oolite' / dolomite / ??????????	
	(K)	Lower Limestone Shales to 50 m	LOWER LIMESTONE SHALES / Upper Old Red Sandstone facies

of shelly oolite and coarser, pisolitic limestone may indicate a shallowing of depositional conditions so that on balance it seems deposition was continuous through to the closing part of the Stage (cycle).

In common with other areas on the North Crop there was probably a break in sedimentation at the beginning of the Asbian Stage (D_1 Zone), following which deposition of oolitic limestones recommenced. A fairly thick and pure oolite sequence, equivalent to the Greenhall Limestone Formation (or in local terms the lower part of the 'Llandyfan Limestone Beds') laid down right across the area, followed by the thin but laterally persistent Honeycombed Sandstone (Formation). Above this, oolite deposition continued with the Penderyn Oolite Formation (still part of the 'Llandyfan Limestone Beds'), which probably continued to the end of the Asbian Stage. The precise line between the Asbian and Brigantian stages has not been mapped in this area and lies within the 'Llandyfan Limestone Beds'. Locally the upper part of this unit is gritty, with discontinuous chert bands, and is assumed to equate with the Penwyllt Limestone Formation, of Brigantian age, further west. No Upper Limestone Shales development has been separately identified in the area, but possibly the beds immediately below the overlying Namurian Basal Grit, which are impure and of mixed lithology, are the local equivalent. Certainly the area is on the extreme limit of the extent of Upper Limestone Shales deposition.

Structurally the Ystradfellte area is much more complex than areas to the west and east. The rock sequence is cut into fairly narrow slices by predominantly NNW–SSE faulting. A strongly developed joint-set parallels this faulting, whilst other joints trend closer to east–west. The regional dip of the sequence is towards the south, being locally slightly variable in both amount and direction due to the structural effects.

Cave development has been both lithologically and structurally controlled. Pant Mawr Pot (SN 8909 1612) on the boundary between the Ystradfellte and Penwyllt areas, has its entrance on the probable unconformity between the Courceyan 'Abercriban Oolite' and the Holkerian Dowlais Limestone ('Cil yr Ychan Limestone'). For a considerable distance the cave follows a significant fault line, gaining depth by trending down-dip. Beyond the limits of exploration the cave must penetrate progressively higher beds of the Dowlais Limestone to reach its resurgence point at Pwll Du (SN 9121 1206) in the Nedd Fechan Valley. At the resurgence a narrow slice of limestone, probably belonging to the Penwyllt Limestone Formation, is present between two faults, with Namurian sandstones to either side. The precise way that the cave drainage which commences in the lowest Holkerian beds reaches the surface in the Brigantian Penwyllt Limestone is unknown, but its route must cross several faults, which have presumably stepped down progressively younger beds along the route. From knowledge of lithological control elsewhere on the North Crop it seems likely that the cave penetrates to appreciable depth within the Dowlais Limestone, probably to hit a deep phreatic zone, before developing phreatic lifts, probably on fault planes, to a stratigraphically and topographically higher bed such as the Greenhall Limestone or Penderyn Oolite, whence down-dip development would again take over. A further fault rise would then be required to complete the

theoretical system and bring the drainage to the resurgence level in the uppermost Dinantian beds.

The picture of the drainage and speleogenesis of the upper reaches of the Nedd Fechan is no clearer. Rather than rising at Pwll Du or one of the many small risings in this part of the valley, it is suspected that water sinking upstream of Pwll y Rhyd (SN 9113 1376) cuts under the surface watershed to resurge in the Mellte Valley to the southeast. Such a route is quite possible in that it would follow the regional dip and avoid the need to cross a large number of faults between Pwll y Rhyd and Pwll Du. The system of Ogof Nedd Fechan (Little Neath River Cave) and Bridge Cave, with its entrances on the east bank of the Nedd Fechan (SN 9118 1420 and SN 9116 1400) probably commences at or near the level of the top of the Dowlais Limestone (Asbian–Holkerian disconformity). As at Ogof Ffynnon Ddu further west, there is a strong possibility that the system follows the highly speleogenic Dowlais Limestone down-dip, possibly stepping up phreatic lifts on several faults to stay in the same bed, before regaining the surface. Without knowing the precise resurgence site it is not possible to predict whether, as in the Penwyllt area, the entire system remains in the Dowlais Limestone, or whether it steps up into the higher oolites which also crop out in the Mellte Valley.

The major cave site in the Mellte Valley, Porth yr Ogof (SN 9281 1241), has a large sink entrance, where the Mellte goes underground, at about the level of the top of the Dowlais Limestone – again the level of the Asbian–Holkerian disconformity. Lower down the valley the river resurges at about the same stratigraphic level, having crossed a significant fault line on route.

Further south, in the Penderyn area, the stratigraphy is very similar. Southwestwards from Penderyn a narrow sliver of Holkerian–Brigantian age beds (Dowlais Limestone–Penwyllt

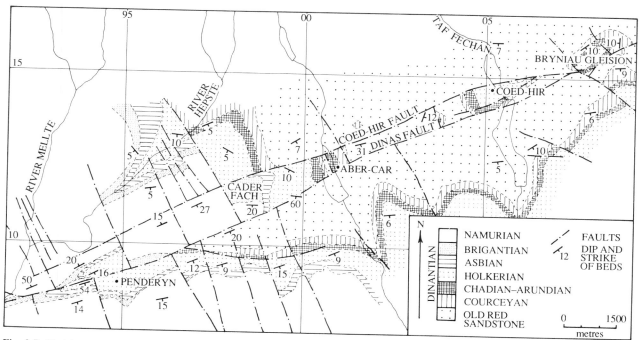

Fig. 2.7. Sketch map of the Dinas and Coed-Hir Fault belt near Penderyn – part of the Vale of Neath Fault system (John Weaver).

Fig. 2.8. A tight fold at Bwa Maen near Pont Nedd Fechan – part of the Vale of Neath Fault system (photo by I. V. Evans, courtesy of the National Museum of Wales).

Table 2.9 *Generalized stratigraphy of Llangattwg area*

Stage	'Zone'	Rock units	Other terms in use
BRIGANTIAN			
ASBIAN			
HOLKERIAN	(S₂)	Dowlais Limestone Formation to 70 m	MAIN LIMESTONE
ARUNDIAN	(C₂S₁)	Llanelly Formation to 20 m	"Calcite–Mudstone Group"
CHADIAN	(?C₁)	Gilwern Oolite to c. 12 m Clydach Beds	'Oolite Group'
COURCEYAN	(?Z to C₁)	Blaen Onneu Oolite / Pwll y Cwm Oolite	
	(K)	Lower Limestone Shales	LOWER LIMESTONE SHALES
			Old Red Sandstone facies

Limestone) is present (Figs. 2.7, 2.8). The main valley here follows a major fracture, the Dinas Fault, which has a similar trend to that of the various Disturbance zones elsewhere in the area, and like them might be an ancient weakness. A number of smaller fractures sub-parallel to the fault line add to the structural complexity. The precise stratigraphical positions of the various caves in the Sychryd–Hepste area are uncertain, but in general terms they all lie close to the top of the Dinantian sequence, close below the basal Namurian unconformity. As elsewhere some development has occurred close to the top of the Dowlais Limestone somewhat lower in the sequence, but the structural complexity is such that a much more detailed study is needed to relate speleogenesis to geology.

The Llangattwg–Clydach Valley area

As elsewhere on the North Crop the earliest Courceyan rocks are of Old Red Sandstone facies, being predominantly of fluviatile origin, and these beds pass upwards into lagoonal and shelf deposits of the Lower Limestone Shales (Fig. 2.3, Table 2.9). The base of the Lower Limestone Shales has been shown to be diachronous, as elsewhere in South Wales (Lovell, 1978) and the unit comprises calcareous mudstones, impure limestones and oolites laid down in a number of transgressive and regressive phases (Burchette, 1981).

Overlying the Lower Limestone Shales is the Abercriban Oolite Group (formerly 'Oolite Group') which is not to be confused with the 'Abercriban Oolite' of the Penwyllt area, although the latter is probably equivalent to one of the formations forming the lower part of the Group. The lowest part of the Abercriban Oolite Group is the Pwll y Cwm Oolite (Formation), a richly bioclastic unit at its base which passes upwards into massive oolitic limestone. At Daren Cilau the unit is about 7 m thick, less than this to the west and thicker, although partly dolomitized, to the east. It is assumed here that the Pwll y Cwm Oolite is of mid- to late Courceyan age (? Z Zone), although some authors (e.g., Wright, 1981, 1982) have assigned it to the early Chadian. Between the Pwll y Cwm Oolite and the overlying Blaen Onneu Oolite (Formation) is a

thin, discontinuous layer of dolomite in the form of lenses and nodules with some shale, possibly indicating a minor cycle boundary. A second dolomitic development occurs higher in the unit. The Blaen Onneu Oolite is of variable lithology and includes sparry bioclastic limestone as well as pure oolites. It can be very dark in colour, although it becomes paler westwards (Raven, 1981) and is generally thickly bedded. The Blaen Onneu Oolite is here attributed to the late Courceyan (? C₁ Zone), although it has also been referred to the mid-Chadian (e.g., Wright, 1981, 1982, 1986).

The overlying Clydach Beds (Formation) may also be of latest Courceyan age, but this thin, laterally persistent unit of sparry limestone, porcellanous limestone and shale, with calcareous sandstone bands, is here assumed to be of Chadian age. There is a well-marked stratigraphic break at the base of the Clydach Beds and a minor disconformity may be present. The formation includes the 'Coral Bed' of George (1954) and is equivalent to the same author's 'Marker Beds'. Above the Clydach Beds is a series of massive, often impure oolitic limestones, which comprises the Gilwern Oolite (Formation). The unit appears to follow the Clydach Beds without sedimentary break and is thus attributed to the Chadian Stage, although other authors have considered it to be of Arundian age (Ramsbottom, 1973; Wright, 1981, 1982). Often the unit is represented only by a thin, rubbly development, becoming thinner to the west, and its upper surface is invariably strongly karstified. The rubbly brecciation, ramifying solutional pipes within the unit and its 'potholed' upper surface probably date from a long period of sub-aerial exposure in late Chadian–early Arundian times. Details of the present form of this ancient karst and its possible mode of formation have been given by Wright (1982) and most of his conclusions seem reasonable, although if any significant cavern formation contributed to collapse and brecciation, a time span greater than 'only a few thousand years' would be required before the deposition of the next higher rock unit. A longer chronological break would make speleogenesis, and collapse, quite feasible.

The lowest beds of the Llanelly Formation (formerly 'Calcite–Mudstone Group') which disconformably overlie the Gilwern Oolite are sub-aerial–fluvial deposits including calcrete palaeosols at the base, conglomerates, sandstones and clays (Wright, 1982, 1986). These beds occupy depressions in the karstic surface of the Gilwern Oolite, and locally penetrate a considerable depth into karstic fissures and solutional tubes. Above these terrestrial deposits, resting on them or on adjacent hummocks of karstified Gilwern Oolite and in places truncating both lithologies, are the remaining members of the Llanelly Formation, which are described in detail by Wright (1981, 1986). The possible significance of this thin (≤ 20 m) essentially non-speleogenic deposit to cavern genesis in the beds above and below will be considered later in this section.

The Llanelly Formation (formerly 'Calcite–Mudstone Group') was assigned to the Upper *Caninia* (C_2S_1) Zone by George (1954), which would correspond to the Arundian Stage in more recent terminology. This age was accepted by Ramsbottom (1973) and Wright (1981) and is retained here, although a Chadian age has also been suggested (George *et al.*, 1976).

A major disconformity separates the Llanelly Formation from the overlying Dowlais Limestone Formation and the break probably represents a significant phase of erosion as well as non-deposition. To the west the Llanelly Formation has been completely removed and the Dowlais Limestone rests on lower beds. The Dowlais Limestone around Llangattwg is thinner than to the west, possibly reaching 70 m; it is also less homogeneous overall and some beds are bituminous. The sequence includes sparry bioclastic limestones and oolites as well as thin intertidal and sub-aerial deposits such as stromatolite beds and seatearths (Wright, 1981, 1986). In this area the Dowlais Limestone represents the deposits of a shallow lagoon, subject to tidal fluctuation and periodic reversion to swamp conditions, implying that the area was on the extreme limit of Holkerian (and possibly even Dinantian) marine deposition. Further west along the North Crop, and on the South Crop, purer, more homogeneous oolite sequences were deposited in somewhat deeper water conditions shoaling away from the lagoonal shelf.

No detailed study of the local structural geology has been carried out, but mapping shows a number of NNW–SSE structures, particularly normal faults. As elsewhere on the North Crop an important joint-set parallels this trend, with a less prominent set approximately at right angles. The regional bedding dip is towards the south or southeast at angles up to 10°.

The relationships between underground drainage, cave formation and geology in the area are somewhat more complex than in areas further west. Water flowing through the major Llangattwg cave systems, Ogof Agen Allwedd, Ogof y Daren Cilau and Ogof Craig a Ffynnon, resurges in the Clydach Valley at Pwll y Cwm, Ogof Capel (Shakespeare's Cave) and Limekiln Springs. These risings and another major spring, Ffynnon Gisfaen, which debouches water from sinks up to 7 km to the west, lie in, and probably close to the base of, the Pwll y Cwm Oolite Formation. Behind the risings the greater part of these major caves so far explored lies within the lower oolites of the Abercriban Oolite Group. The currently active main streams in Agen Allwedd and Craig a Ffynnon flow along passages in the Pwll y Cwm Oolite, while the majority of the abandoned dry passage at higher levels lies in the Blaen Onneu Oolite. There is no evidence of significant primary solutional cave development in the higher part of the Abercriban Oolite Group – the Clydach Beds and the Gilwern Oolite – although these rocks are encountered in the major systems, usually as areas of breakdown. Rarely the still higher Llanelly Formation is seen in passage roofs, and again it is probable that this is due to removal of the beds below rather than by direct involvement of the Llanelly Formation in cavern development. The somewhat limited evidence available indicates that these drainage systems have developed in the lower part of the Abercriban Oolite Group with no direct influence from swallet water sinking into the Dowlais Limestone atop the Llangattwg Escarpment. Whether they were ever fed by major sinks which are no longer active or have been removed by surface modification, or whether the systems were initiated and developed by highly aggressive percolation water penetrating palaeokarstic fissures in the Gilwern Oolite, possibly becoming more aggressive in penetrating the Clydach Beds and eventually finding suitable speleogenic limestone in the Blaen Onneu Oolite, the remarkable narrowing of supposed feeders to the main drainage routes in an 'upstream' direction, strongly supports the idea that the majority of cave formation in the oolites have been solutional rather than due to direct vadose input from the surface. Inlet streams that are currently encountered in these systems are probably late-stage piracies, accidental or underfit to the primary speleogenesis, probably penetrating the Llanelly Formation to Clydach Beds aquiclude through ancient palaeokarstic routes, or very recent upward solutional–collapse weaknesses. The occurrence of current active main streamways in the lower oolite and abandoned systems in the Blaen Onneu Oolite is not coincidental and the probability is that the original systems were confined to the upper formation, hydrologically constrained by dolomitized strata, and it is probable that there were ancient resurgences at, or close above, the Blaen Onneu Oolite–Pwll y Cwm Oolite junction. Much of the lower development in the Pwll y Cwm Oolite has been drained only recently; much in fact is still submerged and some parts are transitional with variations in local weather conditions. Whether the Pwll y Cwm Oolite passage developments took place subsequently to the Blaen Onneu Oolite developments, in response to an alteration of the surface base level, or whether there was simultaneous development above and below the supposed dolomitic aquiclude, is unknown. If both oolites were fed by percolation water from separate fault blocks, simultaneous development and subsequent joining is within the bounds of possibility.

The obvious counterpoint to the above arguments is that a number of major streams sink into the Dowlais Limestone, above the supposed aquiclude of the Llanelly Formation and apparently within the catchment of the three major systems. This water also rises in the Clydach Valley, but the indications are that its underground routes have not yet been encountered in the major cave systems. It would seem that a totally independent drainage system exists, above the currently active systems, in the Dowlais Limestone. That copious quantities of precipitation would sink into the Dowlais Limestone is certain. The upper surface of the formation must have been strongly karstified in the Asbian–Brigantian period of non-deposition, before Namurian beds were deposited on top. In later times when the cover was stripped off the palaeokarst would inevitably be reinvaded. Similarly any groundwater in the Namurian rocks would be funnelled into palaeokarstic fissures beneath and would lead to the development of subsidence or collapse

dolines. If the assumption of an aquiclude at the Llanelly Formation–Clydach Beds level is valid, it is a simple step to assume that a high-level underground drainage system exists above the currently known system in the oolites. There must be leakage from the upper levels through the aquiclude, either by way of palaeokarstic routes or along fault lines, but this does not rule out the other system. No significant part of this upper system has yet been located, but a number of small caves are known. As mentioned the water sinking into the Dowlais Limestone resurges at the same sites as does the water in the major caves. Where or how the two systems join is currently unknown.

Conclusion

The structure and stratigraphy of South Wales, particularly within the rocks of the Carboniferous, are well known. Much works remains, however, to elucidate the links between these factors and cave formation, particularly the importance of disconformities or the potentially acid-yielding deposits associated with them.

References

Burchette, T. P. (1981). The lower limestone shales. In *A Field Guide to the Lower Carboniferous Rocks Near Abergavenny*, ed. V. P. Wright, M. Raven & T. P. Burchette, pp. 13–27. Cardiff: Department of Geology, University College.

Coase, A. & Judson, D. (1977). Dan yr Ogof and its associated caves. *Transactions British Cave Research Association*, 4(1–2), 245–344.

Dixon, E. E. L. & Vaughan, A. (1912). The Carboniferous succession in Gower (Glamorganshire). *Quarterly Journal of the Geological Society of London*, 67, 477–571.

George, T. N. (1954). Pre-seminulan main limestone of the Avonian Series in Breconshire. *Quarterly Journal of the Geological Society of London*, 110, 283–322.

George, T. N. (1970). South Wales. In *British Regional Geology*. London: HMSO.

George, T. N., Johnson, G. A. L., Mitchell, M., Prentice, J. E., Ramsbottom, W. H. C., Sevastopulo, G. D. & Wilson, R. B. (1976). A correlation of Dinantian rocks in the British Isles. *Geological Society of London Special Report*, No. 7, 87 pp.

Holland, C. H., ed. (1978). *A Guide to Stratigraphical Procedure*. Geological Society of London Special Report No. 10, 18 pp.

Lovell, R. W. W. (1978). *Sedimentology of the Upper Old Red Sandstone and Lower Limestone Shale of the South Wales Coalfield*. Unpublished Ph.D. thesis, University of Bristol, 333 pp.

O'Reilly, P. M. (1969). Geology. In *Ogof Ffynnon Ddu*, ed. P. M. O'Reilly, S. E. O'Reilly & C. M. Fairbairn, pp. 17–25. Penwyllt: South Wales Caving Club.

Owen, T. R. (1954). The structure of the Neath disturbance between Bryniau Gleision and Glynneath, South Wales. *Quarterly Journal of the Geological Society of London*, 109, 333–65.

Ramsbottom, W. H. C. (1973). Transgressions and regressions in the Dinantian: a new synthesis of British Dinantian stratigraphy. *Proceedings of Yorkshire Geological Society*, 39, 567–607.

Ramsbottom, W. H. C. (1977). Major cycles of transgression and regression (mesothems) in the Namurian. *Proceedings of Yorkshire Geological Society*, 41, 261–91.

Raven, M. (1981). The oolite group. In *A Field Guide to the Lower Carboniferous Rocks Near Abergavenny*, ed. V. P. Wright, M. Raven & T. P. Burchette, pp. 28–35. Cardiff: Department of Geology, University College.

Weaver, J. D. (1974). Jointing along the Swansea Valley Disturbance between Clydach and Hay-on-Wye, South Wales. *Geological Journal*, 10, 75–86.

Wright, V. P. (1981). The Llanelly Formation and the Dowlais Limestone. In *A Field Guide to the Lower Carboniferous Rocks Near Abergavenny*, ed. V. P. Wright, M. Raven & T. P. Burchette, pp. 36–73. Cardiff: Department of Geology, University College.

Wright, V. P. (1982). The recognition and interpretation of palaeokarsts: two examples from the Lower Carboniferous of South Wales. *Journal of Sedimentary Petrology*, 52, 83–94.

Wright, V. P. (1986). The polyphase karstification of the Carboniferous Limestone in South Wales. In *New Directions in Karst*, ed. K. Paterson & M. M. Sweeting, pp. 569–80. Norwich: Geobooks.

3

Karst geomorphology of South Wales

JOHN CROWTHER

Outcrops of Carboniferous Limestone occur in two relatively narrow bands on either side of the large, east–west synclinal basin of the South Wales Coalfield. The limestones of the North Crop are exposed for a short distance to the west of Pendine (Fig. 3.1), but the main outcrop extends some 110 km from Kidwelly to Abergavenny (Fig. 3.2). For most of its length the Carboniferous Limestone is <200 m thick and has a southerly dip of 10–15°. Locally, the outcrop has been displaced laterally by movements associated with the northeast-trending Cennen, Swansea Valley and Vale of Neath Disturbances and by many northwest faults. Several small outliers are present, the best known providing the very imposing site for Carreg Cennen Castle, 5 km south of Llandeilo. The dip-slope of the Carboniferous Limestone averages about 1.5 km in width and comprises a gently undulating plateau, broken

sporadically by rocky crags and ridges. From Kidwelly to the Loughor the land remains mostly below 200 m O.D., but along much of the rest of the North Crop its altitude rises to between 300 and 550 m. The lower ground supports good quality grassland, whereas the higher land largely comprises rough moorland grazings, with woodland confined to sheltered valleys and depressions. In the far east, between Mynydd Llangynidr and Blorenge, the limestones form an impressive escarpment overlooking the River Usk. Elsewhere, however, there is usually only a low, broken cuesta, although often, as at Pal y Cwrt, southwest of Carreg Cennen, lithological variations within the limestone produce multiple scarp faces.

Exposures of the South Crop occur in southern Pembrokeshire, Gower and the Vale of Glamorgan. In all three areas the limestone is intensively folded, with Devonian Sand-

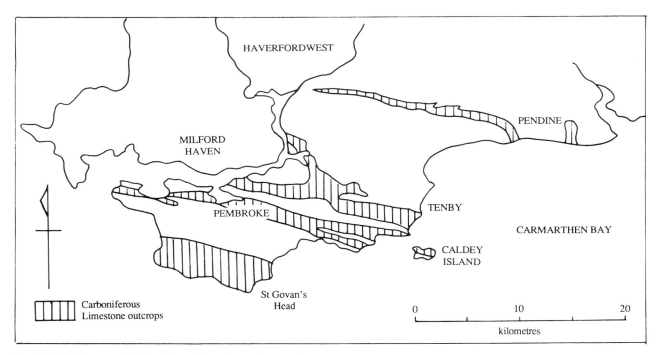

Fig. 3.1. Sketch map of Carboniferous Limestone outcrops of southern Pembrokeshire.

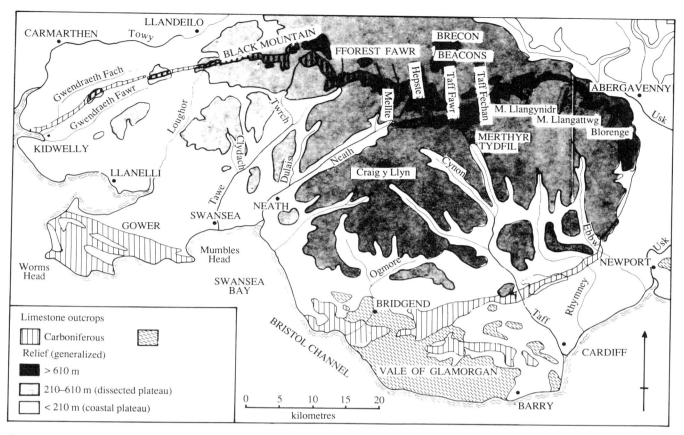

Fig. 3.2. Sketch map of Carboniferous Limestone outcrops and relief of South Wales.

stone outcropping in the anticlinal cores and Carboniferous Millstone Grit preserved in synclines. The limestones are much thicker than those of the North Crop (>1000 m on Gower), but dip much more steeply. With the closer proximity to the sea the relief is lower, being generally <100 m, and erosion is much more advanced. Indeed, the outcrop is breached by Carmarthen and Swansea bays. The low plateau at about 90 m that forms the coastal belt of the Vale of Glamorgan has developed on interbedded limestones and shales of Lower Lias–Rhaetic age (Fig. 3.2). Here, and along much of the South Crop, rainfall ranges from 1000–1500 mm/y and the land is used for both pasture and cereal cultivation, chiefly winter wheat and spring barley.

Compared with most other rock types, limestones are relatively soluble in natural waters. The resulting 'karst' landscape is characterized by thin soils, rocky outcrops, underground drainage and a distinctive suite of solutional forms, ranging in size from small hollows and flutes on bare limestone to much larger depressions ('dolines'). In parts of South Wales, notably on Gower, thick glacial drifts impede, or even preclude, surface karstification. Such features are, however, particularly well displayed along the North Crop to the east of the Loughor, and this area has special geomorphological significance, for three reasons. First, it is the wettest of the major limestone regions in Britain. Locally, on Black Mountain, the annual rainfall is as high as 2300 mm, although this decreases eastwards to about 1250 mm/y on Mynydd Llan-

gattwg and Blorenge. Secondly, because of its narrowness, the evolution of the limestone outcrop has been strongly influenced by adjacent non-carbonate terrain. This is especially true of the Tawe, Neath and Taff catchments, whose headwaters drain the extensive Old Red Sandstone dip-slope of the Black Mountain–Fforest Fawr–Brecon Beacons escarpment before crossing the limestones. Thirdly, the overlying Millstone Grit provides the outstanding and most closely investigated example of 'interstratal karst' to be found anywhere in Britain (Thomas, 1974).

This chapter outlines the origins of the modern karst terrain and examines the factors that influence contemporary geomorphological processes, highlighting in particular those aspects that are relevant to an understanding of cave development and the hydrology and chemistry of groundwaters. It traces events from initial carbonate deposition in Carboniferous times, through the Mesozoic, Tertiary and Pleistocene, to the present Holocene epoch, considering: the pre-glacial evolution of the limestone outcrops as components of the wider Welsh landscape; the stratigraphic evidence for, and interpretation of, the phases of sub-aerial karstification in the pre-glacial period; the effects of Pleistocene glacial and interglacial episodes; and, the principal features of the modern landscape.

Pre-glacial evolution of the Welsh massif

Two features of the Welsh landscape are particularly striking and must be accommodated within any scheme of landform development. First, the ground rises from sea level in a series of steps, that truncates Precambrian, Palaeozoic and Mesozoic rocks with marked discordance. The surfaces fall into two main groups: the Coastal Plateaux, up to an altitude of 210 m; and the Dissected Plateaux, between 210 and 610 m (Fig. 3.2; Brown, 1960). The former are very distinctive, with prominent benches at 61, 122 and 183 m (the 200-, 400- and 600-foot platforms). These features are well displayed along

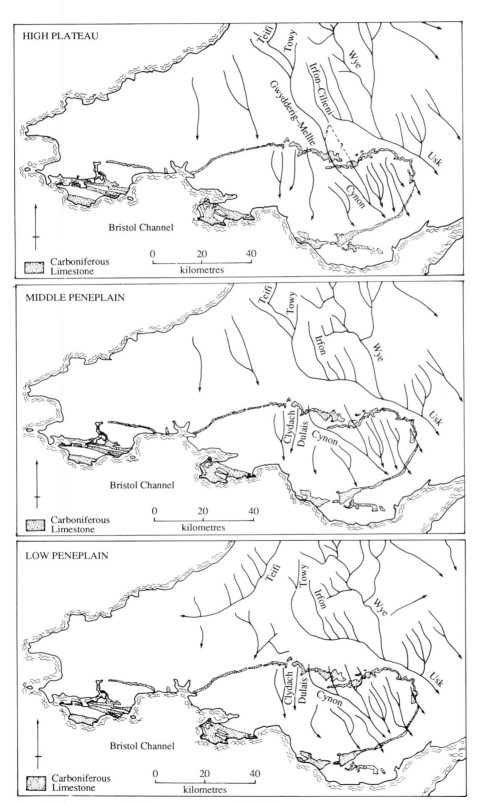

Fig. 3.3. Reconstructed drainage patterns at the High Plateau, Middle Peneplain and Low Peneplain stages (after Brown, 1960).

Table 3.1 *Summary of the pre-Quaternary evolution of the limestone outcrops of South Wales*

Era	Period	Epoch		Age (million yrs BP)	Events
Cenozoic	Tertiary	Neogene	Pliocene	2	Uplift and planation in early Pleistocene/Pliocene–development of Coastal Plateaux. Temperate climatic conditions, but cooling.
			Miocene	7	Uplift and planation–development of Dissected Plateaux and radial drainage network.
		Palaeogene	Oligocene	26	Deep weathering in Palaeogene/early Neogene under tropical–subtropical conditions (possible Tertiary Palaeokarsts).
			Eocene	38	Phases of tectonic uplift and erosion throughout Palaeogene.
			Palaeocene	54	
				65	
Mesozoic	Cretaceous				Cenomanian transgression–extent of chalk cover uncertain in Wales.
	Jurassic			135	Uplift of Welsh Massif (?)
	Triassic			195	Deposition of limestones in Rhaeto/Liassic marine transgression.
				225	Erosion under arid–semi-arid conditions (Mesozoic Palaeokarsts).
Upper Palaeozoic	Permian			280	Hercynian Orogeny–development of synclinal basin of South Wales Coalfield.
	Carboniferous				Surface denudation of limestone prior to deposition of Millstone Grits (Pre-Millstone Grit Palaeokarst). Deposition of limestones, with periodic exposure to subaerial denudation (Carboniferous Limestone Palaeokarsts).
	Devonian			345	
				395	
					Caledonian Orogeny–uplift of St George's Land.

the South Crop of the Carboniferous Limestone, particularly in south Pembrokeshire and on Gower. By comparison, the Dissected Plateaux are less clearly defined, but none the less three main surfaces may be recognized: the 'Low Peneplain' (215–335 m), the 'Middle Peneplain' (365–490 m) and the 'High Plateau' (520–610 m) (Fig. 3.3). The limestones along much of the North Crop lie within the altitudinal ranges of the Low and Middle Peneplains. For example, erosion surfaces at 300–330 m, 360–390 m and 440–470 m have been identified in the upper Tawe valley. The second outstanding feature is the drainage network radiating from Central Wales, which displays almost complete indifference to major folds or faults in crossing the coalfield syncline southwards to the Bristol Channel.

The pre-Tertiary period

The origins of the Welsh massif lie in an ancient landmass, known as St George's Land, that was uplifted during the Caledonian Orogeny (Table 3.1). Sub-aerial erosion, with marginal sedimentary accretion, ensued in the Devonian–Carboniferous period until, by late Westphalian times, only a subdued relief remained and sediment yields finally dwindled (Kelling, 1974). Intense folding and faulting occurred in the south during the Hercynian Orogeny. The complex synclinorium of the South Wales Coalfield developed at this time, but nowhere in Wales was uplift sufficient to produce high mountainous relief (Battiau-Queney, 1984). Following further denudation under arid or semi-arid conditions (Permian–early Triassic), much of the land surface was overlain by Triassic and Liassic deposits in the Rhaeto–Liassic marine transgression.

Somewhat paradoxically, there is less certainty about later

events in the Cretaceous and Tertiary. The traditional view has been that the Welsh massif, except possibly the highest summits, was overlain by chalk during the worldwide Upper Cretaceous (Cenomanian) transgression, before being finally elevated into an asymmetrical dome in the Palaeogene. This hypothesis gained much support, particularly as drainage superimposition from such a cover afforded a plausible explanation of the radial river system (Lake, 1900, 1934; Strahan, 1902; Jones, 1930). Hitherto, however, no Cretaceous rocks, flint pebbles or greensand chert have been found on the present land surface, in solution pockets in the Carboniferous Limestone, or offshore in the Mochras borehole in Tremadoc Bay (Wood & Woodland, 1968; Woodland, 1971). Indeed, Battiau-Queney (1984) has recently suggested that uplift occurred between the Upper Jurassic and Cenomanian transgression and that the Welsh massif remained entirely above sea level during the Upper Cretaceous. Serious doubts have therefore been cast upon the extent and geomorphological significance of the supposed Cretaceous cover, and increasingly the key to the pre-glacial evolution of Wales is thought to lie in an understanding of Tertiary tectonics and sea-level fluctuations.

Palaeogene tectonics and deep weathering

Tectonic instability in the Eocene, as evidenced by synclinal development west of the Llanbedr Fault in Tremadoc Bay, is thought to have led to the removal of much of the Mesozoic cover and exposure of the underlying Palaeozoic basement (George, 1974). Later in the Palaeogene the Welsh hinterland underwent further substantial uplift, resulting in the accumulation of very thick (550 m at Mochras), mid-Oligocene–early Miocene sediments (O'Sullivan & Herbert-Smith, 1979).

The Welsh landmass lay at about 40°N at the beginning of the Tertiary period and continued to drift northwards. Nevertheless, the Palaeogene and early Neogene climate appears to have been tropical or sub-tropical in character (Brown, 1979), and only in mid–late Neogene times was there a gradual reduction in global temperatures. Deep, intensively weathered regoliths, analogous to those of modern tropical landscapes, occur sporadically in Wales and its borderlands. Those on the Carboniferous Limestone outcrop of North Wales, between Llandudno and Mold, are especially significant since they afford some basis for dating (Walsh & Brown, 1971). Here, stratified sediments, containing gibbsite, kaolinite and interbedded lignites, have collapsed into the limestone as a result of solutional subsidence. Although the deposits have yielded no datable material, they are so similar to the Mio–Pliocene 'Pocket Deposits' of the Brassington Formation in Derbyshire (Walsh et al., 1972; Ford, 1977, 1984) in their altitude, composition and mode of occurrence that they are thought to be contemporaneous features. The sediments in North Wales are thus regarded as products of weathering under tropical conditions in the Palaeogene, which were reworked and deposited in a fluvio-lacustrine environment in Mio–Pliocene times, before ultimately being preserved as solution subsidence outliers. Elsewhere, weathered profiles have been found in situ. Along the North

Crop, for example, the Millstone Grit in one quarry section near Pwll Byfre has been weathered to a goethite-rich, ferruginous aggregate in a kaolinite matrix, with much loss of silica (Battiau-Queney, 1984). Near by, on Black Mountain, a regolith comprising almost 30% clay-sized particles has developed from orthoquartzites containing at least 90% quartz. The intensity of weathering displayed in these and similar in situ profiles is thought to be indicative of tropical conditions, and a Palaeogene–early Neogene age is generally assumed.

Neogene tectonics and planation

The Welsh massif was uplifted in early Neogene times, with a downthrow of at least 1350 m along the Llanbedr Fault (Blundell, Davey & Graves, 1971). Deep sediments in St George's Channel and the southern Irish Sea bear witness to the erosion that accompanied uplift. Conventionally, the Plateau surfaces are interpreted as erosional features and their altitudinal concordance is attributed to intermittent uplift. The Coastal Plateaux are widely regarded as wave-cut platforms, probably of late Pliocene–Pleistocene age (George, 1974). By comparison, there is less consensus about the higher Dissected Plateaux. Brown (1960), for example, championed a sub-aerial rather than marine origin in order to account for the local height variations of individual surfaces, the poorly defined junctions between different surfaces, and the absence of beach deposits. According to Brown's synthesis, the land above the 183 m platform has remained above the sea since the end of the Mesozoic era, and the surfaces are erosional forms, possibly analogous to the etchplains of present-day tropical terrains (Walsh & Brown, 1971), that developed during stillstands in the falling sea levels of the Neogene period. A marine hypothesis, first advanced by Ramsay (1846), has also found much support, with discordant notches at the landward edge of several of the Plateaux remnants being interpreted as degraded coastal cliffs. According to George (1974) the entire sequence of surfaces in Wales comprises marine benches, cut during pulsed uplift in Neogene–Pleistocene times. The major Miocene–early Pliocene transgression implicit in the model is partly attributable to global eustatic sea-level fluctuations resulting from sea-floor spreading (Flemming & Roberts, 1973).

In order to account for present-day altitudinal concordance, the corollary of pulsed uplift and erosion (sub-aerial or marine) is that the whole Welsh landmass was uplifted without major deformation. This is a reasonable assumption if the emergence of the massif was largely the result of eustatic sea-level changes, but seems less tenable if regional uplift was the primary cause. Also, the occurrence at low altitudes of in situ deeply weathered profiles cannot be readily accommodated within such a model, particularly if the marine hypothesis is adopted. Thus, if such profiles are of Palaeogene age, then they must have survived a period of submergence and subsequent marine planation, which seems unlikely. On the other hand, they may be products of post-emergence weathering and, hence, younger than has previously been thought. In order to overcome some of these difficulties, Battiau-Queney (1984) has postulated that the present landscape developed in Neogene times by block-faulting and differential uplift of a subdued terrain that had been continuously exposed to weathering under tropical conditions since pre-Cenomanian uplift. The detailed tectonic pattern is uncertain, but is thought to reflect the Caledonian structural lineations of the Lower Palaeozoic basement. In South Wales, for example, she interprets the Black Mountain–Fforest Fawr–Brecon Beacons and Pennant Sandstone escarpments as block-edge forms. However, the absence of major faults and paucity of evidence for large flexures in these areas casts serious doubts on this block-faulting hypothesis, and it seems likely that variations on the erosional model will continue to gain widest support.

Neogene drainage in South Wales

As outlined above, there is little evidence to support the traditional idea of drainage superimposition from a former chalk cover, and the radial river network in Wales is now believed to be linked with the evolution of the Plateaux surfaces in Neogene–early Pleistocene times. Thus, according to the more widely favoured marine planation hypothesis, the drainage system extended seawards and was repeatedly rejuvenated during phases of pulsed emergence (George, 1974). Evidence of rejuvenation is seen in the hillside benches and knick-points of many valleys, and indeed the long profiles of the Twrch, Taff Fechan, Wye and several other rivers in South Wales are cut to base levels of less than 300 m within a few kilometres of their source. The falling water tables that accompanied valley-trenching must have exerted an important control over cave development. Inevitably, most of the limestones in which speleogenesis is likely to have occurred have now been eroded as the escarpments of the North Crop have retreated southwards. In Ogof Ffynnon Ddu, however, shallow phreatic tubes and extensive vadose passages appear to be graded to surfaces at about 400 and 300 m, respectively, and by implication are of Neogene–early Pleistocene age. Battiau-Queney (1984) also speculated that the Agen Allwedd and Craig a Ffynnon Cave systems, located 200 m or more above the Usk River, beneath Mynydd Llangynidr and Llangattwg, may have developed in Neogene times, prior to uplift and incision of the Usk valley. The numerous wind-gaps and misfit streams indicate a complex evolution involving frequent river capture. Figure 3.3 shows reconstructions of the drainage patterns of the High Plateau and Middle and Low Peneplain stages of landform development. Interestingly, the degree of discordance diminishes from the oldest, High Plateau stage to the present day as the initial radial network becomes increasingly constrained by geological factors.

At the High Plateau stage the direction of flow across South Wales was predominantly southeastwards or SSE. The upper Teifi and Towy rivers, for example, formed the headwaters of a formerly extensive Cynon valley system, the connection being via the Llywel wind-gap at the head of the Gwydderig valley, and either the upper Tawe or Llia Mellte valleys. Clearly, therefore, much larger volumes of allogenic water were formerly transmitted through parts of the North Crop of the Carboniferous Limestone, and chemical and mechanical erosion along streambeds are likely to have been greater than at present. Since most other drainage from mid-Wales, including

the Wye, fed into the Usk, most of the remaining valleys on the dip-slope of the modern Black Mountain–Brecon Beacons escarpment, such as the Twrch, Giedd, Taff Fawr and Taff Fechan, are purely local in origin. Indeed, many of the cols along the top of the escarpment, including Storey Arms at the head of the Taff Fawr, that Brown tentatively linked with the Irfon–Cilieni system of mid-Wales, are now simply regarded as products of scarp retreat (Bowen, 1977).

Subsequent adaptation to geological structures has led to a more reticulate pattern. Thus, by the Middle Peneplain stage the Usk had cut back westwards along the strike to capture the Teifi and Towy headwaters from Cynon, only to lose them later as the lower Teifi and Towy systems extended back northeastwards along their major anticlinal axes. While the Teifi, like the Wye, was diverted by Low Peneplain times, diversion of the Towy probably occurred at the time when the 183 m coastal platform was planed (George, 1974). The break-up of the Clydach, Dulais and Cynon systems by the headward extension of the lower Tawe and Neath along the Swansea Valley and Vale of Neath Disturbances took place even later, and is thought to have been completed after uplift (early Pleistocene?) of the 61 m coastal platform (Jones, 1939; North, 1962). Consequent rejuvenation in the headwaters of the Tawe is thought to have been responsible for a third phase in cave development in Ogof Ffynnon Ddu at 250 m O.D. (O'Reilly, 1974).

Pre-glacial palaeokarsts in South Wales

The term 'palaeokarst' refers to karst features and related soils and deposits that developed either at an early stage in the evolution of the present land surface when environmental conditions were different from today ('relict palaeokarst'), or on a former land surface preserved beneath younger rock formations or deposits ('buried palaeokarst'). Palaeokarsts are important in three respects. First, they provide either an indication of the minimum age of the present landscape or, in the case of buried features within a sedimentary sequence, a stratigraphic record of previous periods of sub-aerial exposure. Secondly, karst surfaces are valuable in palaeoenvironmental reconstruction since distinctive suites of micro- and macro-forms develop under different climatic conditions (Sweeting, 1972; Jennings, 1985). It should be noted, however, that karstification is influenced by a diversity of non-climatic factors, especially rock structure and porosity (Brook & Ford, 1978; Brook, 1981). Caution is needed, therefore, when making palaeoclimatic inferences from modern analogues, particularly as quarry exposures often provide only fragmentary cross-sections through surfaces that may in any case have suffered post-burial modification. Finally, palaeokarsts may affect present-day processes and forms. Four main groups of palaeokarst may be identified in South Wales (Fig. 3.4).

Palaeokarst within the Carboniferous Limestone

In common with limestones of similar age elsewhere in Britain (Ramsbottom, 1973), stratigraphic unconformities, resulting from periodic emergence during marine regressions,

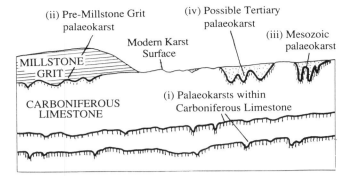

Fig. 3.4. Schematic diagram of palaeokarsts in South Wales, showing ancient karst surfaces developed (i) at intervals during Dinantian sedimentation; (ii) at the end of Dinantian sedimentation before transgression by Millstone Grit deltas; (iii) by solutional weathering before transgression and infilling with late Triassic to early Jurassic sediments; and (iv) by weathering processes in pre-glacial (Tertiary) times.

occur widely throughout the Carboniferous Limestones of South Wales (Riding & Wright, 1981). They are most frequent along the northern, landward edge of the carbonate shelf, and particularly well-developed palaeokarsts and palaeosols are found in scarps and quarries at the eastern end of the North Crop between Penderyn and Abergavenny. Detailed locations and descriptions of key sites are given in Wright (1981). The top of the Gilwern Oolite, for example, displays clear evidence of sub-aerial exposure under humid tropical conditions (Wright, 1982a, 1986). Locally, the surface is riddled by clay-filled pipes, some up to 40 cm wide and more than a metre deep, and bears a close resemblance to the present-day *Kavornossen karren* of certain humid tropical karsts. Elsewhere, weathering has reduced the uppermost few metres of bedrock to a rubbly calcareous clay. Sinuous horizontal fissures within the more massive oolite immediately below the weathered zone are thought to be products of speleogenesis close to the level of the former water table (Raven, 1983; Wright, 1986). Other unconformities indicate that a drier climate also prevailed at times during the Dinantian. Thus, in the floodplain members of the Llanelly Formation overlying the Gilwern Oolite, Wright (1980, 1982b) has identified three calcrete-bearing palaeosol units and a thin xero-rendzina which, by analogy with modern soil distributions, are interpreted as having developed under semi-arid or arid conditions. Detailed fabric studies have shown that the rendzina contains faecal pellets of small arthropods and insect larvae and is locally overlain by a cyanobacterial mat. Further evidence of soil biota is provided by needle-fibre calcite cements and alveolar textures in certain calcretes, which are thought to be associated with fungi (Wright, 1984).

Besides providing valuable insight into the climatic fluctuations and palaeoecology of the Dinantian period, the palaeokarsts may influence modern geomorphic processes and forms. At present their precise role remains to be established, but theoretical considerations suggest several possible effects which merit future research. It is well known, for example, that exposure to meteoric groundwaters is important in lithification and mineral stabilization (Longman, 1980). During diagenesis

cementation increases the density and hardness of carbonate deposits, with a corresponding reduction in porosity and permeability. Vadose cement fabrics have been identified, for instance, within the weathered zone of the Gilwern Oolite (Raven, 1983). In contrast, bedrock permeability is markedly enhanced by the solutional enlargement of joints and bedding planes. Indeed, Ford (1984) suggested that many present-day cave passages in South Wales may owe their origins to intra-Dinantian karstification. The presence of clay-rich palaeokarst and palaeosol beds may be even more significant than limestone diagenesis, for four reasons. First, clay may impede groundwater flow, thereby favouring cave development in overlying beds. Certain major phreatic tubes in the Peak District (Ford, 1984) and Yorkshire Dales (Waltham, 1974) may have been initiated in this manner. Secondly, the characteristically olive-green clays are rich in bases, particularly potassium (Wright, 1982b), and this may affect the solute content of groundwaters. Thirdly, being generally weaker than the adjacent limestones, they cannot support such steep slopes. Differences in slope angle are readily seen in scarps and quarry faces. Finally, clay-rich beds tend to produce deeper soils than the purer limestones, thereby favouring higher rates of chemical weathering.

Pre-Millstone Grit palaeokarst

In the late Dinantian a marine regression advanced westwards across South Wales (Owen, 1974). The limestones in the northeast were eroded considerably before being overlain unconformably by the Millstone Grit (Barclay & Jones, 1978). Compared with the intra-Dinantian palaeokarsts which have been buried and, hence, preserved beneath carbonate-rich strata, those at the very top of the Carboniferous Limestone sequence have been much more vulnerable to later solutional modification by acidic groundwaters percolating through the quartzitic sandstones (i.e. interstratal karstification). Wright (1982a) has suggested a number of criteria that may be used to distinguish sub-aerially formed palaeokarsts from ancient or modern interstratal features. In reality, however, the Carboniferous Limestone–Millstone Grit contact is so poorly exposed that little unequivocal evidence remains of a pre-Namurian karst surface. Indeed, Owen & Jones (1961, 1966) believed that all vestiges of the pre-Millstone Grit surface have been obliterated. Thomas (1973, 1974), on the other hand, argued that the very abruptness of the lithological break within the plane of unconformity in certain quarry sections might indicate the presence of a karst surface prior to deposition of the Millstone Grit. Irrespective, however, of its present state of preservation, it seems likely that surface irregularities in the pre-Namurian palaeokarst would have been important in affecting the distribution of subsequent interstratal solution and cave development (Thomas, 1974).

Mesozoic palaeokarst

By early Triassic times erosion under arid to semi-arid conditions had reduced the Carboniferous Limestones of the South Crop to a series of low hills, flanked by angular screes of insolation-weathering debris. The hills were deeply fissured at the surface and contained sizable underground caverns, some up to 15 m wide (Thomas, 1952; Robinson, 1957). With the advance of the Rhaeto–Liassic sea, the hills were progressively drowned and buried by sediments. While much of the former Mesozoic cover has now been removed, stratified sediments have been preserved in fissures and caves of the limestone outcrops. As in the Mendips, some of the sediments are rich in animal bones and these have provided much insight into the Triassic faunas of Britain (Robinson, 1957, 1971).

Possible Tertiary palaeokarsts

Hitherto, no Mesozoic deposits have been found within the Carboniferous Limestones of the North Crop. This implies that at the time a substantial cover of Upper Carboniferous rocks still overlay the limestones (Wright, 1986) and that there was little or no sub-Triassic karstification. More recent palaeokarsts are exposed, however, in the Vaynor and Cwar yr Ystrad quarries, north of Merthyr Tydfil. In both cases the palaeokarsts occur in beds that nearby are overlain by non-carbonate rocks: the Basal Grit in the former and Garn Caws Sandstone, of the Gilwern Clay Member, in the latter. The karst surface comprises rugged limestone pinnacles, closely resembling many humid tropical karst terrains (Battiau-Queney, 1973, 1985). The intervening pipes and pockets are up to 20 m wide and 10 m deep and their walls display a range of vertical solution channels, which Battiau-Queney (1985) interpreted as being sub-aerial in origin. Alternatively they may be the product of localized groundwater flow within the former grit or sandstone cover, and certainly the possibility of an interstratal origin merits further investigation. Given the high degree of solutional erosion evident in the palaeokarst surface, it seems likely that extensive speleogenesis would also have taken place. Indeed, Battiau-Queney (1985) postulates that the extensive cave systems beneath Mynydd Llangynidr and Llangattwg to the east may be contemporaneous forms.

The pipes and pockets are filled with an unsorted matrix of highly weathered, clay–boulder-sized material. Detailed mineralogical analysis has revealed that this was derived from ferralitic or ferruginous soils, analogous to modern humid tropical soils, which developed on the adjacent grits and sandstones. There is also the suggestion in one exposure of post-fill laterization of the surface horizons, which is indicative of a seasonally humid tropical regime. In contrast to the bone deposits and highly distinctive, red Triassic marls of the Carboniferous Limestone 'islands' of the South Crop, the fill cannot be dated accurately. However, on the grounds that tropical conditions still prevailed after the pockets were filled, Battiau-Queney (1985) speculated that the burial of the karst surface occurred in the Lower Neogene.

Pleistocene evolution

Global temperatures fell gradually in late Neogene times as the warm Tertiary environment gave way to generally cooler conditions in the Quaternary period, which began around 2 million years ago and extends up to the present day. The most striking feature of the Quaternary has been the high magnitude and frequency of climatic fluctuations during what, in geological terms, has been a very short time-span (Table 3.2). Thus, in high- and mid-latitude regions major cold phases ('glacials') have alternated with warm episodes ('interglacials'), some of which were appreciably warmer than today. Conventionally, the period up to the end of the last (Devensian) glacial at about 10 000 years BP is known as the Pleistocene epoch, and the present post-glacial warm phase as the Holocene. Climatic instability in the Quaternary has not only caused fundamental changes in the denudation system, from predominantly glacial or periglacial to one dominated by chemical weathering and fluvial processes, but has also been accompanied by marked eustatic sea-level variations as world ice volumes have waxed and waned. While much insight into the more recent glacial and interglacial episodes can be gained from an analysis of existing landforms and sediment stratigraphies, little terrestrial evidence has survived from early or mid-Pleistocene times. However, oxygen isotope studies of undisturbed ocean sediments have revealed 9 or 10 glacial/interglacial cycles over the last 700 000 years (Shackleton & Opdyke, 1973), and it is thought that there may have been as many as 21 cycles in the full Quaternary period (van Donk, 1976). Shorter or less pronounced cold ('stadials') or warm episodes ('interstadials') are superimposed upon the major oscillations. Three interstadials have been identified, for example, within the Devensian Glacial (Coope, 1975; Lowe & Walker, 1984). Quaternary palaeoclimates are therefore immensely complex and detailed knowledge of the resultant landform development sequence in areas such as South Wales, that have suffered multiple glaciations, is necessarily confined to the late Pleistocene and Holocene.

Pleistocene ice limits

There is evidence of at least two major glaciations in Wales. The earlier of these, which covered the whole of Wales, pre-dates the last (Ipswichian) interglacial and may include either the Anglian or Wolstonian glaciations, although its exact age is unknown (Catt, 1981). Detailed application of absolute dating techniques may augment our knowledge in due course. From the glacial erratic and heavy mineral content of the 'Older Drifts' most of the ice appears to have been derived from source areas in central and southern Wales (Fig. 3.5). However, the ice in Pembrokeshire, western Gower and the Vale of Glamorgan formed part of a very extensive sheet which encroached inland from the southern Irish Sea. The later glaciation was of Devensian age. The onset of the Devensian is difficult to define since temperatures fluctuated quite markedly between 150 000 years BP and the first prolonged cold period that began about 80 000 years BP. What is clear is that the Devensian was, for the most part, characterized by cold continental conditions and, although there is circumstantial evidence for glacier development in highland Britain during the early Devensian (Lowe & Walker, 1984), the most extensive glaciation did not occur until late Devensian times, reaching a maximum about 18 000 years ago. The limit of this last glaciation in South Wales has been the subject of much debate (Bowen, 1981). The latest reconstruction is shown in Figure 3.5. This was originally based upon stratigraphic analysis of glacial, periglacial head and interglacial raised beach deposits, but has subsequently been substantiated by amino-acid dating

Table 3.2 *Summary of environmental change during the Quaternary period in South Wales*

Epoch	Stage	Substage	Age (thousand yrs BP)	Climate–geomorphological processes
Holocene				Present warm 'interglacial' conditions, with active chemical weathering, speleogenesis and speleothem growth. Blanket peat formation during cooler–wetter phase (7600–5000 BP).
Pleistocene			— 10 —	
		Loch Lomond Stadial		Cold periglacial conditions–active slope erosion and sedimentation in caves.
			— 10.8 —	
		Lateglacial Interstadial		Warm.
			— 14 —	
		Late Devensian Glaciation		Cold oceanic climate–glacial maximum *c.* 18 000 BP. Cave sedimentation and dry valley formation during onset and waning of ice.
			— 30 —	
				Cold continental conditions–periglacial processes: frost weathering and solifluction.
			— *c.* 38 —	
		Upton Warren Interstadial		Warm–chemical weathering and speleothem growth.
			— *c.* 42 —	
				Cold continental conditions–periglacial processes: frost weathering and solifluction.
			— 60 —	
		Chelford Interstadial		Warm–chemical weathering and speleothem growth.
			— 65 —	
		Devensian glacial		Cold continental conditions–periglacial processes: frost weathering and solifluction. Possible Early Devensian glaciation. Beginning of Devensian is uncertain.
			— *c.* 100 —	
	Ipswichian Interglacial			Generally warm, up to 3° C higher than present (cold phase from 110 000–105 000 BP)–chemical weathering, speleogenesis and speleothem growth.
			— *c.* 125 —	
	Wolstonian Glacial		— ? —	
	Hoxnian Interglacial			Warm stage of 20 000–25 000 years duration.
			— ? —	
				7 or 8 Glacial–Interglacial cycles.
			— 700 —	
				10 or 11 Glacial–Interglacial cycles
			= 2000 =	

of marine molluscs incorporated within the glacial tills (Andrews, Bowen & Kidson, 1979; Campbell, Andrews & Shakesby, 1982; Davies, 1983). Thus, while doubts remain about the precise ice limit in certain localities, notably in western Gower (Bowen, 1984), the broad picture is well established and it is clear that the whole of the North Crop, the eastern part of the South Crop and northeastern Gower were covered by late Devensian ice. The ice was of Welsh provenance and three drift provinces have been identified: the Central Wales Drift, the Brecknockshire Drift from the Black Mountain–Brecon Beacons dip-slope and, more locally, the Glamorgan Drift from Craig y Llyn. At the glacial maximum an ice sheet some 300 m thick is thought to have blanketed South Wales (Bowen, 1980). However, deglaciation was rapid and much of the area is thought to have been ice-free by 14 000–15 000 years BP.

Glacial erosion and deposition

The valleys of the South Wales Coalfield bear the hallmarks of glacial erosion, being typically U-shaped in cross-section, with hanging tributary valleys. The bedrock floor in the lower sections of the major valleys, now buried beneath

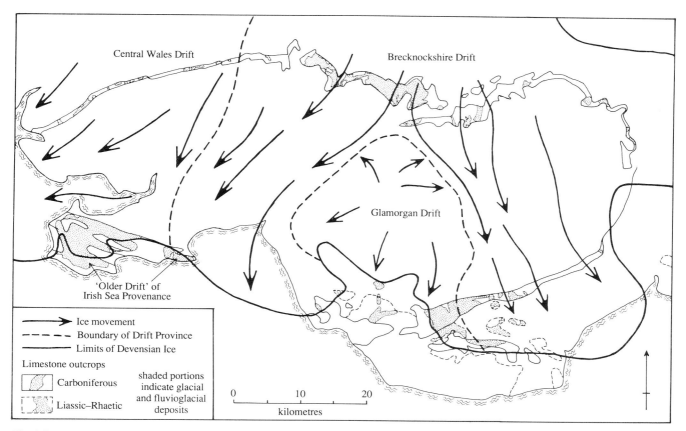

Fig. 3.5. Devensian ice flow directions, drift limits and provenances, and the extent of drift cover on limestone outcrops (after Bowen, 1970, 1981, 1984).

thick post-glacial sediments, has been scooped into deep basins. Thus, one borehole at Resolven (31.5 m O.D.) in the Vale of Neath failed to reach bedrock at −32.5 m O.D., whereas at Clyne, less than 3 km downstream, the rockhead is almost at the surface (Anderson, 1974; Anderson & Owen, 1979), the sea level at the glacial maximum being $c.$ −130 m O.D. Upstream, the valley floors are more or less buried by glacial deposits, although bedrock is frequently exposed along riverbeds in the narrow Carboniferous Limestone reaches of the North Crop. In the absence of an unequivocal, immediately pre-glacial datum level, rates of glacial entrenchment cannot be established with certainty. The mean maximum rate of valley downcutting in the Yorkshire Dales over one or more glacial/interglacial cycles has been estimated from ^{230}Th/^{234}U speleothem dates to be 5–<20 cm/1000 y (Gascoyne, Ford & Schwarcz, 1983; Gascoyne & Ford, 1984), giving a figure of 6–<24 m for the entire Devensian–Holocene period. Interestingly, these figures are of the same order of magnitude as contemporary weathering rates in soil-covered limestone terrain, implying that glacial scour has not been as effective as has often been supposed. Since erosion (glacial and fluvial) is governed by a diversity of local factors, interregional extrapolations must be regarded with caution. If, however, similar entrenchment rates are assumed in South Wales, then the third (250 m) phase of cave development in Ogof Ffynnon Ddu (O'Reilly, 1974), some 70 m above the present valley floor, has a maximum age of about 1.2 million years.

Much of South Wales is covered by glacial and fluvio-glacial deposits (Fig. 3.5). The occurrence of glacial till is especially significant in the karst areas since it mantles terrain that would otherwise have only a thin and patchy soil cover, thereby affecting rock weathering, soil development and limestone hydrology (see below). Agen Allwedd, Ogof Ffynnon Ddu, Ogof y Ci and many other cave systems along the North Crop contain stream-lain sediments, some >5 m thick. Detailed examination of the surface textures of quartz grains from these deposits by scanning electron microscopy has revealed features, such as conchoidal fractures, arc-shaped steps and parallel striations, that result from grinding and sliding within ice bodies (Bull, 1976, 1980). This confirms their fluvio-glacial origin, and they are thought to date from the onset or retreat of the late Devensian glaciers. A later phase of sedimentation in Agen Allwedd resulted in the formation of a finely laminated deposit, known as the 'cap mud', which is interpreted as having developed under cold conditions in the Loch Lomond Stadial ($c.$ 10 800 years BP). Independent evidence of accelerated slope activity at this time is provided by minerogenic sediments in the sequence of late glacial deposits in the Traeth Mawr dead-ice hollow, north of the Brecon Beacons scarp (Walker, 1980, 1982). Elsewhere in South Wales, as in Minchin Hole Cave on Gower which lay beyond the limits of the last glaciation, thermoclastic scree accumulated as a result of cave roof breakdown under periglacial conditions in the Devensian (Sutcliffe & Bowen, 1973; Sutcliffe & Currant,

1984). Generally, therefore, glacial episodes are times of cave infilling and degradation, with many passages, in fact, becoming completely choked by fluvio-glacial and periglacial deposits. In Agen Allwedd, for instance, eight of the entrances that were active during the main phase of sediment in-wash were subsequently blocked (Bull, 1977a). Indeed, one entrance (to the Summertime Series) now lies buried beneath a large periglacial scree.

Dry valley systems

Dry valleys occur quite widely on the Carboniferous Limestone outcrops. Some, such as the Bishopston and Llethrid–Green Cwm systems on Gower and the Mellte near Ystradfellte, form part of larger drainage basins with surface flow in their Millstone Grit or Old Red Sandstone headwaters (Ede, 1975). Others, such as the Pwll Dwfn valley above Dan yr Ogof (Fig. 3.6) are confined within limestone terrain (Coase & Judson, 1977). As in the last example, these valleys are often complex in form, with wide, open sections giving way to steep-sided gorges. They are thought to have developed in glacial times, notably during periods of ice retreat, when infiltration into the limestone was impeded by frozen sub-soils, glacial drift and solifluction deposits, and when the copious volumes of meltwater would have been sufficient to sustain surface flow.

Fig. 3.6. View down the Pwll Dwfn dry valley above Dan yr Ogof; in the distance across the Tawe Valley lies Penwyllt and the slopes above Ogof Ffynnon Ddu (photo by J. Rowland).

Interglacial weathering and speleogenesis

Conditions during glacial episodes were generally unfavourable for chemical weathering and cave development, for two reasons. First, underground flow was reduced. This was partly because of permafrost, but also reflected the relatively dry continental climate that frequently prevailed as, for example, in early and middle Devensian times. Secondly, plant growth and microbial activity virtually cease at very low temperatures and, consequently, the regoliths, glacial deposits and skeletal soils contained little biogenic carbon dioxide. The small volumes of recharge water therefore had a very limited potential to dissolve limestone, as is confirmed by the low solute content of spring waters and glacial meltwaters in modern arctic and alpine karst terrain in North America (Ford, 1971; Cogley, 1972). The warm interglacial environments, on the other hand, supported a complete vegetation cover and a fully active soil biota. Groundwater recharge was therefore chemically aggressive and, given the increased chemical reaction rates at higher temperatures, much of this weathering potential was quickly expended upon contact with limestone. Insect assemblages from the last (Ipswichian) interglacial deposits in southern England, for example, indicate temperatures 3 °C warmer than today (Coope, 1974), implying weathering rates were periodically higher than those of the present day. Furthermore, since some interglacials were of quite long duration, the penultimate (Hoxnian) interglacial being 20000–25000 years (Turner, 1975), extreme caution must be exercised when using modern post-glacial soils and regoliths as interglacial analogues. Indeed, the possibility that some of the deeply weathered profiles reported by Battiau-Queney (1984) may be interglacial, rather than Tertiary, in origin appears to merit further investigation.

The high temperature and weathering potential of karst recharge waters during interglacials also favours the formation of secondary carbonate deposits in caves, and this is highlighted by the age distribution of speleothems in the Mendips (Atkinson et al., 1978) and Yorkshire Dales (Gascoyne et at., 1983; Gascoyne & Ford, 1984). Thus, in the latter area, for which the greatest number of dates is available, there have been three main periods of speleothem growth over the last 350 000 y: Holocene; 90 000–135 000 y BP (Ipswichian Interglacial–early Devensian); and 180 000–320 000 y BP (including Hoxnian Interglacial). Intermittent growth occurred from 30 000 to 80 000 y BP, and this may be linked with climatic amelioration during Devensian interstadials. The last and penultimate (Wolstonian) glaciations stand out as clear hiatuses in the record at 15 000–30 000 and 140 000–165 000 y BP, respectively. Unfortunately, there are few speleothem dates from South Wales, but those available generally conform with the detailed pattern that is emerging from other British limestone areas. For example, one large block of flowstone in Minchin Hole Cave on Gower, which is critical for dating the Patella Beach, has an Ipswichian Interglacial age of 107 000–127 000 y BP (Sutcliffe & Currant, 1984).

The present landscape

Soil and peat cover

The soils of the Carboniferous and Lower Lias–Rhaetic Limestone outcrops are extremely varied (Table 3.3). Some are derived from the underlying bedrock. These are typically <30 cm deep and simply comprise a dark, loamy, humose A horizon, which is occasionally stony towards the base. Locally, as on certain Gower cliff tops, the fine earth fraction (<2 mm diameter) is calcareous and the soil is a neutral to slightly alkaline rendzina (Marian Series). More usually, however, these lithomorphic soils are non-calcareous, slightly acidic rankers. Brown rankers (Crwbin Series) occur widely towards the western end of the North Crop, particularly on the limestone ridges between Crwbin and Meinciau (Clayden & Evans, 1974), whereas humic rankers (Wetton Series) are more

Table 3.3. *Soils of the limestone outcrops*

Table 3.3. *Continued*

Soils derived directly from underlying limestone (lithomorphic)
(1) Rendzina (Calcareous) – Marian Series (Crampton, 1972)
Location: Llansannor, Vale of Glamorgan (SS 987 784)
Slope: 18° Aspect: S Altitude: 50 m
Land use: Ash woodland
Horizons (cm)

0–10 A	Very dark brown (10 YR 2/2) humose loam with limestone fragments; crumb structure; friable; abundant roots; ants present; sharp boundary.
10+	Carboniferous Limestone

(2) Brown ranker (Non-calcareous) – Crwbin Series (Clayden & Evans, 1974)
Location: Greenhall, Four Roads (SN 447 102)
Slope: 15° Aspect: SE Altitude: 175 m
Land use: Ash coppice with some beech and oak; stones and boulders over 20% of surface
Horizons (cm)

L	Thin litter layer of mainly beech leaves
0–2 A	Very dark greyish-brown (10 YR3/2 but darker) humose silt loam; strong fine crumb; very friable; extremely abundant roots; earthworms present; narrow boundary.
2—22 AB	Brown – dark yellowish-brown (7.5–10 YR 4/4) silty clay loam; slightly stony to 9 cm, extremely stony below, tabular Carboniferous limestone; strong fine crumb; very friable; abundant fine roots; sharp boundary.
22+	Carboniferous Limestone

(3) Humic ranker – Wetton Series (Wright, 1981)
Location: Foel Fawr quarry, Llangadog (SN 734 190).
Slope: 1° Aspect: N Altitude: 500 m
Land use: Moorland with abundant sheep's fescue and frequent common bent and larger wild thyme (*Thymus drucei*).
Horizons (cm)

0–14 Ah	Very dark greyish-brown (10 YR 3/2) humose clay loam; stoneless; moist; strongly developed fine granular; low packing density; moderately porous; moderately weak soil strength; non-sticky, slightly plastic; many very fine fibrous roots; non-calcareous abrupt smooth boundary.
14–23 Cu and Ah	Abundant medium and large angular and sub-angular limestone stones with a very dark greyish-brown (10 YR 3/2) humose clay loam matrix; many very fine fibrous roots.
23+	Limestone bedrock

Drift-derived soils
(4) Brown earth – Lulsgate Series (Clayden & Evans, 1974)
Location: St Hilary Down, ESE of Cowbridge (ST 014 737)
Slope: 1° Aspect: SSE Altitude: 120 m
Land use: Ley pasture
Horizons (cm)

0–13 Ap	Brown – dark brown (7.5 YR 4/3) silt loam; friable; crumb structure; abundant roots, particularly 0–5 cm; a few limestone fragments; earthworms; merging boundary.
13–25 B1	Brown – dark brown (7.5 YR 4/4) silt loam; friable; fine sub-angular blocky structure; some roots, with organic matter in channels; earthworms; a few limestone fragments; merging boundary.

25–38 B2	Reddish-brown (5 YR 4/3) silt loam; friable; crumb structure; some roots, with organic matter in vacated channels; a few limestone fragments; sharp boundary.
38+	Limestone

(5) Stagnohumic gley – Wenallt enclosed phase (Wright, 1981)
Location: Pant y Dderwen, Llangadog* (SN 707 207)
Slope: 1° Aspect: SSW Altitude: 215 m
Land use: Rough grazing dominated by purple moor grass.
Horizons (cm)

0–31 Oh	Dark reddish-brown (5 YR 2/2) semi-fibrous peat; stoneless; moist; moderately developed fine sub-angular blocky; many fine fibrous roots; gradual smooth boundary.
31–63 Eg	Light grey – grey (5 YR 6/1) clay loam with many prominent fine strong brown (7.5 YR 5/8) mottles; few medium sub-rounded soft weathered sandstone fragments; very moist; weakly developed adherent coarse prismatic; medium packing density; moderately porous; moderately sticky, very plastic; common fine fibrous roots; clear smooth boundary.
63–80 BCg	Reddish-brown (2.5 YR 5/4) clay loam with common distinct medium light grey to grey (5 YR 6/1) and prominent strong brown (7.5 YR 5/8) mottles following root channels; common small sub-rounded soft weathered sandstone fragments; very moist; high packing density; slightly porous; moderately sticky, very plastic; clear smooth boundary.
80–100 Cu	Reddish-brown (2.5 YR 4/4) sandy silt loam; common medium sub-rounded sandstone fragments; moist; massive; high packing density; slightly porous; fine macropores; moderately strong soil strength; moderately sticky, very plastic.

(6) Ferric stagnopodzol – Lydcott bouldery phase (Wright, 1981)
Location: Truman, Llangadog* (SN 745 203)
Slope: 16° Aspect: NNW Altitude: 400 m
Land use: Moorland with abundant bilberry and frequent wavy hair-grass and mat-grass.
Horizons (cm)

0–4 Oh	Black (5 YR 2/1) amorphous peat; moist; moderately developed fine granular; many fine fibrous roots; sharp smooth boundary.
4–10 Ah	Very dark grey (5 YR 3/1) humose sandy loam; a few small sub-angular and tabular red sandstone fragments; moist; weakly developed adherent fine sub-angular blocky; low packing density; extremely porous; very weak soil strength; very weak ped strength; non-sticky, non-plastic; many fine fibrous roots; abrupt wavy boundary.
10–18 Eag	Pinkish-grey (7.5 YR 6/2) sandy loam; common mainly medium sub-angular and tabular red sandstone fragments; moist; massive; low packing density; extremely porous; very weak soil strength; non-sticky, non-plastic; common fine fibrous roots; abrupt wavy boundary.
18–21 Bw(g)	Reddish-brown (5 YR 4/4) clay loam; common distinct fine pinkish-grey (7.5 YR 6/2) and common fine yellowish-red (5 YR 4/6) mottles; common mainly medium subangular and tabular red sandstone fragments; moist; weakly developed adherent fine

Table 3.3. *Continued*

	angular blocky; low packing density; moderately porous; very weak soil strength; very weak ped strength; slightly sticky, moderately plastic; common fine fibrous roots; abrupt smooth boundary.
21–46 Bs	Yellowish-red (5 YR 4/8) clay loam; common, medium sub-angular and tabular red sandstone fragments; moist; moderately developed fine angular blocky; low packing density; moderately porous; very weak soil strength; very weak ped strength; slightly sticky, moderately plastic; common fine fibrous roots; clear smooth boundary.
46–80 Cu	Reddish brown (2.5 YR 4/4) clay loam; abundant very large sub-angular and tabular red sandstone fragments; very fine; few fine fibrous roots.

* 5 and 6 are not located on limestones but the soils are very similar to those on adjacent limestone outcrops.

common in the wetter, central part around Ystradfellte and north of Merthyr Tydfil (Rudeforth *et al.*, 1984). The majority of soils, however, are formed from drift deposits. Brown earths are dominant where the cover is thin (<60 cm). While being only rarely calcareous in the fine earth fraction, these soils are strongly affected by the underlying bedrock and are characteristically well drained and neutral in reaction, often with limestone rubble near their base. They include soils of Ston Easton, Lulsgate and Radyr Series on Gower and on both the Carboniferous and Mesozoic Limestones in the Vale of Glamorgan (Crampton, 1972). The deeper drifts mostly originate from adjacent lithologies and the bedrock influence is correspondingly reduced. Thus, acidic, surface-water gleys with peaty topsoils, have developed on the Old Red Sandstone and Namurian Shale drifts that mantle extensive tracts of the North and South Crops, respectively. On the southern flank of the Black Mountain, for example, stagnohumic gley soils (Wenallt association) are dominant, with ferric stagnopodzols (Lydcott association) on some of the more permeable drifts (Rudeforth *et al.*, 1984). Deep blanket peats are found in certain topogenous bogs. For example, Waen Fignen Felen in the Dan yr Ogof catchment contains over 4 m of peat, much of which accumulated between 7600 and 5000 y BP (Smith & Cloutman, 1984).

The weathering potential of soil water is greater than that of rainwater for two reasons. First, soil has a much higher carbon dioxide concentration (typically, 0.5–2.5% in Britain) than the atmosphere (0.034%). Secondly, organic acids are present in water draining humic topsoils and peats. These react directly with limestone, thereby imparting a high initial aggressiveness to recharge waters. Aggressiveness values in excess of 50 ppm $CaCO_3$ have been recorded, for example, in water issuing from Waen Fignen Felen (Ede, 1972). Furthermore, organic materials such as humic acid are readily oxidized in vadose stream water, releasing carbon dioxide. Oxidation studies along the Ogof Ffynnon Ddu streamway have demonstrated the importance of this mechanism in enhancing underground solution (Bray, 1972, 1977; Bray & O'Reilly, 1974).

The nature of the surface cover is important in two further respects. First, where the soil–drift is calcareous or contains limestone rubble, much of the solutional potential is expended before the water reaches the bedrock, and the resultant groundwater is therefore relatively ineffective in cave development. Thus, in laboratory simulation studies of percolation through different Gower soils, a 25 cm profile containing limestone fragments (Radyr Series), yielded water with over 300 ppm $CaCO_3$, whereas a limestone-free, peaty-gley soil (Hirwaun Series) produced about 10 ppm $CaCO_3$ (Groom & Ede, 1972). Secondly, the permeability of the cover is significant in affecting the spatial distribution of groundwater recharge. Inputs from beneath a highly permeable cover will generally be of a diffuse type, with chemical weathering being distributed fairly evenly near the top of the limestone. In contrast, an impermeable drift protects the underlying rock from solution (Thomas, 1970) and there is considerable lateral movement of water as surface run-off and throughflow. Under these circumstances, groundwater input is focused at discrete points where the cover is thinner or more permeable, and underground flow is predominantly of the conduit type, which favours cave development.

Limestone pavements

In contrast with limestone areas in other parts of Britain, such as northern England and western Ireland, pavements form a relatively minor element of the karst landscape of South Wales, occupying approximately 73 ha (Thomas, 1970). They are virtually confined to the Lower Holkerian and Asbian stages of the Carboniferous Limestone and are most extensive where the dip is lowest, as on the spur between the Mellte and

Fig. 3.7. Limestone pavement above Ogof Ffynnon Ddu (photo by C. Howes).

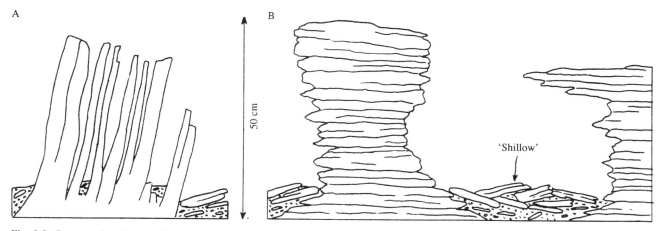

Fig. 3.8. Cross-sections through limestone pavements on the North Crop showing the effects of (*a*) sub-vertical, and (*b*) sub-horizontal fracture and bedding planes (after Thomas, 1970).

Hepste valleys. Although often slightly elevated above the surrounding drifts and occasionally bounded by scars, the pavements themselves are unimpressive and degraded in appearance. Well-developed solutional features, including runnels and hollows (Williams, 1963; Sweeting, 1974), are restricted to the most massive beds, and there is much flaggy rubble ('shillow') (Fig. 3.7). Clint blocks, for example, account on average for only 24.2% of the surface of 15 pavements between Penwyllt and Ystradfellte, compared with 36.0 and 39.8%, respectively, for shillow and turf. Because of close joint spacing the clints are small, averaging only 88 cm in length and 49 cm in width, and the intervening grykes are mostly <20 cm deep (Lewis, 1983). Since local variations in bedrock chemistry are quite small, the micromorphology of rock exposures is largely controlled by physical factors. In particular, minor fracture planes act as foci for solutional erosion and frost shattering (Fig. 3.8). These are either sub-parallel with the main bedding, reflecting the original (often highly irregular) current bedding, or sub-vertical, the frequency of both types commonly exceeding 30/m.

Limestone dolines

Extensive tracts of drift-covered limestone along the North Crop are pock-marked with closed depressions ('dolines'). Overall, they number as many as 80 000 (Thomas, 1974) and on low-gradient moorland slopes, such as those east of Penwyllt on the Tawe–upper Neath interfluve, occur in concentrations of up to 75/ha. About three-quarters of the area is dominated by small features (<10 m diameter), typically 2–5 m in diameter and 1–2 m deep (Fig. 3.9). Groups of larger dolines are mostly confined near the junction with the overlying Millstone Grit. The dolines are almost invariably circular in plan and funnel-like in cross-section, the depth:diameter ratio generally ranging from 1:3 to 1:6. The edges of the depressions are so clearly defined that a post-glacial origin is normally assumed. Thomas (1963) calculated the average volume of doline craters in two sampling areas to be 3212 m³/ha, which amounts to an average of 32.1 mm/1000 y of solutional lowering over the Holocene period (Table 3.4).

Some dolines may simply be products of solutional enlarge-

Fig. 3.9. Some of the many dolines in the limestone outcrop above Ogof Ffynnon Ddu (photo by C. Howes).

Table 3.4. *Doline volumes and subsidence rates in 4.5 ha blocks of drift-covered Carboniferous Limestone* (after Thomas, 1963)

Site	Grid reference of central point	Volume of doline craters (m³/ha)	Average rate of solutional lowering in sample block[a] (mm/1000 yrs)
Blaen Clydach	SN 758 185	3023	30.2
Twyn Walter	SN 825 173	3401	34.0

[a]The dolines are assumed to be of Holocene age.

ment at joint intersections, whereas others may be solely the result of collapse into underlying caves. Usually, however, both processes have been instrumental in doline development and such cases represent extreme members of what may be regarded as a solution–collapse continuum. In parts of the North Crop surface features are closely associated with underground passages (Fig. 3.10). There are, for example, more than 50 dolines in a 1000 × 200 m strip directly overlying Sections I and II of the Dan yr Ogof system. The largest of these dolines (The Crater), which is >100 m wide, lies directly

Fig. 3.10. The Waen Fignen Felen sink north of Dan yr Ogof; a wet weather feeder of water into the northern reaches of the cave system (photo by J. Rowland).

Table 3.5. *Comparison of weathering rates on bare limestone pavements in South Wales with those elsewhere in the British Isles*

Region	Weathering rate (mm/1000 yrs)	Source
South Wales:		
Coral colonies (*Lithostrotion martini*)	Negligible	Thomas (1970)
Breccias – pseudo-breccias	25–35	"
Coarse oolites–pisolites	50–60	"
Fine-grained oolites	65–75	"
Northwest Yorkshire	30–50	Sweeting (1966)
County Clare, Galway, Leitrim (Ireland)	Negligible–50	Williams (1966) Trudgill (1985)

Table 3.6. *Comparison of rates of solution in South Wales with other Carboniferous Limestone outcrops in the British Isles*

South Wales:		
Mellte catchment	16	Groom & Williams (1965)
Nedd Fechan catchment	18	Williams (1963)
Burry catchment	36	Chambers (1983)
Mendip Hills	81(23–102)[a]	Drew (1967), Newson (1970) Atkinson (1971)
Derbyshire	79(25–193)	Pitty (1968)
Northwest Yorkshire	83	Sweeting (1966)
County Clare, Ireland	55	Williams (1968)

[a] Figures in parentheses indicate range recorded over several catchments.

above the Long Crawl and the Platten Hall–Shower Aven boulder chokes (Coase & Judson, 1977). In other areas, the doline distribution bears little or no relationship with known underground systems, although this in itself does not preclude a collapse origin, since there are probably extensive networks of inaccessible passages within the Carboniferous Limestone. Irrespective of their origin, once dolines have formed, subsequent growth is self-enhanced, as surface run-off, through-flow and subcutaneous flow drain towards the existing sinks. This not only concentrates solutional activity beneath the base of depressions, but may also initiate collapse as the strength of the bedrock progressively diminishes. Although some of the hollows contain small ponds or peat deposits, the majority are free-draining and water is transmitted rapidly underground along well-defined routeways. Ede (1972), for instance, recorded a flow-through time of 6 d from the base of a depression above Dan yr Ogof to the resurgence, the vertical and horizontal distances of flow being 175 and 300 m, respectively. This compares with more than two weeks over corresponding distances of 60 and 245 m in the Mendip Hills (Drew, 1968).

Chemical denudation rates

Solution rates on limestone pavements in South Wales are similar to those reported elsewhere in Britain (Table 3.5). Thus, while evidence of glacial polishing is still preserved on certain coralliferous limestone outcrops, estimates of solutional lowering normally range from 25 to 75 mm/1000 y, depending on the nature of the bedrock (Thomas, 1970). Despite the high rainfall, catchment-wide denudation rates calculated from solute yields tend, in contrast, to be lower than on the other major Carboniferous Limestone outcrops (Table 3.6). Rates are particularly low along the North Crop, as little as 16 mm/1000 y in the Mellte Basin (Groom & Williams, 1965), and even in the Burry Catchment on Gower denudation amounts to only 35 mm/1000 y (Chambers, 1983). There are two reasons for this. First, the thick veneer of drift forms an effective seal over a significant proportion of the underlying

bedrock surface. Secondly, the rapid transmission of groundwater through the limestone limits the duration of water–rock contact. Often, therefore, the solutional potential is not fully expended. Some underground seepages in Ogof Ffynnon Ddu, for example, have a hardness of <50 ppm $CaCO_3$ (Ede, 1972). The predominance of conduit flow is largely a consequence of groundwater recharge being concentrated at discrete points. This is partly attributable to the drift cover and dolines, as discussed above, but in addition reflects the influence of adjacent lithologies. The latter effect is particularly important along the North Crop, for here the limestones receive not only surface run-off from the Old Red Sandstone dip-slope, but also seepage waters through the overlying Basal Grit. Indeed, much of the openness of the master joints in the present Carboniferous Limestone outcrop may be inherited from beneath the former Grit cover.

Interstratal karst

Where limestone is covered by another lithology, sub-surface karstification may induce deformation and collapse in the overlying strata. Landscapes affected in this way are termed 'interstratal karsts'. The Millstone Grit Series of the North Crop, particularly along the 60 km section between the western end of Black Mountain and Blorenge, provides the outstanding British example. Its unprecedented development in South Wales is attributable to two factors. First, in contrast to other parts of Britain, there is a very abrupt lithological break between the Carboniferous Limestone and Basal Grit. Secondly, the regional gradient of the broad dip-slope interfluves is only slightly less than the dip of strata and, consequently, the Grit remains quite thin over a considerable distance. Clearly, solutional erosion within the limestone is greatest along the northern margin of the interstratal karst tract where groundwater recharge occurs freely through the thin and broken cover, and the hydraulic gradient towards the floor of adjacent valleys maintains a substantial underground flow. The potential for groundwater flow and, hence, solutional activity diminishes down-dip as the hydraulic gradient falls. Figure 3.11 illustrates the nature and disposition of the principal landform elements along an idealized section through the interstratal karst.

Down-dip, where the cover is thickest, collapse into solution cavities in the Carboniferous Limestone is generally confined within the lower part of the Grit sequence, and there is little or no surface deformation. However, repeated collapses may over time cause sagging, without rupture, in the uppermost beds, leading to the formation of structural basins. These are typically broad, shallow features with depth:diameter ratios of between 1:40 and 1:60. Two such depressions to the west of Twrch Fechan on Black Mountain are some 200–300 m long and 100 m wide, and have developed through a 120–150 m cover (Thomas, 1974). Many of the larger basins, including that of Llyn y Garn Fawr on Mynydd Llangynidr (Fig. 3.12) contain permanent pools.

As the cover thins up the dip-slope of the Millstone Grit, collapse is more readily transmitted to the surface, resulting in the development of 'collapse dolines'. These are undoubtedly the most striking features of the interstratal karst and occur extensively where the Grit is <30 m thick (Fig. 3.13). They are significantly larger than their counterparts on the Carboniferous Limestone, with almost half the doline zone being dominated by depressions >30 m wide (Thomas, 1954). While some are secondary forms within structural basins or are aligned along the floor of minor dry valleys, most occur as isolated forms on the gently undulating dip-slope. The most spectacular assemblage occurs on Mynydd Llangynidr where over 600 dolines are concentrated within an area of no more than 10 km² (Fig. 3.12). One exceptionally large crater, located close to the down-dip limit of the field (Grit cover 30–35 m thick) is 55 m wide and 17 m deep. The dolines appear to be Holocene in age and their characteristically symmetrical cross-sections suggest that individual depressions are the products of a single collapse. Over large areas the amount of subsidence is equivalent to an overall surface lowering of 1 m (100 mm/1000 y) and, locally, as in one sampling block on Mynydd Llangynidr, the figure is almost 3 m (Table 3.7) (Thomas, 1954). Of 437 collapse dolines investigated by Thomas (1974), nearly three-quarters are funnel-like (depth:diameter ratio *c*. 1:3), their sides being typically 28–32° and veneered by loose Grit blocks and clay-rich head. The remainder are either well- or bowl-shaped, with depth:diameter ratios of *c*. 1:3 and 1:10, respectively. The former are distinguished by their steep, often rocky walls and irregular floors, whereas the latter are bounded by gentle slopes and have flattish floors.

Erosion of the underlying Carboniferous Limestone is at an advanced stage along the northern edge of the Millstone Grit and for the most part the Grit forms only a minor and discontinuous escarpment. Analogy with the present landscape suggests that interstratal karstification would have been initiated at a time when the Grit escarpment lay at least 1–2 km

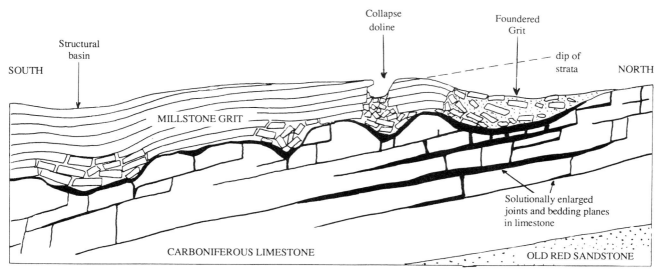

Fig. 3.11. Schematic diagram of the nature and disposition of the main landform elements in interstratal karst terrain where the limestone is covered by the Millstone Grit (after Thomas, 1974).

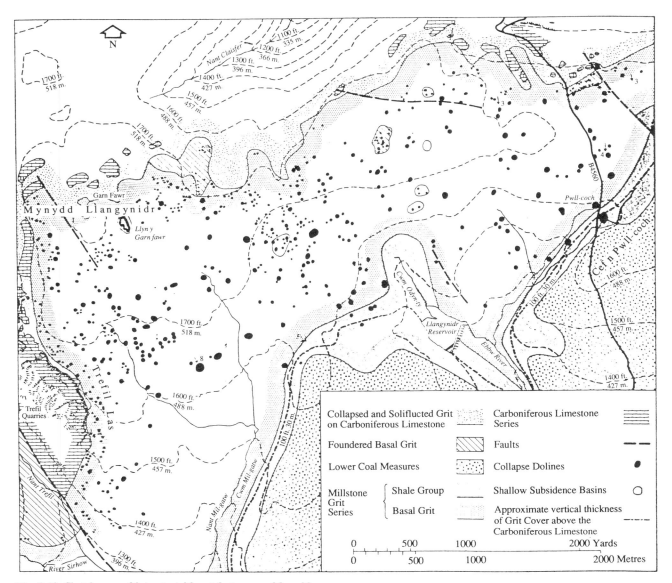

Fig. 3.12. Sketch map of interstratal karst features on Mynydd Llangynidr (reproduced from Thomas, 1974, Fig. 12, with permission).

Fig. 3.13. Collapse structures in the margin of the Millstone Grit above Dan yr Ogof; the collapse in the foreground lies directly above Gerrard Platten Hall in Dan yr Ogof II (photo by J. Rowland).

Table 3.7. *Collapse doline volumes and subsidence rates in 4.5–6.25 ha blocks in interstratal karst terrains* (after Thomas, 1963, 1974)

Site	Grid reference of central point	Volume of collapse doline craters (m³/ha)	Average rate of subsidence in sample block[a] (mm/1000 yrs)
M. Llangynidr		28 800	288
Upper Tawe–Nedd Fechan interfluve	SN 889 147	12 280	123
" "	SN 895 141	11 146	111
" "	SN 893 140	7 746	77
" "	SN 890 138	2 645	26

[a] The collapse dolines are assumed to be of Holocene age.

Table 3.8. *Solutional subsidence in 4.5 ha blocks of foundered Grit and Shales* (after Thomas, 1963)

Site	Grid reference of central point	Volume of foundered Grit–Shale (m³/ha)	Average depth of subsidence in sampling block (m)
Clogan Mawr	SN 712 188	132 250	183
Blaen Clydach	SN 755 188	28 339	213
Giedd	SN 812 184	93 520	107
Pwll Byfre	SN 876 166	69 904	168

further north. Clearly, therefore, this zone is complex, with several generations of subsidence and collapse forms superimposed. In certain locations, often extending over 2–3 km, the Millstone Grit is so distorted that few vestiges of its original structure remain. These areas of solution subsidence are mapped as 'foundered Basal Grits' (Fig. 3.12). The most impressive example is, in fact, a 360 × 200 m outlier of the Millstone Grit shales at Clogau Mawr on the northern flank of Black Mountain. Here the subsidence amounts to almost 200 m (Table 3.8). Calculations based on post-glacial subsidence rates in modern collapse doline fields give a minimum age of 2 million years. However, allowing for greatly reduced rates of chemical erosion during glacial episodes and at the time when the area lay near the down-dip extremity of interstratal karstification, then 10–15 million years is probably closer to the mark (Thomas, 1963).

As with the limestone dolines, surface subsidence or collapse features in the Millstone Grit act as foci for groundwater recharge (Newson, 1970), and their development is self-enhancing. A peat cover is often present and, locally, the resultant acidic run-off (pH, 3.8–4.2) has eroded vertical shafts in the underlying Carboniferous Limestone (Burke & Bird, 1966). The initial development of the internally drained depressions is less well understood. In certain locations, as above Ogof Ffynnon Ddu I and Dan yr Ogof II, collapse dolines are closely aligned above major underground passages, and their origin is readily inferred. Indeed, direct sediment links have been established between Pwll-Gwy-Rhoc and Waen Rudd dolines on Mynydd Llangattwg and boulder chokes in Agen Allwedd (Bull, 1977b). Over large tracts of the interstratal karst, however, there is no clear relationship between surface subsidence and collapse phenomena and known cave systems. Ogof Ffynnon Ddu II and III, for example, penetrate some distance beneath the Basal Grit cover with no discernible surface manifestation. The dearth of explored caves on Black Mountain is particularly striking. One explanation of this apparent anomaly may be that much of the solutional erosion responsible for the present interstratal karst terrain occurred under phreatic conditions at the very top of the Carboniferous Limestone sequence, possibly guided by the relief of the pre-Grit palaeokarst surface. Pwll y Pasg in Agen Allwedd may have originated in this way (Bull, 1977b), but clearly the majority of such passages will either have been filled by collapse debris or are simply inaccessible from modern vadose systems which are generally located some depth below the Carboniferous Limestone–Millstone Grit unconformity. Certainly, the effects of river rejuvenation and falling water tables upon interstratal karstification merit further investigation.

References

Anderson, J. G. C. (1974). The buried channels, rock-floors and rock-basins, and overlying deposits of the South Wales valleys from Wye to Neath. *Proceedings of the South Wales Institute of Engineers*, **88**, 3–17.

Anderson, J. G. C. & Owen, T. R. (1979). The late Quaternary history of the Neath and Afan valleys, South Wales. *Proceedings of the Geologists' Association*, **90**, 203–11.

Andrews, J. T., Bowen, D. Q. & Kidson, C. (1979). Amino acid ratios and the correlation of raised beach deposits in south-west England and Wales. *Nature, London*, **281**, 556–8.

Atkinson, T. C. (1971). *Hydrology and erosion in a limestone terrain*. Unpublished Ph.D. thesis, University of Bristol.

Atkinson, T. C., Harmon, R. S., Smart, P. L. & Waltham, C. A. (1978). Palaeoclimatic and geomorphic implications of ^{230}Th/^{234}U dates on speleothems from Britain. *Nature, London*, **272**, 24–8.

Barclay, W. J. & Jones, D. G. (1978). Recent boreholes in the attenuated Carboniferous strata of the Blaenavon–Pontypool area, Gwent. *Bulletin of the Geological Survey of Great Britain*, **67**, 17pp.

Battiau-Queney, Y. (1973). Mise en évidence d'un karst tropical fossile au Pays-de-Galles. *Norois*, **77**, 136–40.

Battiau-Queney, Y. (1984). The pre-glacial evolution of Wales. *Earth Surface Processes and Landforms*, **9**, 229–52.

Battiau-Queney, Y. (1985). Buried palaeokarst features in South Wales: examples from Vaynor and Cwar yr Ystrad quarries, near Merthyr Tydfil. In *New Directions in Karst*, ed. K. Paterson & M. M. Sweeting, pp. 551–67. Norwich: Geobooks.

Blundell, D. J., Davey, F. J. & Graves, L. J. (1971). Geophysical surveys over the south Irish Sea and Nymphe Bank. *Journal of the Geological Society*, **127**, 339–75.

Bowen, D. Q. (1970). South-east and central South Wales. In *The Glaciations of Wales and Adjoining Regions*, ed. C. A. Lewis, pp. 197–227. London: Longman.

Bowen, D. Q. (1977). The land of Wales. In *Wales: A New Study*, ed. D. Thomas, pp. 11–35. Newton Abbot: David & Charles.

Bowen, D. Q. (1980). *The Llanelli Landscape: the Geology and Geomorphology of the Country Around Llanelli*. Llanelli: Llanelli Borough Council, 280 pp.

Bowen, D. Q. (1981). Sheet 1.3. In *National Atlas of Wales*, ed. H. Carter & H. M. Griffiths. Cardiff: University of Wales Press.

Bowen, D. Q. (1984). Introduction. In *Wales: Gower, Preseli and Fforest Fawr*, ed. D. Q. Bowen & A. Henry, pp. 1–17. Quaternary Research Association Field Guide.

Bray, L. G. (1972). Preliminary oxidation studies on some cave waters from South Wales. *Transactions of the Cave Research Group of Great Britain*, **14**, 59–66.

Bray, L. G. (1977). The role of organic matter in limestone solution in the Ogof Ffynnon Ddu streamway. *Proceedings of the 7th International Speleological Congress*, 65–8. Sheffield.

Bray, L. G. & O'Reilly, P. M. (1974). Preliminary oxidation studies on some waters from the Ogof Ffynnon Ddu system, Breconshire. *Transactions of the British Cave Research Association*, **1**, 75–85.

Brook, D. A. (1981). The limestone pavements of Nahanni: an example of micro-scale labyrinth karst. *South African Geographical Journal*, **63**, 35–46.

Brook, G. A. & Ford, D. C. (1978). The origin of labyrinth and tower karst and the conditions necessary for their development. *Nature, London*, **275**, 493–6.

Brown, E. H. (1960). *The Relief and Drainage of Wales*. Cardiff: University of Wales Press.

Brown, E. H. (1979). The shape of Britain. *Transactions of the Institute of British Geographers, NS*, **4**, 449–62.

Bull, P. A. (1976). An electron microscope study of cave sediments from Agen Allwedd, Powys. *Transactions of the British Cave Research Association*, **3**, 7–14.

Bull, P. A. (1977a). Cave sediment studies in South Wales: towards a reconstruction of a Welsh palaeoclimate by means of the scanning electron microscope. *Studies in Speleology*, **3**, 13–24.

Bull, P. A. (1977b). Boulder chokes and doline relationships. *Proceedings of the 7th International Speleological Congress*, 93–6. Sheffield.

Bull, P. A. (1980). Towards a reconstruction of timescales and palaeoenvironments from cave sediment studies. In *Timescales in Geomorphology*, ed. R. A. Cullingford, D. A. Davidson & J. Lewin, pp. 177–89. Chichester & New York: Wiley.

Burke, A. R. & Bird, P. F. (1966). A new mechanism for the formation of vertical shafts in Carboniferous Limestone. *Nature, London*, **210**, 831–2.

Campbell, S., Andrews, J. T. & Shakesby, R. A. (1982). Amino acid evidence for Devensian ice, west Gower, South Wales. *Nature, London*, **300**, 249–51.

Catt, J. A. (1981). British pre-Devensian glaciations. In *The Quaternary in Britain*, ed. J. Neale & J. R. Flenley, pp. 9–19. Oxford: Pergamon.

Chambers, W. (1983). Denudation rates in the River Burry catchment, Gower, Glamorgan. *Transactions of the British Cave Research Association*, **10**, 181–7.

Clayden, B. & Evans, G. D. (1974). Soils in Dyfed, I: Sheet SN41 (Llangendeirne). *Soil Survey Record*, **20**.

Coase, A. C. & Judson, D. M. (1977). Dan yr Ogof and its associated caves. *Transactions of the British Cave Research Association*, **4**, 245–344.

Cogley, J. G. (1972). Processes of solution in an arctic limestone terrain. In *Polar Geomorphology*, ed. R. J. Price & D. E. Sugden, pp. 201–11. Institute of British Geographers, Special Publication, No. 4.

Coope, G. R. (1974). Interglacial Coleoptera from Bobbitshole, Ipswich. *Quarterly Journal of the Geological Society of London*, **130**, 333–40.

Coope, G. R. (1975). Climatic fluctuations in north-west Europe since the last interglacial, indicated by fossil assemblages of Coleoptera. In *Ice Ages: Ancient and Modern*, ed. A. E. Wright & F. Moseley, pp. 153–68. Liverpool: Seel House Press.

Crampton, C. B. (1972). Soils of the Vale of Glamorgan. *Memoir of the Soil Survey of Great Britain*. 87 pp.

Davies, K. H. (1983). Amino acid analysis of marine molluscs from the Gower Peninsula. *Nature, London*, **302**, 137–9.

Drew, D. P. (1967). *Aspects of the limestone hydrology of the Mendip Hills*. Unpublished Ph.D. thesis, University of Bristol.

Drew, D. P. (1968). Tracing percolation waters in karst areas. *Transactions of the Cave Research Group of Great Britain*, **10**, 107–14.

Ede, D. P. (1972). *An investigation into some factors influencing the hardness of streams and springs on limestone with particular reference to South Wales*. Unpublished Ph.D. thesis, University of Wales, Swansea.

Ede, D. P. (1975). Limestone drainage systems. *Journal of Hydrology*, **27**, 297–318.

Flemming, N. C. & Roberts, D. G. (1973). Tectono-eustatic changes in sea level and sea floor spreading. *Nature, London*, **243**, 19–22.

Ford, D. C. (1971). Characteristics of limestone solution in the southern Rocky Mountains and Selkirk Mountains, Alberta and British Columbia. *Canadian Journal of Earth Sciences*, **8**, 585–608.

Ford, T. D. (1977). Sands and solution in the Tertiary era. In *Limestones and Caves of the Peak District*, ed. T. D. Ford, pp. 79–96. Norwich: Geobooks.

Ford, T. D. ((1984). Palaeokarsts in Britain. *Transactions of the British Cave Research Association*, **11**, 246–64.

Gascoyne, M. & Ford, D. C. (1984). Uranium series dating of speleothems, Part II. Results from the Yorkshire Dales and implications for cave development and Quaternary climates. *Transactions of the British Cave Research Association*, **11**, 65–85.

Gascoyne, M., Ford, D. C. & Schwarcz, H. P. (1983). Rates of cave development in the Yorkshire Dales from speleothem age data. *Earth Surface Processes and Landforms*, **8**, 557–68.

George, T. N. (1974). The Cenozoic evolution of Wales. In *The Upper Palaeozoic and Post-Palaeozoic Rocks of Wales*, ed. T. R. Owen, pp. 341–71. Cardiff: University of Wales Press.

Groom, G. E. & Ede, D. P. (1972). Laboratory simulation of limestone solution. *Transactions of the Cave Research Group of Great Britain*, **14**, 89–95.

Groom, G. E. & Williams, V. H. (1965). The solution of limestone in South Wales. *Geographical Journal*, **131**, 37–41.

Jennings, J. N. (1985). *Karst Geomorphology*. Oxford: Blackwell.

Jones, O. T. (1930). Some episodes in the geological history of the Bristol Channel region. *Report of the Proceedings of the British Association for the Advancement of Science*, 57–82.

Jones, R. O. (1939). The development of the Tawe drainage. *Proceedings of the Geologists' Association*, **44**, 305–21.

Kelling, G. (1974). Upper Carboniferous sedimentation in South Wales. In *The Upper Palaeozoic and Post-Palaeozoic Rocks of Wales*, ed. T. R. Owen, pp. 185–224. Cardiff: University of Wales Press.

Lake, P. (1900). Bala Lake and the river systems of North Wales. *Geological Magazine*, **7**, 204–15, 241–5.

Lake, P. (1934). The rivers of Wales and their connection with the Thames. *Science Progress*, **113**, 25–39.

Lewis, K. (1983). A morphometric and geological study of limestone pavements in South Wales. *Transactions of the British Cave Research Association*, **10**, 199–204.

Longman, M. W. (1980). Carbonate diagenetic textures from near-surface diagenetic environments. *Bulletin of the American Association of Petroleum Geologists*, **64**, 461–87.

Lowe, J. J. & Walker, M. J. C. (1984). *Reconstructing Quaternary Environments*. London: Longman.

Newson, M. D. (1970). *Studies of chemical and mechanical erosion by streams in limestone terrains*. Unpublished Ph.D. thesis, University of Bristol.

North, F. J. (1962). *The River Scenery at the Head of the Vale of Neath*. Cardiff: National Museum of Wales Press.

O'Reilly, P. M. (1974). Morphology and hydrology of the Ogof Ffynnon Ddu karst area. *South Wales Caving Club Newsletter*, **76**, 6–17.

O'Sullivan, K. N. & Herbert-Smith, M. (1979). The Tertiary rocks of the Llanbedr (Mochras Farm) borehole. *Institute of Geological Sciences Report*, **No. 78/24**.

Owen, T. R. (1974). The Variscan Orogeny in Wales. In *The Upper Palaezoic and Post-Palaeozoic Rocks of Wales*, ed. T. R. Owen, pp. 285–94. Cardiff: University of Wales Press.

Owen, T. R. & Jones, D. G. (1961). The nature of the Millstone Grit–Carboniferous Limestone junction of a part of the South Wales coalfield. *Proceedings of the Geologists' Association*, **72**, 239–49.

Owen, T. R. & Jones, D. G. (1966). The Millstone Grit succession between Brynmawr and Blorenge, South Wales and its relationship to the Carboniferous Limestone. *Proceedings of the Geologists' Association*, **77**, 187–98.

Pitty, A. F. (1968). The scale and significance of solutional loss from the limestone tract of the southern Pennines. *Proceedings of the Geologists' Association*, **79**, 153–77.

Ramsay, A. C. (1846). The denudation of South Wales and the adjacent English counties. *Memoir of the Geological Survey of Great Britain*, **No. 1**.

Ramsbottom, W. H. C. (1973). Transgressions and regressions in the Dinantian: a new synthesis of British Dinantian stratigraphy. *Proceedings of the Yorkshire Geological Society*, **39**, 567–607.

Raven, M. (1983). *The diagenesis of the Oolite Group between Blaen*

Onneu and Pwll Ddu, Lower Carboniferous, South Wales. Unpublished Ph.D. thesis, University of Nottingham.

Riding, R. & Wright, V. P. (1981). Palaeosols and tidal flat/lagoonal sequences on a Carboniferous carbonate shelf: sedimentary associations and triple disconformities. *Journal of Sedimentary Petrology*, **51**, 1323–39.

Robinson, P. L. (1957). The Mesozoic fissures of the Bristol Channel area and their invertebrate faunas. *Journal of the Linnean Society (Zoology)*, **43**, 260–82.

Robinson, P. L. (1971). A problem of faunal succession on Permo-Triassic continents. *Palaeontology*, **14**, 131–53.

Rudeforth, C. C., Hartnup, R., Lea, J. W., Thompson, T. R. E. & Wright, P. S. (1984). Soils and their use in Wales. *Soil Survey Bulletin*, **11**, 336.

Shackleton, N. J. & Opdyke, N. D. (1973). Oxygen isotope and palaeomagnetic evidence for early Northern Hemisphere glaciations. *Nature, London*, **261**, 547–50.

Smith, A. G. & Cloutman, E. W. (1984). Waen-Fignen-Felen. In *Wales: Gower, Preseli, Fforest Fawr*, ed. D. Q. Bowen & A. Henry, pp. 83–90. Quaternary Research Association Field Guide.

Strahan, A. (1902). On the origin of the river system of South Wales and its connection with that of the Severn and Thames. *Quarterly Journal of the Geological Society of London*, **58**, 207–25.

Sutcliffe, A. J. & Bowen, D. Q. (1973). A preliminary report on excavations in Minchin Hole, April–May 1973. *Newsletter of the William Pengelly Cave Studies Trust*, **21**, 12–25.

Sutcliffe, A. J. & Currant, A. P. (1984). Minchin Hole Cave. In *Wales: Gower, Preseli, Fforest Fawr*, ed. D. Q. Bowen & A. Henry, pp. 33–7. Quaternary Research Association Field Guide.

Sweeting, M. M. (1966). The weathering of limestones with particular reference to the Carboniferous Limestone of northern England. In *Essays in Geomorphology*, ed. G. H. Dury, pp. 177–210. London: Heinemann.

Sweeting, M. M. (1972). *Karst Landforms*. London: Macmillan.

Sweeting, M. M. (1974). Karst geomorphology of north-west England. In *Limestones and Caves of North-West England*, ed. A. C. Waltham, pp. 46–78. Newton Abbot: David & Charles.

Thomas, T. M. (1952). Notes on the structure and occurrence of littoral Trias in the Vale of Glamorgan. *Geological Magazine*, **89**, 153–62.

Thomas, T. M. (1954). Swallow holes on the Millstone Grit and Carboniferous Limestone of the South Wales coalfield. *Geographical Journal*, **120**, 468–75.

Thomas, T. M. (1963). Solution subsidence in south-east Carmarthenshire and south-west Breconshire. *Transactions of the Institute of British Geographers*, **33**, 45–60.

Thomas, T. M. (1970). The limestone pavements of the North Crop of the South Wales coalfield. *Transactions of the Institute of British Geographers*, **50**, 87–105.

Thomas, T. M. (1973). Solution subsidence mechanisms and end products in south-east Breconshire. *Transactions of the Institute of British Geographers*, **60**, 69–86.

Thomas, T. M. (1974). The South Wales interstratal karst. *Transactions of the British Cave Research Association*, **1**, 131–52.

Trudgill, S. T. (1985). *Limestone Geomorphology*. London: Longman.

Turner, C. (1975). The correlation and duration of Middle Pleistocene interglacial periods in northwest Europe. In *After the Australopithecines*, ed. K. W. Butzer & G. D. Isaac, pp. 259–308. The Hague: Mouton.

van Donk, J. (1976). An ^{18}O record of the Atlantic Ocean for the entire Pleistocene. *Memoirs of the Geological Society of America*, **145**, 147–64.

Walker, M. J. C. (1980). Late Glacial history of the Brecon Beacons, South Wales. *Nature, London*, **287**, 133–5.

Walker, M. J. C. (1982). The late Glacial and early Flandrian deposits at Traeth Mawr, Brecon Beacons, South Wales. *New Phytologist*, **91**, 147–65.

Walsh, P. T., Boulter, M. C., Ijtaba, M. & Urbani, D. M. (1972). The preservation of the Neogene Brassington Formation of the southern Pennines and its bearing on the evolution of upland Britain. *Quarterly Journal of the Geological Society of London*, **128**, 519–59.

Walsh, P. T. & Brown, E. H. (1971). Solution subsidence outliers containing probable Tertiary sediments in north-east Wales. *Geological Journal*, **7**, 299–320.

Waltham, A. C. (1974). The geomorphology of the caves of north-west England. In *Limestones and Caves of North-West England*, ed. A. C. Waltham, pp. 79–105. Newton Abbot: David & Charles.

Williams, P. W. (1966). Limestone pavements with special reference to western Ireland. *Transactions of the Institute of British Geographers*, **40**, 155–72.

Williams, P. W. (1968). An evaluation of the rate and distribution of limestone solution and deposition in the River Fergus Basin, Western Ireland. In *Contributions to the Study of Karst*, pp. 1–40. Australian National University, Department of Geography, Publication No. G/5.

Williams, V. H. (1963). *A study of the solutional processes and phenomena in limestone with special reference to the North Avonian outcrop of the South Wales syncline*. Unpublished Ph.D. thesis, University of Wales, Swansea.

Wood, A. & Woodland, A. W. (1968). Borehole at Mochras, west of Llanbedr, Merioneth. *Nature, London*, **219**, 1352–5.

Woodland, A. W., ed. (1971). The Llanbedr (Mochras Farm) borehole. *Institute of Geological Sciences Report*, **71/18**, 115 pp.

Wright, V. P. (1980). Climatic fluctuations in the Lower Carboniferous. *Naturwissenschaften*, **67**, 252–3.

Wright, V. P. (1981). *Stratigraphy and sedimentology of the Llanelly Formation between Penderyn and Blorenge, South Wales*. Unpublished Ph.D. thesis, University of Wales.

Wright, V. P. (1982a). The recognition and interpretation of palaeokarsts: two examples from the Lower Carboniferous of South Wales. *Journal of Sedimentary Petrology*, **52**, 83–94.

Wright, V. P. (1982b). Calcrete palaeosols from the Lower Carboniferous Llanelly Formation, South Wales. *Sedimentary Geology*, **33**, 1–33.

Wright, V. P. (1984). The significance of needle-fibre calcite in a Lower Carboniferous palaeosol. *Geological Journal*, **19**, 23–32.

Wright, V. P. (1986). The polyphase karstification of the Carboniferous Limestone in South Wales. In *New Directions in Karst*, ed. K. Paterson & M. M. Sweeting, pp. 569–80. Norwich: Geobooks.

4

The hydrology of the limestone outcrop north of the Coalfield

W. GASCOINE

The Northern Outcrop of Carboniferous Limestone is a strip 3–5 km wide, stretching from Pontypool in the east to Ammanford in the west. The limestone forms high ground north of the towns in the Heads of the Valleys region and is part of the catchment for the rivers flowing south to Swansea, Cardiff and Newport. An unpublished report to the Welsh Water Authority (1980) has been the only previous survey of the whole area.

In this chapter, the Northern Outcrop is subdivided into segments in which caves and hydrology are considered individually from east to west (Fig. 4.1). Throughout the chapter dye-tracing refers to tests to trace the underground waterflow from sinks to resurgences by means of the insertion of the non-toxic dyes fluorescein (green) or rhodamine B (pink). The quantities used are kept small so as not to colour the rivers downstream of resurgences, and detection is by means of absorbent charcoal detectors placed in resurgences with subsequent laboratory measurement of the amount of dye present. Alternatively, the otherwise unknown links between sinks and resurgences can be traced by inserting dyed *Lycopodium* spores into the sinks and catching them in plankton nets placed in resurgences. This latter method has the advantage that several sinks can be tested at once using differently coloured spores.

The Afon Lwyd Valley

The Afon Lwyd River rises above Blaenavon and flows southwards through Pontypool joining the River Usk near Caerleon. Limestone outcrops on the western slopes of the valley for most of its upper reaches, crossing to the eastern

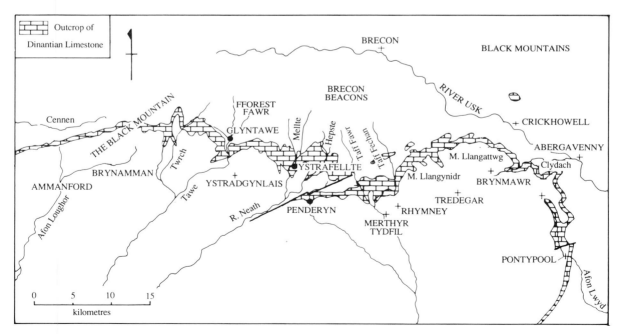

Fig. 4.1. Sketch map of the Carboniferous Limestone outcrop along the northern margin of the South Wales coalfield, in relation to the drainage pattern.

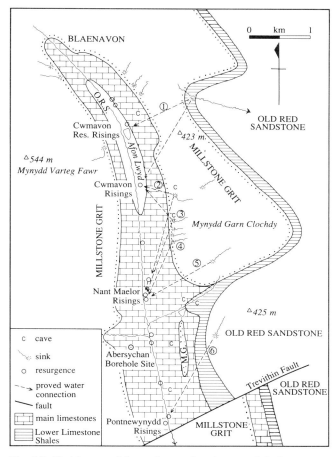

Fig. 4.2. Sketch map of the geology and underground drainage pattern of the Afon Lwyd.

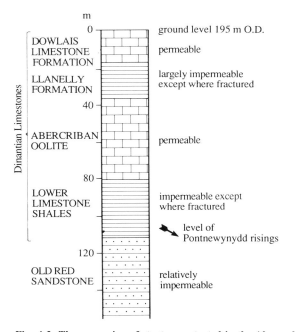

Fig. 4.3. The succession of strata penetrated in the Abersychan borehole (SO 269 033).

slopes north of Pontypool, where it is offset by the Trevithin Fault. The limestone is partly capped by Millstone Grit of variable thickness so that it outcrops only on the lower flanks of the Afon Lwyd Valley (Fig. 4.2.) and on the western flank of the Vale of Usk on the other side of the intervening ridge. Caves occur along the valley sides with approximately 400 m of known passage and several resurgences at or just above river level. No significant stream caves are known and hydrological studies have been concerned with tracing the many streams which sink along the gritstone–limestone contact high on the hill to reappear at the various resurgences in the valley floor. Figure 4.3 illustrates a representative geological succession, based on a borehole near Abersychan. In general, the underground flow follows the dip of the strata, cutting down through the sequence to resurge at or just above the contact of the limestone with the underlying Old Red Sandstone.

The sinks

A series of sinks (Table 4.1) exists in the Afon Lwyd Valley, mainly near the limestone contact where the Millstone Grit cap is only a few metres thick. Most of them are only active in wet weather, but a few are constantly active and in times of high rainfall take a considerable flow from the surrounding hillsides. On the western flank of the valley, two sinks near the head of the valley take small streams continuously and a third, further south, takes water after heavy rain. None have yet been dye-tested or dug to ascertain their potential as surface-feeders for caves or resurgences. Many sinks are present on the eastern flank. Some are named, associated with caves and with their resurgences proved by dye-testing, others have yet to be studied in detail. Virtually all are catalogued in the Cambrian Cave Registry.

The caves

Of about 15 main caves in the valley, the longest, at over 130 m, is at Jackdaw Quarry. The passages are narrow, with a vertical rift type of development higher in the limestone and a

Table 4.1 *The Eastern Valley*

	Grid reference	Altitude (m)	Length (m)
Sinks and caves			
Cuckoo Pot	SO 2774 0805	404	
Garn Clochdy Sink	SO 2828 0530	415	
Jackdaw Quarry Cave	SO 2660 0683	335	122
Jackdaw Quarry Pot	SO 2660 0682	335	21
Ogof Cwmavon	SO 2715 0742	336	91
Ogof Dial Pridd	SO 2772 0606	335	9
Ogof Wenog	SO 2777 0580	335	14
Pwll Gwyn	SO 2800 0660	415	
Pen Lasgarn Sink	SO 2810 0360	308	
Resurgences			
Cwmavon Reservoir Risings	SO 2692 0724	267	
Cwmavon Risings	SO 2702 0620	227	
Nant Maelor Risings	SO 2710 0470	195	
Pontnewynydd Risings	SO 2745 0190	120	

bedding-plane form in the strata near or at the contact with the Lower Limestone Shales. All the sinks have been formed at the Millstone Grit–limestone contact, and none is accessible for more than a few metres. Larger caves must exist, at least near Blaenavon where a borehole at Pen Ffordd Goch (SO 2553 1058) penetrated a 4 m cavity at approximately 120 m depth in the upper part of the Gilwern Oolite Limestone below the Llanelly Formation.

The resurgences: traces and feeders

Many resurgences are present at or close to river level in the Afon Lwyd Valley, and several are substantial (Table 4.2). Cwmavon and Nant Maelor Risings yield constant streams while Pontnewynydd Risings issue over 6 million l d⁻¹ from a large gravel bed on the west bank of the river. The reservoir at Cwmavon is fed by an intermittent spring at its northern end, with a large flow after wet weather.

The results of water tracing by dye-tests (numbers refer to trace-lines on Fig. 4.2)

1   Water which sinks at Cuckoo Pot, lying in the Lower Limestone Shales on the eastern scarp of Mynydd Garn Glochdy above the Usk Valley, has been traced to Cwmavon Reservoir Risings (a strong trace) and Cwmavon Risings (a weaker trace). Both of these resurgences are on the other side of the hill and showed positive in <24 h.

2   Water sinking at the cave above Cwmavon, called Dial Pridd, has been traced in approximately 4 h to Cwmavon Risings, with a much weaker trace to Nant Maelor Risings.

3   The sink and cave at Hafod Wenog has been traced to Cwmavon Risings and Nant Maelor Risings in 4 h.

4   An unnamed sink south of Hafod Wenog has been traced to Nant Maelor Risings in 5 h.

5   Garn Clochdy Sink, a very large depression with multiple swallets, has been traced to Nant Maelor Risings in <24 h.

6   The sink at Pen Lasgarn Farm has been traced to small resurgences on the river bank opposite Pontnewynydd Risings in <48 h; Pontnewynydd Risings did not register any dye.

All of these tracing experiments used fluorescein or related dyes in 50–100 g aliquots. The dyes were adsorbed on activated charcoal and the tests were carried out in falling water conditions after periods of heavy rain.

Dye tests indicate that much of the flow within the limestone on the eastern flanks of the valley occurs at the contact between the Pwll y Cwm Oolite limestones and the Lower Limestone Shales. Most of the resurgences are at this level or within the shales, and groups of them may be interconnected. On present evidence, the potential for large cave systems is limited. On the western flank of the valley, strata dip away from the river, presenting a better prospect for stream caves and a cave system running parallel to the Afon Lwyd on its western side feeding Pontnewynydd Risings may be present.

Llangattwg and Mynydd Llangynidr

Mynydd Llangattwg and Mynydd Llangynidr are high points on a continuous stretch of moorland north of Brynmawr, Ebbw Vale and Tredegar, in Gwent. In the north they comprise limestone scarps up to 200 m high running east–west and forming the southern flank of the Vale of Usk. The high moorlands are dissected in the east by the Clydach Gorge and partially cut in the west by the Trefil Valley, both valleys being the main outlets for water flowing within the limestone. Details of the strata beneath Mynydd Llangattwg and Mynydd Llangynidr are shown on the borehole data in Figures 4.4 and 4.5; the boreholes being at opposite ends of the moorland at Clydach and Trefil. Also noted are the main resurgences in the valleys and the levels where they occur with respect to the limestone succession. The area is similar to the Afon Lwyd in that much of the limestone is overlain by the Millstone Grit. Many sinks and dolines occur near its northern edge where the gritstone cap is <15 m thick. Numerous caves are present under the moorland, with entrances in the Clydach Gorge, the

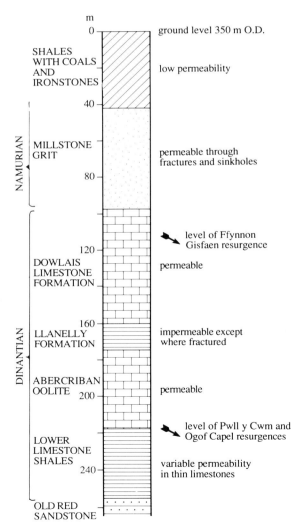

Fig. 4.4. The succession of strata penetrated in the Clydach borehole (SO 197 122).

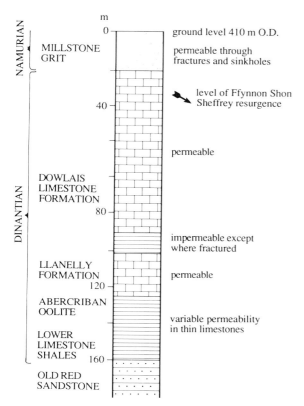

Fig. 4.5. The succession of strata penetrated in the Trefil borehole (SO 124 125).

limestone scarp, the Trefil Valley and on the surface of the moor.

Llangattwg

Figure 4.6 shows the cave locations and hydrological connections. The main resurgences in the Clydach Gorge are considered here with their respective water catchments. Surface-feeders which supply Agen Allwedd are also described.

Ogof Capel and the Limekiln Springs   Ogof Capel is a resurgence cave approximately 500 m long, located at the base of a cliff above the north bank of the Afon Clydach just downstream of the large waterfall at Pwll y Cwm. The water issues from a sump which is fed by two streams emerging from boulder chokes. The Limekiln Springs are a series of small resurgences near the old limekilns on the side of the road at Blackrock near the Rock and Fountain Inn. The discovery of Ogof Craig a Ffynnon, in 1976, stimulated dye-testing. Using this technique part of the catchment area for these springs has been identified. All the streams encountered in Ogof Craig a Ffynnon have been tested (Table 4.2) and, with one exception, have been found to resurge at Ogof Capel and the Limekiln Springs. The times taken for the water to appear at the resurgences vary from <6 to >24 h. Little is known about the courses of the more distant streams except that they feed the main stream in the entrance passages of the cave (Gascoine, 1979, 1980).

Only one surface-feeder, from Pwll Coedog, has been linked to Ogof Capel and a hydrological study of this cave and its feeders has yet to be carried out. During heavy rain, the

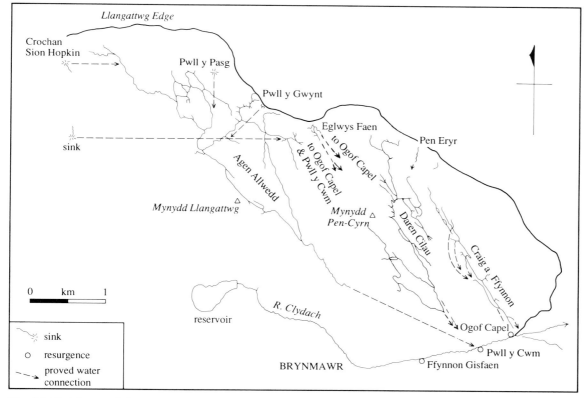

Fig. 4.6. Sketch map of the caves and underground drainage pattern beneath Mynydd Llangattwg.

Table 4.2. *Water-tracing in Ogof Craig a Ffynnon area*

Dye in NW inlet sump OCAF, SO 2185 1307.	Positive trace at Limekiln Springs, SO 2185 1274, and Ogof Capel, SO 2165 1257, in 4–6 h.
Dye in stream below pot in Hall of Mountain King OCAF, SO 2188 1316.	Visual trace after 6 h at Rock and Fountain Springs, SO 2203 1283.
Dye in Blaen Elin stream OCAF, SO 2166 1355.	Positive trace to Limekiln Springs in under 24 h, visual in NW inlet after 6 h.
Dye in Lower Series Stream OCAF, SO 2140 1400.	Positive trace to Limekiln Springs and Ogof Capel in under 24 h; Blaen Elin remained negative.
Dye in Helictite Passage OCAF, SO 2140 1405.	Positive trace to Ogof Capel in 25–30 h, Blaen Elin negative.
Dye in Promised Land OCAF, SO 2135 1405.	Visual trace at Limekiln Springs and Ogof Capel after 26 h.
Dye in Pwll Coedog, SO 2141 1327.	Positive trace at Ogof Capel in 24 h.

volume of water issuing from Ogof Capel and the Limekiln Springs indicates that the risings have an extensive catchment area. The tendency to flood pulse over a short period, when the other Clydach resurgences are much less responsive to heavy rain, suggests the presence of an unknown swallet or a flood overflow passage to Ogof Craig a Ffynnon from another of the Mynydd Llangattwg caves.

All the passages in Ogof Craig a Ffynnon occur in the Gilwern Oolite below the Llanelly Formation. The uppermost passages touch the underside of this formation at intervals but nowhere do they extend upwards into the Dowlais Limestone Formation. The stream passages in the cave are generally situated in the lowest oolite (the Pwll y Cwm Oolite) and the resurgences, although up to 0.8 km apart, all occur in the Pwll y Cwm Oolite, or in a shale bed at its upper limit. Downstream sections of Agen Allwedd and its resurgences at Pwll y Cwm also lie within this bed. There is thus strong evidence that considerable waterflow occurs in the lower part of the Oolites over a large area of Mynydd Llangattwg. This water resurges at points where the Afon Clydach, cutting down through the limestones, has intercepted pre-existing conduits.

The Pwll y Cwm resurgences   In the Afon Clydach a few metres upstream of the waterfall at Pwll y Cwm, near the confluence of a stream draining northwards from the Coal Measures, there are several springs in the riverbed and the resurgence cave, Elm Hole. One of the riverbed springs is a flooded pothole 6 m deep and 2 m in diameter blocked by boulders and river silt. Other springs emerge from small cracks in the limestone pavement which forms the bed of the Afon Clydach at this point. Elm Hole is a 76 m long cave which terminates in a 20 m deep flooded rift. Both the cave and the riverbed pot have been explored by divers but no major extensions have been found.

Since the early 1960s, several dye-tests have been carried out in Agen Allwedd. To prove the link with resurgences at Pwll y Cwm, 6 kg of Durazol Orange were used. As part of a more thorough hydrological study of Agen Allwedd, this dye-testing

was repeated and times of 36–48 h were recorded for water to travel from the known cave, through at least four sumps, to the resurgences. In the late 1970s and in 1980–1, the results of dye-tracing in Ogof Craig a Ffynnon stimulated the search for feeders to this cave. Ogof Daren Cilau, Ogof Pen Eryr and Llangattwg Swallet were dye-tested and the first two proved positive at Pwll y Cwm, using only 1 kg of fluorescein (Fig. 4.6).

There is obviously considerable flow southeast from the northern scarp of the hill to the resurgences at Pwll y Cwm. Ogof Agen Allwedd, Ogof Daren Cilau and Ogof Pen Eryr, together with the Pwll y Cwm resurgences, are all in the Pwll y Cwm Oolite, so similar conclusions to those concerning Craig a Ffynnon hydrology may apply here. The strength of the traces from Daren Cilau and Llangattwg Swallet indicates that it is unlikely that their water passes through any known passages in Agen Allwedd since dilution in four very extensive sumps would have been too great to allow the visual confirmation that was obtained.

Surface-feeders to Agen Allwedd   Dye-tests have confirmed that several sinks on Mynydd Llangattwg are feeders to the Agen Allwedd cave system (Fig. 4.6).

Pwll y Pasg Cave (SO 1820 1610), on the surface of the moor, has been dye-tested to Coal Cellar Passage in Agen Allwedd in <4 h.

Pwll y Gwynt Cave (SO 187 157), high in the limestone scarp, has been traced to Second Boulder Choke, Agen Allwedd, in 4 h.

The sink and dig on the moor at Crochan Sion Hopkin (SO 1651 1628) has been traced to Remembrance Series, Agen Allwedd, in <2 h.

A sink on the moor (SO 1660 1536) has been traced to Ace of Spades Inlet, Agen Allwedd, in approximately 48 h.

These tests indicate waterflow in the limestone sequence of Mynydd Llangattwg above the Llanelly Formation. They show, in all but the Crochan Sion Hopkin test, that the flow passes over known cave passages in Agen Allwedd entering the cave only where fractures in the Llanelly Formation shales allow leakage. The presence of manganese- and iron-rich deposits near the points where the streams enter the cave suggests that the flow occurs near the contact between the Dowlais Limestone and the Millstone Grit. The tests to Ffynnon Gisfaen, described later in this chapter, tend to confirm this view.

Ogof Daren Cilau   The discovery of the large extensions to the cave Ogof Daren Cilau (see Fig. 4.6) have allowed several more dye-tests to be carried out underground, adding extra information on the drainage pattern in this mountain. Three surface tests have also been carried out.

In Ogof Daren Cilau, large mature passages trending southeast parallel to Ogof Craig a Ffynnon almost reach the Clydach Gorge near the Pwll y Cwm resurgences. Within these passages the following dye-tests have been done (up to July 1985): (1) the stream in the Ogof Daren Cilau entrance passage has been dye-traced throughout the cave to the resurgences at Pwll y Cwm; (2) the stream in Ogof Pen Eryr has been dye-traced to the Daren Cilau extensions and through to Pwll y Cwm; (3) the stream sinking at Llangattwg Swallet has been shown to enter Daren Cilau at a sump near its southern end – 200 g of dye took 25 h to appear; and (4) a dye-trace (using 1 kg of dye) in Agen Allwedd has shown Daren Cilau *not* to intercept Agen Allwedd water, at least in the cave explored so far.

Surface traces, conducted in 1985, showed the following: (1) a sink at SO 2090 1410, almost directly above the Epocalypse chokes in Daren Cilau, has been traced to Craig a Ffynnon without passing through known parts of Daren Cilau; (2) a second sink at SO 2058 1455, almost directly over the ladder pitches in Daren Cilau, has also been traced to Craig a Ffynnon direct; and (3) a third sink at SO 2015 1493, west of most of the Daren Cilau passages, has also been dye-tested but the dye did not appear in Daren Cilau. These surface traces indicate that Craig a Ffynnon will ultimately be shown to trend westwards either under or over Ogof Daren Cilau and they may go some way to explain why the cave contains such large volumes of floodwater on occasions.

A further series of dye-tests, done in autumn 1985 and spring 1986, in Eglwys Faen have shed further light on the underground waterflow in Llangattwg. These traces were conducted during two periods of very heavy rain when most cave streams were in full flood. In St Patrick's Passage 600 g of dyed *Lycopodium* spores were put into streams and 1 kg of spores was introduced into the Inner Choke stream of Eglwys Faen. The results were as follows: (1) the St Patrick's streams were found to resurge at Ogof Capel after a period of 3 d, two separate tests having been carried out on two streams sinking at different locations in St Patrick's Passage; and (2) the Inner Choke stream, which sinks in boulders at the base of a boulder choke, was found to resurge at Pwll y Cwm and Ogof Capel after a period of 3 d and at Craig a Ffynnon after 14 d – following a further resurge flood. These traces add further weight to the conclusions that Pwll y Cwm, Ogof Capel and the Craig a Ffynnon resurgence at the limekilns take water from a very large area of Mynydd Llangattwg and are hydrologically linked in times of flood. They also indicate that the caves discovered so far are likely to be part of a single system and that further large phreatic passages may exist under the hill.

Ffynnon Gisfaen   In the Clydach Gorge, 5 m below the top of the limestone sequence, is a series of three resurgences collectively called Ffynnon Gisfaen. Two are on the north bank of the river enclosed by brick and concrete structures and the third is on the south bank surrounded by a stone trough. The springs at Ffynnon Gisfaen supply over five million litres of drinking water per day to the town of Brynmawr.

Water-tracing experiments were carried out in 1979–80 to ascertain the catchment of Ffynnon Gisfaen and, because the resurgences lie so high in the limestone sequence, Craig a Ffynnon and Agen Allwedd were not considered to be likely sources. Surface sinks were therefore examined north and west of the resurgences and tests were carried out using *Lycopodium* spores and plankton nets to avoid visible pollution from dyestuffs. The use of spores also enabled tests to be carried out over larger distances (>5 km) than would have

been feasible with dyes. The traces were carried out simultaneously using *Lycopodium* dyed four different colours so the input of each sink to the resurgence could be recorded. The most distant sink was 6 km from the resurgence and over 260 m higher in altitude. Ffynnon Gisfaen south was sampled after 24 h and again after 50 h so times of travel are not particularly significant; however, each trace was clearly defined, several spores from each sink being identified. The results of the *Lycopodium* traces were as follows: (1) Pwll Coch, a large doline and sink at SO 165 150 and an altitude of 486 m was treated and after 50 h Ffynnon Gisfaen south spring (SO 2074 1240), altitude 277 m, registered a positive trace (distance 4.5 km); (2) a sink at SO 153 158 (538 m altitude) was traced to Ffynnon Gisfaen south in 50 h (distance 5.5 km); (3) a sink at SO 1425 1560 (545 m altitude) was traced to Ffynnon Gisfaen south in 50 h (distance 6 km); and (4) a sink in a Water Authority tunnel in Cwm Carno, Ebbw Vale at SO 1562 1423 (369 m altitude) was traced to Ffynnon Gisfaen south in 24 h (distance 5 km).

The position of the Ffynnon Gisfaen springs being so close to the top of the Dowlais Limestone Formation implies that a substantial flow of water southwards from the western end of Mynydd Llangattwg is present at the contact between the Millstone Grit and the limestones or in cavities near the top of the Dowlais Limestone. Such cavities can be seen on parts of the mountain and represent the upper levels of some of the caves whose entrances are on the surface of the moor, e.g., Ogof Cynnes (SO 1408 1540), Pwll y Pasg (SO 1820 1610) and Blaen Onneu Pot (SO 1584 1623). The visible cavities are mostly bedding planes with gritstone roofs but some have rifts in their floors going deep into the Dowlais Limestone. Often manganese- and iron-rich deposits coat the walls of the passages. The sink at SO 1425 1560 lies on Mynydd Llangynidr and represents the most westerly point yet traced to Ffynnon Gisfaen (Fig. 4.7). Indeed, traces a little further west of this sink indicated flows in the opposite direction, towards Ffynnon Shon Sheffrey, indicating the presence of a hitherto unknown watershed beneath Mynydd Llangynidr.

Mynydd Llangynidr

Ffynnon Shon Sheffrey   At the northern end of Shon Sheffrey's reservoir, near the village of Trevil north of Tredegar, is a large resurgence in a stone enclosure, called Ffynnon Shon Sheffrey (Fig. 4.7). The water issues from a large sand- and gravel-filled depression on the east bank of the Nant Trefil. The flow of over six million litres a day feeds directly into the reservoir and is used to supply the towns in the nearby Heads of the Valleys Region. White cave trout have been seen on occasions in the resurgence. Some 50 m upstream of Ffynnon Shon Sheffrey on the same bank of the Nant Trefil is a flood resurgence in a large pile of boulders. In times of high rainfall, Ffynnon Shon Sheffrey and its flood overflow are a very impressive sight with enormous volumes of water issuing from both sources. The position of this spring near the top of the limestone sequence (Fig. 4.5) but down-dip from the high ground on Mynydd Llangynidr (over 550 m) makes Ffynnon Shon Sheffrey an interesting site for hydrological investigation.

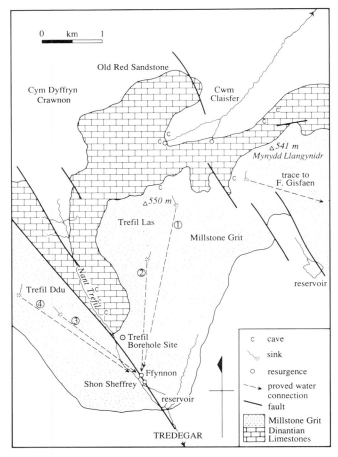

Fig. 4.7. Sketch map of the geology and underground drainage pattern of the Trefil area, north of Tredegar.

Table 4.3. *Summary of the main features of the Mynydd Llangynidr area (Cambrian Cave Registry)*

	Grid reference	Altitude (m)	Length (m)
Caves			
Chartists Cave	SO 128 152	533	671
Crescent Cave	SO 129 151	533	15
Ogof Cynnes	SO 1408 1540	524	914
Ogof Fach Trefil	SO 0940 1460	495	
Cwar yr Ystrad Caves	SO 088 145	533	204
Ogof Ap Robert	SO 099 134	518	31
Pwll Pirs	SO 1006 1370	540	20
Resurgences			
Claisfer Resurgence	SO 1290 1601	454	
Ffynnon Cae Rhos	SO 1368 1608	454	
Ffynnon Shon Sheffrey	SO 1265 1188	355	
Ogof Purgad	SO 1046 1536	396	122
Ogof Blaen Crawnon	SO 0950 1494	442	91

on the east side of the valley did not encounter any significant cavities or aquifers before it reached the Old Red Sandstone. Small caves occur throughout this area but, since only Ogof Ap Robert contains an appreciable stream, it is not feasible to carry out water-traces and thus the level at which water flows in the limestone remains unknown.

The two streams in Cwm Dyffryn Crawnon and Cwm Claisfer, which flow northwards to join the River Usk, both originate from limestone resurgences with substantial flows. No positive trace has been obtained at either rising from any sinks on Mynydd Llangynidr or the Trefil Hills. Consequently, the hydrology of the northern flanks of the outcrop in this area remains largely unknown.

The Rhymney Valley
This shallow, high-level valley lies approximately 5 km west of Trefil and has a similar structure to the Trefil Valley, with many sinks on the hill flanks and a large resurgence at river level near the southern end (Fig. 4.8). The Blaen Rhymney stream collects water from widespread surface peat deposits and from a large limestone resurgence in its bed. A borehole sunk to a depth of 55 m feeds limestone water to the reservoir at Blaen Rhymney and to the towns lower down the valley. Figure 4.8 also shows two of the spore traces which gave positive results at Ffynnon Shon Sheffrey and the position of Pontsticill Risings in the next valley to the west, the Taff Fechan.

The Blaen Rhymney stream is a promising site for hydrological study, with a large resurgence in a rift in the riverbed and, some way upstream, a large sand-filled depression on the west bank from which a copious amount of water issues during high rainfall. The upper valley also probably has the greatest concentration of dolines and sinks on the Northern Outcrop.

After a period of heavy rain, sinks at SO 0855 1310 (570 m altitude), SO 0965 1245 (527 m altitude) and SO 0988 1220 (515 m altitude) (Fig. 4.8) were treated with a dyed *Lycopodium*-spore slurry and a plankton net was placed in the

In 1979–80, with encouragement from the Welsh Water Authority, a series of multi-coloured *Lycopodium* spore tests were made on Mynydd Llangynidr, Trefil Las and Trefil Ddu, with a net placed in the main outflow from Ffynnon Shon Sheffrey (Fig. 4.7). The results of these traces were as follows: (1) spores introduced into a sink near Crescent Cave (SO 127 147; 545 m altitude) were recovered in Shon Sheffrey (SO 1265 1188; 345 m altitude) after 48 h; (2) a sink at SO 127 129 (511 m altitude) was traced to Shon Sheffrey after 48 h; (3) spores introduced into Wet Sink (SO 110 132; 498 m altitude) on the western side of the Nant Trefil Valley on Trefil Ddu reached Shon Sheffrey after 48 h; and (4) the stream cave Ogof Ap Robert (SO 099 134; 523 m altitude) was traced to Shon Sheffrey in 48 h.

Ffynnon Shon Sheffrey lies on a large fault which runs southeast, parallel to the Nant Trefil Valley, passing beneath the northern edge of the South Wales Coalfield near Tredegar. These tests may indicate that the fault captures water from the area around this valley delivering it to the surface where the limestone reaches the water table as it plunges under the Millstone Grit at the southern end of the valley. An interesting feature is the proximity of the sink giving the easternmost trace to Shon Sheffrey and that giving the westernmost trace to Ffynnon Gisfaen, <1 km apart on the largely featureless surface of Mynydd Llangynidr (Table 4.3). It is also of interest that a borehole drilled just upstream of Ffynnon Shon Sheffrey

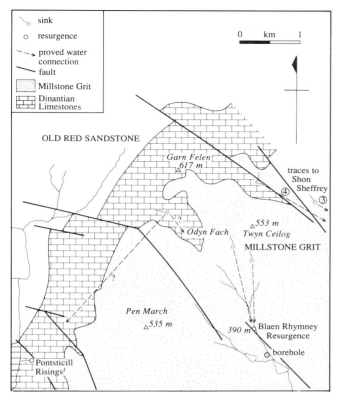

Fig. 4.8. Sketch map of the geology and underground drainage pattern of the area around Pontsticill and Blaen Rhymney.

Blaen Rhymney stream just below the risings at SO 1018 1050 (370 m altitude). Another net was placed over the outflow pipe from the borehole at SO 1040 0995 (340 m altitude) which flows in all but severe drought conditions. The latter two traces were positive in the stream net after only 24 h, whereas the trace from the northernmost sink, at SO 0855 1310, failed to show in the net after 7 d. The net over the borehole overflow also failed to give a positive result.

The riverbed resurgence (SO 1995 1082; 396 m altitude) lies in the Millstone Grit just above the top of the limestone, as do the two sinks which produced positive results in the capture net. The rapid throughput of the spores suggest good conduit flow but whether or not the flow is in mature caves cannot be ascertained. Water must traverse part of the limestone sequence since the resurgence has the chemical composition characteristic of a limestone spring. The sink near the head of the valley to the north may be a feeder for the Pontsticill Risings in the adjacent valley at SO 062 105 (312 m altitude). This spring was not netted on the occasion of the spore test and further confirmatory work is needed. The failure to detect spores at the borehole spring also requires explanation.

No caves of any significance are known in the Blaen Rhymney Valley, although digs have been undertaken in several of the sinks and the flood resurgence. Thus there is little firm evidence of this valley's speleological potential.

Between Rhymney and Glyntawe water generally drains from the Old Red Sandstone of the Brecon Beacons on to north of the limestone outcrop and passes southwards through this rock as it flows south. The southerly dip of the strata means that many rivers enter the base of the limestone at its contact

with the Old Red Sandstone and resurge at its top, the contact with the Millstone Grit. Hydrological study of this area of the North Outcrop has thus been limited, large resurgences or sinks being few beyond those where the main rivers leave or enter caves.

The Taff Fechan and Taff Fawr Valleys

North of Merthyr Tydfil in Mid Glamorgan, two rivers flow southwards from the Brecon Beacons to join just above the town, at Cefn Coed (Fig. 4.9). Each cuts through the limestone outcrop but neither passes underground, although they could have done so in the past. Only the Nant y Glais, a stream which enters the Taff Fechan near Vaynor, currently flows for any distance underground and this is the only stream which has been the subject of water-tracing. A sink at SO 0415 1140 (362 m altitude) called the 'Old Quarry Sink' was dyed in the late 1960s to try and prove a link with Ogof Rhyd Sych in the Nant y Glais Gorge. The test was negative and it is now thought that the sink may feed risings below Vaynor Church (see below). The majority of the water flowing in Ogof Rhyd Sych is from Nant y Glais and is captured upstream of the cave. Nant y Glais feeds and resurges from most of the caves in its valley providing an interesting example of a limestone river gorge in the making. The locations of the main caves and resurgences in the Taff Fawr and Taff Fechan valleys are shown in Figure 4.9 and are detailed below with some additions to complete the picture.

The caves

Ogof y Ci (SO 0404 1047; 300 m altitude; length 548 m). This is a stream cave which in normal weather conditions carries the whole of the Nant Glais water. One major inlet from an unknown source is present in the cave. An adjacent quarry covers the area where likely sinks for this inlet might be expected and although one dye-test has been carried out, it was not successful.

Ogof Rhyd Sych (SO 0412 1021; 292 m altitude; length 914 m). This is the major stream cave in the Nant Glais Valley, carrying the entire stream in normal conditions. It lies downstream of Ogof y Ci on the opposite side of the valley and the majority of its flow is water that resurges from Ogof y Ci to resink into Rhyd Sych.

Ogof Pysgodyn Gwyn (SO 0416 1016; 286 m altitude). This is a small, largely flooded cave near the entrance to Ogof Rhyd Sych on the opposite side of a small limestone gorge. Ogof Pysgodyn Gwyn has been proved by dye-tracing to be an active oxbow cave carrying some of the Nant y Glais water sinking some 50 m upstream.

Ogof Pont y Meirw (SO 0250 0840; 240 m altitude; length 38 m). This is a resurgence cave in the west bank of the Afon Taff Fawr, upstream of Cefn Coed Risings. It was once thought to be the resurgence for Pant Sychbant, high on the hill to the west, but this assumption has been proved incorrect (see the Cwm Cadlan section).

The resurgences

Pontsticill Risings (SO 062 105; 312 m altitude). This is a large resurgence from a collapse in a field close to the road

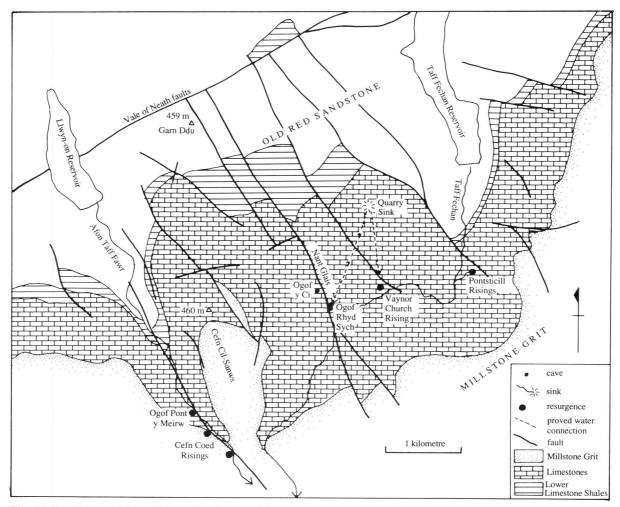

Fig. 4.9. Sketch map of the geology and underground drainage pattern of the Taff Fechan, Taff Fawr and Nant y Glais area.

leading to Pontsticill Reservoir. An attempt to locate its feeders was made in 1980 as part of the Rhymney Valley tests, but was not successful.

Vaynor Church Risings (SO 0490 1021; 246 m altitude). These risings cover a large area of the west bank of the Afon Taff Fechan and may be associated with the Old Quarry Sink to the northwest but this is still unconfirmed. Large amounts of tufa are present in these risings, which are the only ones on the North Outcrop yielding tufa in any quantity.

Cefn Coed Risings (SO 0305 0775; 216 m altitude). These risings lie on the east bank of the Taff Fawr and the flow emerges at three points on a large pebble bank. Collectively, they represent the largest resurgence in the area and knowledge of the source of their water would give valuable clues to the hydrology and cave potential of both the Taff Fawr and Taff Fechan area.

Other sites of hydrological interest include: Ogof Robin Goch (SO 0392 1075; 312 m altitude; length 107 m); Taff Fawr Caves (SO 0270 0810; 225 m altitude); and Morlais Castle Resurgence (SO 048 099).

Cwm Cadlan and Penderyn

Cwm Cadlan is a minor valley running eastwards from high ground lying west of the Taff Fawr (Fig. 4.10). The

limestone outcrop lies on its southern slopes and contains several sinks and dolines. These include a large swallet cave called Ogof Fawr (SN 9853 0961; 326 m altitude) which takes the constant Nant Sychbant stream, and the swallet Ogof Fechan (SN 9707 0972; 297 m altitude). Several other dolines and sinks are present on both sides of Cwm Cadlan.

Dye-testing has shown that water sinking at Ogof Fawr resurges at Llygad Cynon near the village of Penderyn (SN 9524 0774; 219 m altitude), nearly 5 km distant (Fig. 4.10). Llygad Cynon is a substantial resurgence and is the source of the River Cynon which flows through Aberdare to join the Taff near Pontypridd. A borehole near the resurgence located an underground 'lake' at a depth of 55 m, whence about five million litres of water per day is extracted to supply a reservoir above Hirwaun to the south.

Llygad Cynon is the only major resurgence in the Penderyn area but in times of high rainfall a substantial flood rising flows from three adjacent shakeholes in a dry valley (SN 9529 9784; 221 m altitude). It is possible that this resurgence represents water which sinks in all of Cwm Cadlan. West of Cwm Cadlan the limestone outcrop narrows and terminates at the Sychryd Gorge north of Glynneath. North of Penderyn and Cwm Cadlan, the main outcrop continues west and is separated from the Penderyn Oolite by the Vale of Neath Disturbance. The

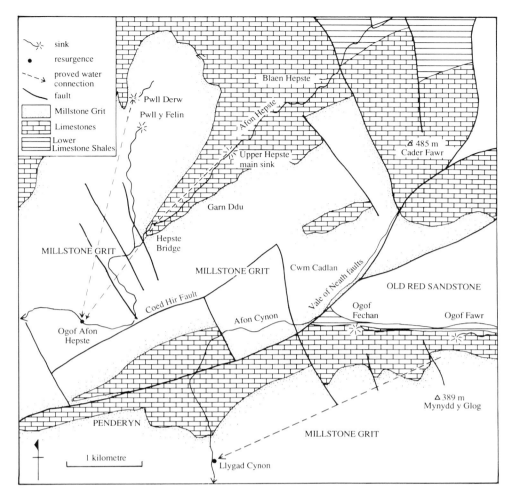

Fig. 4.10. Sketch map of the geology and underground drainage pattern of the area around the Hepste and Cynon Valleys, Penderyn.

Sychryd Gorge contains some fine cave remnants, sinks and resurgences in steeply inclined limestone and the Afon Sychryd flows partly on the surface and partly in a series of parallel cave passages beneath the gorge. Some of the main features of the gorge are mentioned below.

Wills Hole (SN 9147 0802; 102 m altitude; length 393 m). This is a stream cave which carries water some 15 m below and parallel to the floor of the gorge. Large amounts of silt washed into the cave by the stream cause its point of entry, its route through the cave and its sump exit to vary with flow.

Ogof Bwa Maen (SN 9145 0804; 93 m altitude; length 91 m). This is an abandoned stream cave on the opposite side of the gorge from Wills Hole lying on one of the Neath Disturbance faults.

Ogof Pont Sychryd (SN 9109 0791; 83 m altitude; length 225 m). This is the resurgence cave at the foot of the Sychryd Gorge where the water flows into the Afon Mellte. It has been proved by dye-testing to carry drainage from Wills Hole.

The Afon Hepste area

From its source high on Fan Fawr, a peak in the Brecon Beacons, the Afon Hepste flows south and southwest to eventually join the Afon Mellte (Fig. 4.10). For 8 km it flows at or just below the top of the Carboniferous Limestone, much of its route being underground. There are several sinks and

resurgences in the riverbed, and most of the upper ones are closely connected, the river only appearing for a few metres before sinking again. Those of note are listed in Table 4.4.

The stream caves contain a total of 880 m of passage and are fed largely by river water. Most of the passages are only accessible to divers, except in severe drought conditions. Some

Table 4.4 *Sinks and resurgences of the Hepste area*

	Grid reference	Altitude (m)	Length (m)
Nant Cwrier Bridge Rising	SN 9747 1396	341	
Tir yr Onnen Sink	SN 9647 1286	316	
Upper Hepste Resurgence Cave	SN 9632 1286	314	
Blaen Hepste Hole	SN 9615 1281	312	213
Upper Hepste Main Sink	SN 9541 1208	298	
Hepste Fawr Sinks	SN 9518 1189	293	
Ogof Glan Hepste	SN 9519 1182	291	46
Tucks Rift	SN 9518 1185	293	55
Hepste Bridge Sinks	SN 9475 1132	280	
Ogof Afon Hepste	SN 9383 0968	234	488
Tarddiant Hepste Sink	SN 9367 0970	226	
Ogof Tarddiant Hepste	SN 9366 0970	224	
Hepste Main Resurgence	SN 9360 0973	219	76

feeder streams to the Afon Hepste sink directly into the caves while others flow on the abandoned riverbed for varying distances before sinking. The river caves are seldom more than a few metres beneath the surface, having developed in general at the contact between the limestone and the Millstone Grit. Using dye-testing, the Upper Hepste Main Sink (SN 9541 1208) has been traced to the Hepste Main Resurgence (SN 9360 0973) in under 24 h, and Pwll Derw, a large stream sink (SN 9413 1235; 344 m altitude) containing 15 m of cave passage, has been traced to the Hepste Main Resurgence. More dye-tracing is required to obtain a fuller picture of the hydrology of this valley.

The Afon Mellte area

This river flows southwards 3–4 km west of the Afon Hepste, rising at the head of two valleys, one adjacent to Fan Fawr, lying on either side of the peak Fan Dringarth. The Afon Mellte is a surface stream until it reaches the limestone at the village of Ystradfellte. It sinks at the Main Mellte Sink (or Church Sink) (SN 9315 1332; 255 m altitude) and from this point flows in the cave Porth yr Ogof, partly accessible only to cave divers (Fig. 4.11). There are five resurgences from Porth yr Ogof spanning 30 m (SN 9378 1247; 230 m altitude) (Lloyd, 1979). South of this point the Afon Mellte remains on the surface, eventually joining the Hepste and Nedd Fechan to become the River Neath. In high water conditions the flow can

exceed the capacity of the Main Sink and the stream runs on the surface to the entrance of Porth yr Ogof (SN 9281 1241; 230 m altitude). Several dye-tests in the Mellte Valley have identified minor feeders to the Porth yr Ogof system. These include: (a) Cwm Porth Farm Well Sink (SN 9298 1213; 280 m altitude) feeds water into Porth yr Ogof cave in the area of Sump Ten; (b) Ogof Ffynnon (SN 9340 1202; 320 m altitude; length 100 m), a stream cave, has been traced to Cwm Porth Inlet in Porth yr Ogof; (c) Waterfall Cave (SN 9287 1188), a stream cave, which under flood conditions feeds water to Sumps Four and Seven in Porth yr Ogof; and (d) Church Sink (Main Mellte Sink) (SN 9315 1332) has been traced to the uppermost entrance of Porth yr Ogof in 2.25 h. A resurgence at Ffynnon Garreg Fawr (SN 9379 1387; 302 m altitude) is interesting from a hydrological standpoint, but access to the cave is difficult and hence no dye-tracing or chemical analyses have yet been done.

The Nedd Fechan area

The Afon Nedd Fechan rises between Fan Nedd and Fan Gihirych to the west of the Afon Mellte. It flows on the surface over Old Red Sandstone and sinks into tight bedding planes and fissures at the foot of a small limestone cliff (SN 9117 1404; 303 m altitude). During high flow conditions, this sink can overflow to a series of sinks downstream and in very high floods, the Afon Nedd Fechan continues on the surface as far

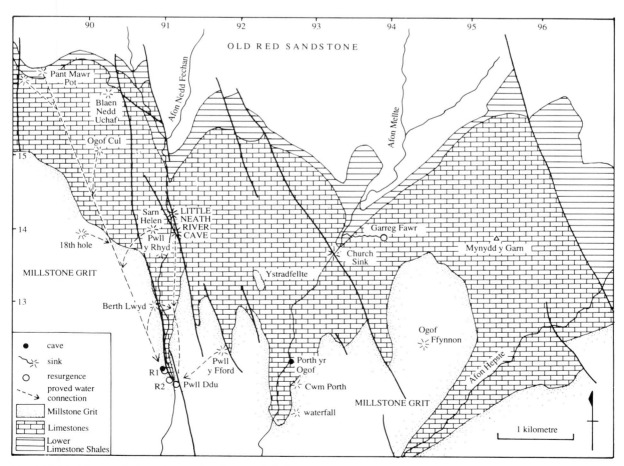

Fig. 4.11. Sketch map of the geology and underground drainage pattern of the Mellte and Nedd Fechan Valleys.

as Pwll y Rhyd, a large cavity in the riverbed (SN 9114 1375; 294 m altitude). In these conditions, the stream resurges from White Lady Cave (SN 9110 1367; 288 m altitude) continuing its course downstream to join the Afon Mellte near Pont Nedd Fechan north of Glynneath.

The majority of the water under all conditions flows in the Bridge Cave and Little Neath River Cave (= Ogof Afon Nedd Fechan) systems (Standing, Newson & Wilkins, 1971). Dye-tracing has shown that these streams resurge at two points on the river bank; a deep pool, called Pwll Ddu (SN 9121 1206; 215 m altitude), and a resurgence, called R2, below a gritstone cliff opposite Pwll Ddu (SN 9119 1207; 215 m altitude). Details of the dye-tracing and some of the speleological sites in this valley are given below and also in Figure 4.11: fluorescein placed in Bridge Cave (SN 9116 1400; 308 m altitude) was detected at Pwll Ddu and R2 in low water conditions after 36 h; a large wet-weather sink called Pwll y Fford (SN 9154 1267; 289 m altitude) has been traced to Little Neath River Cave between Sumps Four and Five; and Berthlwyd Swallet (SN 9132 1315; 295 m altitude), a large rock-filled sink, has been traced to Foot and Mouth Passage in Little Neath River Cave between Sumps Four and Five.

On the western side of the Afon Nedd Fechan is a stretch of moorland containing many large collapse features, sinks and dolines. Pant Mawr Pot, with its impressive shaft entrance (SN 8909 1612; 436 m altitude), contains 1151 m of fine mature cave passage and a substantial stream which has been dye-traced to the R1 resurgence in the Nedd Fechan Valley just upstream of Pwll Ddu and R2. At the riverbed resurgence (SN 9120 1226; 227 m altitude), water issues from a gravel bank over a distance of 6 m. In midwinter 1984 a series of *Lycopodium* spore traces in surface sinks to the east of Pant Mawr Pothole and to the north of Little Neath River Cave provided further information about the underground drainage to the Pwll Ddu, R1 and R2 resurgences (see Fig. 4.11). These tests were carried out in times of high waterflow and the results are summarized here. The sinks tested are listed in Table 4.5. Most of the sinks gave positive traces in 24–48 h to the resurgence called R1, previously identified as that issuing water from Pant Mawr Pothole. The exception was Ogof Cul which drained to R2, a resurgence below R1 on the same bank of the river and opposite Pwll Ddu, the resurgence for Little Neath River Cave. This trace was repeated to verify the rather surprising result and detectors were placed in Little Neath River Cave to see if it featured in the drainage of water from Ogof Cul (both dye and spores were used in this test). The underground detectors did not show traces of any dye and R2 was confirmed

as the resurgence for water from Ogof Cul. This result suggests that waterflow may be at two levels within the limestones of this area, with Ogof Cul water passing over or under the Pant Mawr–R1 water.

In December 1985 the Ogof Cul dye-test was repeated under high but more constant water conditions and the Ogof Cul stream was proved to feed Little Neath River Cave at the Blaen Nedd Isaf inlet just beyond the entrance crawls and then travel to the R2 resurgence through the cave. An interesting result of tracing carried out in December 1985 was from *Lycopodium* spores introduced into a cave called Hole by the Wall (SN 9209 1351). This cave lies between the Mellte and Nedd Fechan valleys in a large closed depression carrying an active stream in wet weather. The spores were detected in both the Mellte and Nedd Fechan rivers at points downstream of Pwll Ddu and Porth yr Ogof, respectively, in <24 h; the highest concentration being detected in the Nedd Fechan net.

Two small features in the Blaen Nedd Uchaf area at the northernmost edge of the limestone outcrop have been linked by dye-tracing. Water takes only minutes to flow from Blaen Nedd Uchaf Swallet (SN 9043 1550; 388 m altitude) to Blaen Nedd Uchaf Resurgence (SN 9056 1540; 367 m altitude). The high ground called Pant Mawr to the west of the Nedd Fechan Valley contains many sites of interest indicative of cave systems. Dye-tracing could yield valuable information on other caves feeding water to the valley and on the eastern limit of the Ogof Ffynnon Ddu drainage system just over the surface watershed to the west, in the Glyntawe Valley.

Glyntawe and the Black Mountain (Mynydd Ddu)

The Afon Tawe starts as a series of streams on the slopes of Fan Gihirych and Fan Hir, eventually flowing southwards to Swansea. The streams all collect on the Old Red Sandstone uplands of the Brecon Beacons and the river is augmented after a few kilometres by large resurgences where the River Tawe cuts through the limestones of the Northern Outcrop. Two major cave systems (Ogof Ffynnon Ddu and Dan yr Ogof) are found behind the two main resurgences at Glyntawe, and they have been the focus of much exploration and scientific research. Ogof Ffynnon Ddu has been explored and surveyed from the resurgence to a point almost beneath the main sink (Pwll Byfre) and several minor feeders have been identified. By comparison the hydrology of Dan yr Ogof is less well understood.

Lycopodium-spore tracing has shed considerable light on the hydrology of the whole of the Black Mountain region in recent years, from Dan yr Ogof in the east to Llygad Llwchwr in the west. The results are detailed below and in Figure 4.12.

The Ogof Ffynnon Ddu catchment

Ogof Ffynnon Ddu comprises 45 km of passage with a vertical range of over 350 m extending from SN 8750 1640 (458 m altitude) to its resurgence at SN 848 153 (204 m altitude) (Smart & Smith, 1976). The results of dye-testing are as follows: (a) dye placed in Pwll Byfre (SN 8745 1660; 488 m altitude), a major sink at the foot of a gritstone cliff, resurges at Ffynnon Ddu within 48 h; (b) Horseless Carriage Dig (SN 8561 1573; 335 m altitude) has been traced to the main stream

Table 4.5. *Sinks in the Nedd Fechan catchment*

	Grid reference	Altitude (m)
Ogof Cul	SN 9017 1510	400
Pant Mawr Sink	SN 8909 1622	436
18th Hole	SN 8991 1383	400
Sarn Helen Sink	SN 9076 1395	327
Pwll Derwen	SN 9087 1294	331

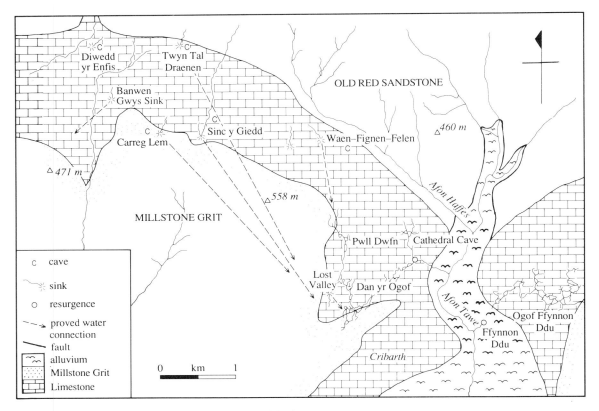

Fig. 4.12. Sketch map of the geology and underground drainage pattern of the Tawe Valley area.

in the Cwm Dwr Series of the cave after an unknown time; and (c) dyes introduced into two normally dry dolines at SN 859 169 and SN 859 160 were seen in the Cwm Dwr stream after 20 h.

The Dan yr Ogof catchment

Dan yr Ogof (SN 8381 1600; 216 m altitude), contains 14.4 km of surveyed passage and is the source of the Afon Llynfell which flows a few hundred metres to join the Afon Tawe near Craig y Nos Castle (Coase & Judson, 1977; Gascoine, 1983). The cave is developed in the Dowlais Limestones beneath the Honeycombed Sandstone (Fig. 4.13). The Sandstone is visible in the roof and in avens throughout the system. Sinks are plentiful in the area to the west and north of the known cave and, even in dry weather, many take small streams.

Dye-traces to date: the first two dye-traces to identify major stream inflows to the Dan yr Ogof system were carried out in 1970. Sinc y Giedd (SN 8101 1784; 434 m altitude), and a sink at the edge of a peat bog, Waen Fignen Felen (SN 826 177; 438 m altitude), were traced to the Dan yr Ogof system, the dye taking 36 h and 24 h, respectively, to reach the resurgence. In 1982, traces using *Lycopodium* spores identified three other feeders to the Dan yr Ogof system: (a) spores put into a sink at Twyn Tal Draenen (SN 807 191; 511 m altitude) were recovered in the Dan yr Ogof resurgence after 48 h, having travelled 5 km; (b) a sink on Carreg Lem (SN 805 178; 461 m altitude) proved positive after 48 h; and (c) a sink at SN 826 156 (415 m altitude) in the Lost Valley area above Dan yr Ogof

near Pwll y Wydden was traced to the resurgence in 48 h. In the same *Lycopodium*-tracing exercise two other sinks gave negative results: (d) spores placed in a 300 m long cave sink called Ogof Diwedd yr Enfis (SN 796 191; 423 m altitude) did not appear at Dan yr Ogof, even though nets remained in place for 8 and 15 d; and (e) spores placed in a sink near Banwen Gwys (SN 798 184; 431 m altitude) failed to appear in Dan yr Ogof.

These tests indicate that the Dan yr Ogof drainage system has a western hydrological limit in the Carreg Lem area. Its northern extent is controlled by the limestone outcrop in the Tal Draenen area and its southern limit is the Cribarth Disturbance, a complex fault-belt trending northeast–southwest to the south of the resurgence and cave entrance. The limestone in the Lost Valley dips steeply to the north whereas the general dip in Dan yr Ogof is shallow and to the south.

Adjacent to Dan yr Ogof lie three other caves of note, one sink and two resurgences:

Cathedral (= Tunnel) Cave (SN 8388 1610; 236 m altitude) contains 2200 m of passage and several small streams, thought to be mostly percolation water.

Pwll Dwfn (SN 834 165; 399 m altitude) contains 97 m of passage, mostly with vertical development, and lies above and to the east of the Great North Road area of Dan yr Ogof. In the late 1960s, its water was traced to the Dan yr Ogof resurgence in 91 h.

Hospital Cave (SN 8394 1522; 274 m altitude), with 122 m of passage, lies to the south of Dan yr Ogof.

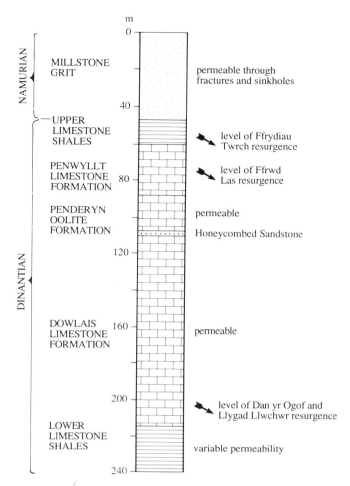

Fig. 4.13. The succession of strata in the Dan yr Ogof area of the Tawe Valley.

Pwll y Wydden (SN 8305 1574; 396 m altitude) is a large sink which was traced to the Dan yr Ogof resurgence in the late 1960s. No flow time was recorded.
Gwyn Arms Rising (SN 846 166; 207 m altitude).
Craig y Nos Hospital Rising (SN 8440 1560; 195 m altitude).

Neither of these two risings has been confirmed as a resurgence for any known caves in the Glyntawe area, although watches were kept on both risings when dye-tests were made in Ogof Ffynnon Ddu.

The Twrch Valley

Seven kilometres to the west of the River Tawe, the Afon Twrch has cut a shallow valley into the Black Mountain uplands (Fig. 4.14). It begins as two streams near the summit of Carreg yr Ogof (590 m altitude) and flows southwards, cutting through the limestone outcrop, to join the Afon Tawe near Ystradgynlais.

The Twrch Valley is sub-parallel to two major faults which have displaced the limestone outcrop 1 km to the north. On

one of these faults, at a point where the overlying Millstone Grit appears, lie two substantial resurgences. On the east bank of the Afon Twrch is Ffrwd Las, a multiple resurgence issuing from gravel- and sand-filled depressions and from a small cave passage which is totally flooded. Above the west bank of the Afon Twrch, 60 m higher, lies Ffrydiau Twrch, a large resurgence flowing from the top of a high boulder scree (Hartwright, 1974). Ffrwd Las is at SN 7739 1638 (290 m altitude) and the water issues from the Penwyllt Limestones. Ffrydiau Twrch is at SN 7704 1623 (351 m altitude) with water draining from the Upper Limestone Shales (see Fig. 4.14).

Results of water-tracing

When spore tests were being undertaken on the Dan yr Ogof catchment, a plankton net was also placed in Ffrwd Las, as it was felt that sinks beyond the western limit of the Dan yr Ogof catchment could feed Ffrwd Las. Previous work, in the early 1970s, had identified one feeder to Ffrwd Las. Fluorescein placed in a sink near the summit of Carreg yr Ogof (SN 782 213; 551 m altitude) appeared in Ffrwd Las resurgence after a few days. This sink, at the southern edge of a peat-filled depression, is over 5 km from Ffrwd Las and >260 m higher. The test was repeated using *Lycopodium* spores and after 4 d the net at Ffrwd Las gave positive results.

The sinks at Diwedd yr Enfis and Banwen Gwys were considered to be likely candidates as feeders to Ffrwd Las, especially when they failed to give positive indications at Dan yr Ogof. Despite repeat tracing, Diwedd yr Enfis failed to register a positive connection to either Dan yr Ogof or Ffrwd Las. Banwen Gwys has not been tested a second time, and this leaves a puzzle in the central area of the Black Mountain requiring further work, perhaps to locate as yet undiscovered resurgences in the bed of the Afon Twrch or the Gwys Fawr and Gwys Fach to the east.

Just downstream of Ffrwd Las and above the opposite bank of the Twrch lies Ffrydiau Twrch which gives an almost constant flow of water of over six million litres per day. Some dye-traces were carried out in the early 1970s, following the discovery of Ogof Pwll Swnd, and two positive results were obtained: (a) Ogof Pwll Swnd (SN 7606 1844; 579 m altitude), with 914 m of passage, contains a small stream which was traced to Ffrydiau Twrch; and (b) Llynfell Sink (SN 7620 1750; 457 m altitude), a large sink in a quarried area, was also traced to Ffrydiau Twrch in around 24 h. The Llynfell Sink test was repeated using *Lycopodium* spores in 1982 and a positive trace to a net in Ffrydiau Twrch was obtained in under 4 d. Also in 1982, a sink on Moel Gornach (SN 748 182; 492 m altitude), to the west of Ogof Pwll Swnd, was treated with *Lycopodium* spores in an attempt to delimit the western edge of the catchment of Ffrydiau Twrch. The spores did not appear at the resurgence or in the resurgence cave Llygad Llwchwr.

The volume and constancy of the flow from Ffrydiau Twrch must indicate a considerable area of catchment but the tracing tests failed to give a clear picture of the cave potential in this area of the Black Mountain. Many discoveries possibly await any caver who pursues a detailed attack on the many sinks and dolines in this part of the Northern Outcrop.

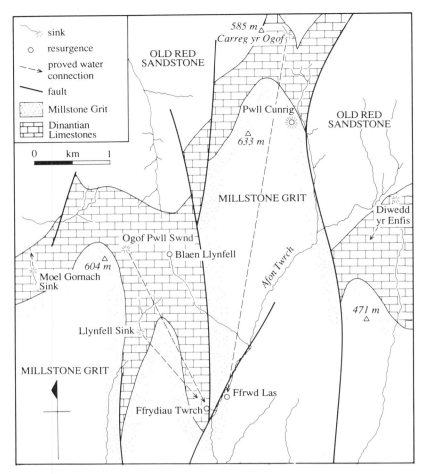

Fig. 4.14. Sketch map of the geology and underground drainage pattern of the Afon Twrch area.

The western area of the Black Mountain

The Llygad Llwchwr catchment

To the west of the Twrch, the Black Mountain uplands stretch unbroken for over 10 km from Brynamman in the south to Llangadog in the north, eventually falling to the valleys of the rivers Cennen, Llwchwr and Towy (Machin, 1974).

Two large resurgences lie at the western end of the mountain:

Llygad Llwchwr (SN 6688 1780; 229 m altitude) where the mature River Llwchwr flows out of 305 m of fine stream cave containing large chambers and deep canal passage (Fig. 4.15).

Llandyfan Church Risings (SN 6477 1712; 152 m altitude), a resurgence enclosed by the Welsh Water Authority for supply near the village of Llandyfan.

Llygad Llwchwr has long been a site of interest to cavers. Being similar to the Dan yr Ogof resurgence, it has always given the impression that a large cave system may extend beyond the flooded passages under Black Mountain. The water issues from the Dowlais Limestones in an area where the dip is steeply inclined to the south.

Until recently no dye-traces have been attempted to Llygad

Llwchwr because of the involvement of the Water Authority and the enormous dilutions that would be experienced by dye emerging from such a system. However, using the *Lycopodium*-spore technique, two successful traces were made to the resurgence in 1982: (a) a sink called Pwll Cwm Sych (SN 6907 1835; 341 m altitude), a large peaty depression on the northern flank of Banwen Gwythwch, was treated with spores and the net in Llygad Llwchwr showed positive after 4 d; and (b) more significantly, a sink called Sinc Ger y Ffordd alongside the Brynamman–Llangadog Road (A4069) (SN 7317 1892; 472 m altitude) containing a few metres of rather unstable passage was spore-traced to Llygad Llwchwr in less than 7 d. The distance of over 7 km from the sink to the resurgence and the presence of several faults between the two sites, makes this trace highly significant to the hydrology of the Black Mountain. The test results have since been augmented using fluorescein dye placed in a stream in the cave Ogof Foel Fawr (SN 7342 1875; 427 m altitude). Ogof Foel Fawr is to the east of Sinc Ger y Ffordd, above and just to the south of Herberts Quarry, i.e., at a greater distance from and a higher altitude above Llygad Llwchwr.

The Black Mountain area between Dan yr Ogof and Llygad Llwchwr gives an unbroken band of Carboniferous Limestone upland bounded by Millstone Grit to the south. The region

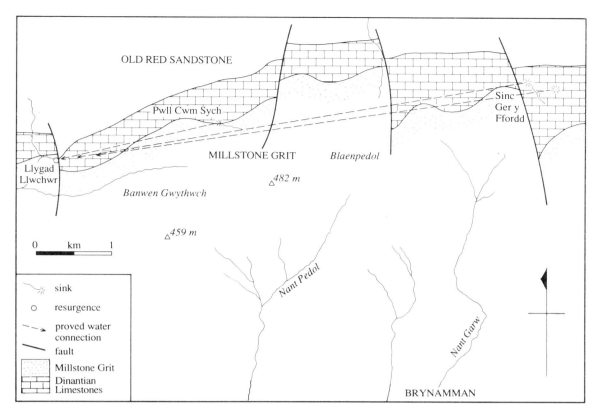

Fig. 4.15. Sketch map of the geology and underground drainage feeding Llygad Llwchwr, north of Brynamman.

contains sinks and dolines in an area spanning 17 km east–west and 3 km north–south. Considering the thickness, dip and character of the limestones, the only substantial cave, Dan yr Ogof, must represent a small fraction of the potential caves in the area. Resurgences like Ffrwd Las, Ffrydiau Twrch and

Llygad Llwchwr must be outlets for major cave systems and, like Mynydd Llangattwg and Llangynidr at the eastern end of the limestone outcrop, this western area has the potential for several cave systems of international importance.

References

Coase, A. C. & Judson, D. M. (1977). Dan yr Ogof and associated caves. *Transactions of the British Cave Research Association*, **4(1–2)**.

Gascoine, W. (1979). Ogof Craig a Ffynnon: an attempt to trace its water. *Y Ddraig Goch*, **6**, 90–4.

Gascoine, W. (1980). Continued water tracing on Mynydd Llangattwg and Mynydd Llangynidr. *Y Ddraig Goch*, **7**, 57–64.

Gascoine, W. (1983). A hydrological study of the Dan yr Ogof and Ffrwd Las resurgences. *South Wales Caving Club Newsletter*, **97**, 9–14.

Hartwright, P. (1974). Pwll Swnd–Ffrydiau Twrch connection proved. *Hereford Caving Club Newsletter*, **28**, 22–3.

Lloyd, O. C. (1979). The Hepste river caves and a study of the Hepste–Mellte area. *Proceedings of Bristol University Spelaeological Society*, **15(2)**, 107–27.

Machin, W. R. (1974). Underground drainage of the Carmarthenshire Black Mountain limestone outcrops. *Hereford Caving Club Newsletter*, **28**, 15–21.

Smart, P. & Smith, R. (1976). Water-tracing in Ogof Ffynnon Ddu. *Y Ddraig Goch*, **3**, 45–50.

Standing, P. A., Newson, M. D. & Wilkins, A. G. (1971). Little Neath River Cave. *Proceedings of Bristol University Spelaeological Society*, **12(3)**, 303–25.

Welsh Water Authority (1980). *Southeast Wales Groundwater Study*. Report by Howard Humphreys & Partners.

5

Cave biology in South Wales

G. T. JEFFERSON

Although there had been some sporadic collecting as early as the latter half of the last century, the systematic study of cave biology in Britain started only in the 1930s. It was, indeed, not until 1938 that any attempt was made to examine the fauna of a wide range of British caves; this was the time when E. A. Glennie, home on leave from India, started, together with his niece, Mary Hazelton, the serious collection of underground fauna in all the major British caving areas. South Wales, at that time, hardly ranked as one of these but, nevertheless, Glennie did go caving in the area and even undertook some collecting (Railton, 1969); this seems not to have been very successful as his fauna records for 1938 include none from South Wales. The area was, however, starting to interest cavers, and in August 1939 the British Speleological Association held its annual conference in Swansea. Mary Hazelton took the opportunity to collect in several caves, and it was in Dan yr Ogof that she made one of the outstanding discoveries of the early days of British biospeleology. This was the finding, for the first time in a British cave, of the blind white crustacean, *Asellus* (now *Proasellus*) *cavaticus*, a truly underground animal known from some European caves (Fig. 5.1); the very few previous British records of it had been from wells and similar situations.

Within a month of the BSA conference war broke out, and for six years there was little opportunity for caving, but in South Wales the stage was set for great developments, and with the coming of peace many cavers were attracted to the area. The biological side was not neglected; in 1945 Don Coase was collecting in South Wales, and by the following year so were several others. It was Harold Davies who, on 19 April 1946, found the 'well-shrimp', *Niphargus*, for the first time in a British cave. This was again in Dan yr Ogof, but on the very next day Brigadier Glennie collected one in Porth yr Ogof; both specimens were identified as *Niphargus? aquilex*, but were probably *Niphargus fontanus*.

This basically important work of collecting specimens, so that they could be sent away to experts for identification and recording, was undertaken by a number of enthusiastic cavers, both in the previously known caves and in the new ones as they were discovered. It has continued, although more recently at a somewhat reduced rate and on a more selective basis, ever

Fig. 5.1. *Proasellus cavaticus* – an aquatic cave isopod (Crustacea) widely distributed in South Wales, particularly in the stream caves of the North Crop. Scale bar = 1 mm (photo by M. Ware).

since. Indeed, South Wales is represented in every issue of British cave fauna records published, under a variety of titles, by the Cave Research Group (Hazelton, 1955 *et seq.*). The major part of the biological work done in South Wales caves, as in those of other parts of Britain, has been of this basic faunistic type, building up a picture of the range of organisms occurring in the various caves (Hazelton, 1971, 1972). It is an essential part of cave biology, and provides a foundation for ecological and other types of studies which have begun in more recent years. This is not to say that faunistic work is finished. Much remains to be done and South Wales caves have continued to have their share of exciting finds of creatures new to Britain, or, in some cases, new to science (Hazelton, 1974).

This chapter deals mainly with animals, but it is interesting to note that the study of the flora of caves in South Wales has had a somewhat different history. It began rather later with the pioneering work of A. Mason-Williams & K. Benson-Evans (1958) which, although resulting in bacterial and botanical flora lists for a number of caves, involved a distinctly ecological approach from the start. More attention was also paid to the cave threshold than has been the case in zoological work until quite recently. Subsequent studies, some of which will be mentioned later, have, in the main, been confined to the microbiological side of the flora, except for some interesting observations on the ferns growing in cave entrances (Lloyd, 1977). Very limited research has been carried out in North Wales, and the results are summarized in Appendix 5.1. A full list of fauna recorded so far is given in Appendices 5.2 and 5.3.

The ecological classification of cave animals

It was realized very early in the history of biospeleology that cave animals could be put into several different categories on a basis of the use they make of the cave. Some come and go, spending only part of their lives in the cave. Others, although occurring elsewhere, are able to live in caves permanently, while others yet again are only to be found underground. The usefulness of such a broadly ecological classification has led to the almost universal adoption of the long-established Schiner–Racovitza system, so named after its originator (Schiner, 1854) and subsequent modifier (Racovitza, 1907). This recognizes three main categories which can be defined along the following lines:

> Trogloxenes – Species which may be found in caves, but do not complete their life cycles there. They may be accidental or habitual trogloxenes. The former come into the cave accidentally, survive for varying lengths of time and may even complete their development, but are unable to establish further generations there. Habitual trogloxenes – bats are a familiar example – move in and out of the cave, using it for only part of the time, perhaps as a roost or for hibernation.
>
> Troglophiles – Species which live successfully underground, and form permanently established populations there, but also occur in suitable habitats elsewhere.
>
> Troglobites – Species found only in underground habitats, and, in nature, unable to survive elsewhere.

It is also possible to put the animals into categories based on the kind of habitat they occupy within the cave. There are, of course, many different habitats which can be recognized, but in broad terms the animals are either aquatic or terrestrial, and occur in either the cave threshold or the dark zone, and sometimes in both. Whether an animal is aquatic or terrestrial depends on whether it normally lives in or out of water; in caves high atmospheric humidities make this distinction less rigorous than in many other situations, and animals are sometimes found in, as it were, the wrong medium, but there is rarely much difficulty in deciding which is the normal one. The threshold is usually defined as extending from the cave entrance to the furthest point to which daylight ever penetrates. It frequently has well-marked gradients in its physical characteristics such as light, temperature and humidity, and these show some daily and seasonal variation, especially near the entrance; the dark zone, on the other hand, has no natural light, and in its deeper parts, the other physical characteristics tend also to be remarkably uniform and constant.

The threshold fauna

The threshold fauna has been rather neglected in South Wales as in other parts of the country, although it is now attracting greater interest (Jefferson, 1982, 1983). Of the more obvious threshold trogloxenes, bats also penetrate into the dark zone. They are rare in South Wales caves, but can occasionally be seen roosting or hibernating underground in most parts of the region. Most sightings are of lesser horseshoe bats (*Rhinolophus hipposideros*), the greater horseshoe (*Rhinolophus ferrumequinum*) now being very rare in this area. A threshold trogloxene which does not penetrate into the dark zone, nor even into the deeper parts of the threshold, is the rock dove (*Columba livia*). These birds regularly nest in numbers in several of the coastal caves of Gower; in this part of the country they have presumably reverted from feral pigeons.

The various invertebrates, many of them insects, which use cave thresholds merely as overwintering sites, are also trogloxenes. Among these, and more conspicuous than most, are two fairly large moths, the tissue moth (*Triphosa dubitata* (L.)) and the herald moth (*Scoliopteryx libatrix* (L.)) (Fig. 5.2a,b). These species are to be seen, from autumn to late spring, settled on the walls or roofs of cave thresholds in many parts of South Wales, as they are in most cave areas of Britain and some other temperate regions. The common mosquito (*Culex pipiens*) also overwinters in cave thresholds and large numbers can be seen in many Welsh caves from September to April. These are females of the autumn generation which have already mated but which will not lay their eggs until after leaving the cave in the spring.

A few threshold trogloxenes use the cave for aestivation, presumably avoiding the heat and dryness outside – at any rate they appear in caves during the summer. *Stenophylax permistus*, a limnephilid caddisfly, seems to be such a case in South Wales. The adults appear fairly regularly during the summer months in caves situated, as many are, in the vicinity of streams suitable for the larvae. These adults may be accidentals, but there is no obvious reason why this species should be more prone to being stranded in caves than other caddisflies. It appears (e.g., Dobat, 1973; Bouvet, 1975, 1976) that in Europe adults of *S. permistus* and a few related species, on emerging from the pupa, enter caves and mate there. The females appear to need a period of quiescence before development of their eggs can be completed and they may stay in the cave for over two months before leaving to oviposit in a suitable stream. It would be interesting to know whether their behaviour in Britain is similar. *S. permistus* is not the only summer visitor to cave thresholds; the cranefly *Limonia nubeculosa* is another and this can be seen in many South Wales caves, sometimes in considerable numbers.

Many of the threshold troglophiles, especially the Diptera, also occur in the dark zone, but the 'cave spider' (*Meta menardi*) and the related *Meta merianae*, are outstandingly characteristic of the threshold. Both are widely distributed in South Wales, and in some caves, for instance parts of Porth yr Ogof, are very numerous. These spiders have been studied at

(a)

(b)

Fig. 5.2. Moths; (a) the tissue moth (*Triphosia dubitata*); and (b) the herald moth (*Scoliopterix libatrix*) are often seen over-wintering in cave thresholds (photo (a) by P. R. J. Chapman and (b) by C. Howes).

University College, Cardiff, and this work has confirmed the findings of Dresco-Derouet (1960) that *M. menardi* has a low metabolic rate, as shown by its oxygen consumption, compared with that of many surface-dwelling spiders. The basal metabolic rate of *M. merianae* is similar to that of *M. menardi*, but is distinctly higher in juveniles of both species. A low metabolic rate is thought to be an adaptation, or in this case a pre-adaptation, to life in caves, where food tends to be sparse.

A detailed study by J. N. Goodwin (pers. comm.) of the distribution of the two species in various parts of Porth yr Ogof, a cave with many thresholds, has shown, as has long been suspected, a marked zonation. *M. merianae* is almost always nearer to an entrance, or at any rate is in a higher light intensity, than *M. menardi* which extends at least to the extreme depths of the threshold. These spiders appear to feed on virtually any available prey, and succeed in catching in their webs both flying and crawling insects. They have been observed feeding on, among other things, woodlice and millipedes.

The dark zone fauna

The fauna of the dark zone of South Wales caves has received more attention than that of the threshold, but the amount of work done in different parts of the region has varied greatly. It is not perhaps surprising that most has been done in the great caves of the North Crop, but other areas trail far behind in terms of the amount of information available. Their caves are fewer, smaller and on the whole less interesting than those of the North Crop, but they would repay a little more attention from the cave biologist. Of these areas the Forest of Dean–Wye Valley, particularly Otter Hole, has fared best, followed by Gower. Records are sparse for the South Crop, and almost non-existent for the far west.

Although bats may penetrate into the dark zone, the majority of trogloxenes found there are accidentals. This applies particularly to aquatic animals carried underground by sinking streams. Most of the species present immediately upstream of a swallet can usually be found, naturally in reduced and variable numbers, for some distance below it. Many of these accidental trogloxenes are the larvae or nymphs of flying insects, and some eventually complete their development and emerge as adults. These can often be seen in stream passages, such as those of Ogof y Ci, Porth yr Ogof or Ogof Nedd Fechan, sometimes flying, but more often immobile on the walls. The groups most commonly represented are mayflies, stoneflies, caddisflies and various kinds of midge.

Vast numbers of 'shrimps' have been seen in the Otter Hole streamway; these were probably *Praunus flexuosus* which had swum upstream. This essentially marine form has been recorded from the entrance series of this cave (Chapman, 1979). Fish may also enter underground stretches of rivers or streams, sometimes almost certainly through the sink, but in other cases probably by swimming up through the resurgence. Trout are the fish most often seen in South Wales caves, but there are others; *Pomatoschistus* (probably *P. microps*) which is an estuarine goby, has been caught in Otter Hole, and *Cottus gobio*, the freshwater bullhead or miller's thumb, can be found among stones in the river in Porth yr Ogof. It seems to be little affected by the darkness, even its colour being substantially normal.

This, however, is not the case with trout seen underground; they appear pale, and are often reported as 'white fish', although in the hand they usually belie the first impression and prove to be not quite completely white. They are common brown trout (*Salmo trutta*), but in a remarkably blanched condition (Fig. 5.3). The dark spots, which in any case tend to

Fig. 5.3. A blanched trout (*Salmo trutta*). This specimen shows signs of emaciation, but some trout living in caves appear to be well fed (photo by G. T. Jefferson).

be rather variable in trout, are reduced to almost nothing, although the row of orange–red spots along the lateral line is often retained. Whether the blanching is entirely the result of normal melanophore action – the withdrawal of pigment granules into the centres of the chromatophore cells – or due to any actual loss of melanin pigment is not known, but such fish darken rapidly if brought into daylight (Proudlove, 1982).

Similar underground fish occur in other parts of the country (e.g., Dixon, 1974), but they seem to be particularly well represented along the North Crop limestone of South Wales, having been seen in the Glais Valley caves, Porth yr Ogof, Ogof Nedd Fechan, Cwm Pwll y Rhyd/White Lady, Ogof Ffynnon Ddu III and probably other caves as well. The status of these underground trout – whether they are trogloxenes or troglophiles – is not settled. They give the impression of being able to maintain themselves, and even of being well established. Food is available, particularly in the form of invertebrates carried underground by the stream, and presumably the trout feed, even though they normally rely on sight when catching their prey. One such trout kept in the laboratory in darkness, was provided with *Gammarus* from time to time, and although the feeding behaviour of the fish could obviously not be observed, the *Gammarus* seemed always to be taken.

There is as yet no real evidence that trout living in caves, well established though individuals may appear to be, belong to permanent populations reproducing and maintaining themselves underground. In the absence of such evidence there is no reason to consider them to be troglophiles, and it seems probable that they are trogloxenes. Even so it would be interesting to know whether these fish move in and out of the cave, presumably becoming darker when outside. This sort of movement could take place in some South Wales caves, but seems virtually impossible in others; in these latter cases, if the trout are trogloxenes, they must presumably be purely accidental ones.

The dark zone troglophiles call for little comment; most of the species, ranging from flatworms to mites, which are recognized as troglophilic in Britain, have been recorded from South Wales. Mention must nevertheless be made of the very common cave gnat (*Speolepta leptogaster*) which is particularly numerous and widespread in this region. Like many troglophiles, *Speolepta* can be found in the deep threshold as well as in the dark zone, but it is only rarely seen outside caves, and some consider it to be a troglobite. The thin translucent larvae, up to approximately 14 mm in length, live on damp cave walls, and are usually thought to feed on micro-organisms and fungal material (Matile, 1962), but there have been claims (see Thinès & Tercafs, 1972) that they are really predators (Fig. 5.4). In South Wales at least this seems unlikely. Observations on larvae in several caves, and the examination of gut contents, have produced no evidence of such feeding. The larval eyes of *Speolepta* are sometimes de-pigmented (Schmitz, 1913), but this appears to be a very rare phenomenon (Jefferson, 1981) and has not been observed in specimens from South Wales.

An animal not previously recorded from British caves also warrants a mention as it may be troglophilic. This is *Aeolosoma hemprichi*, an aquatic oligochaete worm, less than

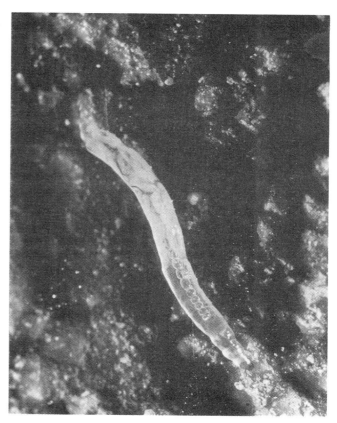

Fig. 5.4. A larva of the cave gnat (*Speolepta leptogaster*); the head capsule, towards the bottom of the picture, is opaque, but the rest of the body is transparent and bubbles can be seen in the gut (photo by P. R. J. Chapman).

1 mm long and eyeless. It is transparent and colourless except for the contents of scattered epidermal glands which embellish it with bright orange–red spots. These worms were found, in 1972, to be present in some numbers in a small drip-fed pool in Ogof Ffynnon Ddu II, living on, and in, a layer of silt where they seemed to be feeding on bacteria or detritus. *Aeolosoma* is able to reproduce asexually by a process of fission, and the population could have built up from a single individual which happened to reach the isolated pool, possibly in an encysted state. If this is in fact what happened, the species would qualify for the status of troglophile.

Troglobites are, in general, confined to the dark zone, with an occasional appearance in the deepest parts of the threshold. They are, however, sometimes swept to the surface accidentally, and in times of flood may appear briefly in springs. Some aquatic troglobites also live in phreatic waters, and are not uncommon in wells, particularly in the chalk areas of southern England. Certain species of *Niphargus*, indeed, are commonly called 'well-shrimps'. There has been very little biological investigation of wells in South Wales, and almost all records are from caves, with just a few from springs and interstitial ground waters. All known British troglobites belong to the phylum Arthropoda, but are distributed among three of its four major classes; the Crustacea, Insecta and Arachnida. South Wales has representatives of all three classes among its troglobites.

Jeannel (1959) noted that troglobites may be, relatively speaking, old or new – the so-called palaeotroglobites and neotroglobites, respectively. The former have been long established as troglobites, having probably gone underground in Tertiary times. They are usually modified in various ways, and tend to be relicts or survivors of rather ancient groups. Neotroglobites, on the other hand, are recent cave animals, little modified, and in most cases closely related to existing surface forms. Many of them seem to have gone underground since the last retreat of the Pleistocene glaciers. In South Wales, as in the rest of Britain, it is only a few aquatic forms, all crustaceans, which are undoubted palaeotroglobites. Other troglobitic animals, both terrestrial and aquatic, are present but these are probably neotroglobites in most cases.

Most of the British troglobitic crustaceans belong to the sub-class Malacostraca, but the Ostracoda includes many interstitial forms which live in the water-filled spaces between grains of sand or gravel, and some of those which penetrate into phreatic waters appear in caves. Ostracods have been found in several South Wales caves, but their status is far from clear. One which is present in some numbers in Ogof Ffynnon Ddu I was identified by the late Professor Munro-Fox as *Cypridopsis subterranea*, and this can provisionally be considered a troglobite. There are a few records of it from springs, but their occurrence on the surface may have been accidental.

Among the Malacostraca the syncarid *Bathynella* is certainly an ancient relict form. It occurs in some Yorkshire caves (see Dixon, 1974), but seems to be essentially an interstitial animal inhabiting rather shallow phreatic waters, in which situation it is quite widespread. It has not been found in Wales, but this may merely mean that it has not been looked for in the right places. The isopod *Proasellus cavaticus*, better known under its old name, *Asellus cavaticus*, is a palaeotroglobite. It is blind and de-pigmented, and seems to have been an underground animal for a very long time. Since it occasionally appears in wells and springs, it must occur in phreatic waters, but it is typically an animal of caves where it is most often to be found in the film of water flowing over stalagmite slopes or on stones in streams.

The main areas of distribution of *P. cavaticus* in Britain, and the places where it occurs in caves, are the Mendip Hills, Somerset, and South Wales. There appear to be differences between specimens from the two areas and the suggestion that they represent different sub-species might be justified. South Wales specimens consistently grow larger than those from Mendip (Hazelton, 1975) and recent observations (Jefferson, unpublished) suggest that they produce more young per brood. *P. cavaticus* is widely distributed in caves along the North Crop limestone of South Wales. It also occurs in Lesser Garth Cave on the South Crop and in Otter Hole near Chepstow where the specimens seem to be of the South Wales type. It has not been found elsewhere in the region except for one record, apparently from the top of Barland Sink in Gower. This is interesting because Guzzle Hole, lower down the valley, yielded what were apparently aberrant *P. meridianus*, a related but not troglobitic species. Clearly the asellids of the underground stream in the Bishopston Valley warrant further investigation.

The remaining five malacostracan troglobites of Britain are all amphipods. *Crangonyx subterraneus* appears to inhabit interstitial waters, and has been found in some numbers in wells, gravel beds and the like in southern England. It has, however, been recorded twice from caves, and one of these was in South Wales, namely Ogof Pant Canol, which is part of the Ogof Ffynnon Ddu system. The other was Gough's Cave in Mendip. The habitat of *Niphargus aquilex* is similar, and this species is widely distributed in interstitial waters, particularly those of gravel beds, in many parts of Wales and the southern half of England. It has been found in the gravel of a number of South Wales rivers including some which are quite heavily iron-polluted. *N. aquilex* is occasionally reported from caves, but in South Wales the only firm record is from the Red Chamber in Gower. The two dubious records mentioned earlier from Dan yr Ogof and Porth yr Ogof were probably not of this species, although *N. aquilex* does occur interstitially in the river downstream from Porth yr Ogof.

Niphargus kochianus inhabits rather deeper phreatic waters, being typically a 'well-shrimp' of the limestone and chalk areas of southern England, excluding Devon. It occurs, although not commonly, in some Somerset and Bristol caves, but is rather surprisingly absent, or at least has not yet been recorded, from any location in Wales. This is all the more remarkable because this is the species, even though usually considered a different sub-species, which occurs in Ireland.

The largest British species of *Niphargus*, *N. fontanus*, seems to extend more deeply into phreatic waters than any of the others (Fig. 5.5). It has been recorded from wells in much the same parts of southern England as *N. kochianus*, but it is characteristically a cave animal. In this habitat it occurs, like *Proasellus cavaticus*, in South Wales and Mendip. Within South Wales it is widely distributed throughout most of the

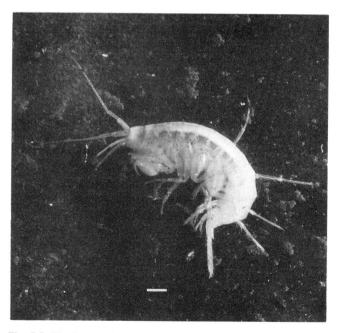

Fig. 5.5. The largest of the British cave 'shrimps' (*Niphargus fontanus*), is an amphipod crustacean quite common in South Wales caves. Scale bar = 1 mm (photo by G. T. Jefferson).

North Crop limestone with only two records from elsewhere in the region – from one of the Clearwell Mines in the Forest of Dean,and from Otter Hole. In caves *N. fontanus* is usually seen in gour pools. The remaining troglobitic British amphipod, *Niphargellus glenniei*, does not occur in Wales, its total known world range being confined to Devon.

The terrestrial troglobites of Britain are nearly all neotroglobites, and most, but not all, of them are closely related to species living in soil, the habitat from which many such cave-dwelling forms were no doubt derived. Of the insects in the British cave fauna some dipteran flies come close to qualifying as troglobites and a beetle, a species of *Aloconota*, has, in Britain, only been found in Otter Hole (Chapman, 1979). It is, however, only to a few species of collembolans (springtails) that the status of troglobite can really be accorded, and then only with some diffidence owing to the difficult taxonomy of the group and the very limited information about the habits and ecology of many of its species. There are perhaps four British Collembola which can provisionally be regarded as troglobites, and so far three of these have been recorded from South Wales. These are *Onychiurus schoetti*, which is widely recorded from caves in most parts of the region, and *Onychiurus dunarius* which has been found in Ogof Rhyd Sych and Ogof y Ci. The third species, *Pseudosinella dobati*, is known to occur in a few South Wales caves including Otter Hole and Ogof Ffynnon Ddu. Several of the other collembolans recorded may be troglobitic in some parts of their ranges. *Arrhopalites pygmaeus*, for instance, is listed by Dobat (1975) as a troglobite in Swabia, southern Germany, and by Turquin (1972) in the Jura méridional in southeastern France.

The troglobitic arachnids of Britain comprise one spider and some half-dozen mites, although, as with the Collembola and for similar reasons, the status of some of the latter can only be provisional. The troglobitic spider, *Porrhomma rosenhaueri*, is only about 2 mm long. It is known from several European caves but in the United Kingdom has only been found in South Wales, and then only in two caves – Lesser Garth and Ogof y Ci – in both of which it seems to be well established. There are, however, a few cave records from Ireland (New Mitchelstown Cave and Fisherstreet Pot). The genus includes several species which are troglophilic in Britain, some of them rarely seen on the surface, but only *P. rosenhaueri* is generally agreed to be troglobitic. It is pale in colour and its eyes are either vestigial or absent (Locket & Millidge, 1953). A wide range of mites has been recorded from South Wales caves but these include very few species thought to be troglobitic. Indeed, apart from a slightly doubtful record of *Rhagidia longipes*, there is only one. This record, however, is particularly interesting. It is of a mite, new to science, found by W. G. R. Maxwell in Ogof Cynnes, and described by Dr Turk (1972) under the name of *Rhagidia odontochela*. There is some evidence (Turk, 1972) that it belongs to an archaic group of mites and may be a troglobitic species of long standing. Among the mites from Ogof Ffynnon Ddu are two species, *Alicorhagia clavipilus* and *Rhagidia punkva*, which are new to Britain (A. Baker, pers. comm.). The status of these mites, whether troglobitic or otherwise, is not clear at present.

The cave flora

The important earlier work in South Wales by Mason-Williams & Benson-Evans (1958, 1967) has already been mentioned. It involved 13 caves, with all but one (Lesser Garth Cave) on the North Crop. It was essentially a microbiological study by the first author and a botanical one by the second. Green plants in thresholds were found to show a marked zonation controlled by light intensity, but also affected by the nature of the substrate. In the dark zone such plants can only be accidentals, but they were represented by bryophyte and algal spores together with a few seeds of higher plants and the occasional hopeless etiolated seedlings. A list of the fungi and moulds found in the dark zone was given, but, in the main, the microbiological work concerned the occurrence and distribution of bacterial populations in various habitats within the caves. Many were found, and they included both heterotrophic and autotrophic species.

Mason-Williams (also publishing as Edington) continued to work on various aspects of cave microbiology, and produced, either individually or in collaboration, several further papers, most of which had some particular relevance to South Wales (e.g., Mason-Williams, 1966, 1967, 1969a,b; Jones & Edington, 1968; Edington, 1977). Her work on 'moon milk' showing the part played in its formation and deposition by microorganisms, particularly the bacterium *Macromonas* sp. (probably *bipunctata*) and apparently also an alga, was done on material from Ogof Ffynnon Ddu I and some caves in the Nedd Fechan Valley (Mason-Williams, 1959, 1961). An investigation into the nature of 'wall fungus' by Mason-Williams & Holland (1967) indicated that these curious encrusting growths consist of a complex bacterium on the borderline between bacteria and fungi (a species of *Streptomyces*), possibly associated with a fungus (*Fusarium* sp.). The material for this work came from Ogof Ffynnon Ddu I and Dan yr Ogof.

The most comprehensive set of British cave flora records to be published to date is that of Cubbon (1970, 1972). This covers the entire country, but the fact that the earlier part is heavily weighted with South Wales records is indicative of the relative neglect of this field in many other cave areas until recently. The numerous species recorded include some that correspond to accidental trogloxenes, and others to troglophiles, but no plants or bacteria have yet been found in Wales, or indeed elsewhere in Britain, which correspond to troglobites in being confined to the underground habitat.

Perhaps the most surprising, although not particularly significant, recent development in South Wales cave botany has been the discovery of fully grown flowering plants, even producing inflorescence shoots, in the dark zone of a part of Porth yr Ogof (Jefferson, unpublished). This may seem to be biologically nonsensical, but the key to the enigma is that the species in question is the toothwort (*Lathraea squamaria*). Although this is a flowering plant, it is not green and does not photosynthesize. It is completely parasitic, deriving its whole nourishment from holophytic green plants. If it is objected that there can be no photosynthesizing plants in the dark zone to act as hosts, this is, of course, true, but the toothwort normally occurs in the soil, deriving its food from tree roots, and sending its flowering shoot up to the surface. In this case the limestone is sufficiently fissured for some tree roots to reach the roof of the cave, and the toothwort is attached to these. Its flowering shoots, unlike the tree roots, cannot penetrate the limestone to reach the surface, and presumably no seed can be set, but vegetatively the plants seem to be thriving. This is, of course, a special case where the plants, in a sense, are having solar energy and nutrients piped directly down to them through the tree's vascular system from the leaf canopy above.

Ecology

Two factors above all others are significant in the ecology of the dark zone of caves – the absence of light, and the remarkable constancy of the environment. Temperature, for instance, stays substantially constant at 8 or 9 °C in the deeper parts of North Crop caves in South Wales, although it may vary somewhat in the vicinity of active stream passages. The lack of light means that the cave community is devoid of a direct source of energy which elsewhere is provided by the sunlight taken up by photosynthesizing plants. This means that the food chain must, in general, start outside the cave.

There can be no doubt about the reliance of many cave communities upon *exogenous* energy coming underground in the form of organic matter originating on the surface. The two main ways in which such organic material reaches the dark zone are: by leaves, twigs or other debris being carried in by water, especially during floods; and in the material, especially faecal droppings, left by trogloxenes which feed outside, but roost in the cave. The masses of guano deposited by bats are an important example of the latter, and in some parts of the world they support very rich cave communities. Animals may eat some of the organic matter directly as when predators devour stream invertebrates carried in as accidental trogloxenes, but more often there is partial decomposition by moulds and bacteria to produce the detritus on which many cave animals feed and in their turn provide food for carnivores.

The effect of a plentiful supply of exogenous organic matter can be seen in the case of Porth yr Ogof. This cave carries a sizeable river – the Mellte – and much of it is subject to flooding. It also has many entrances, open avens and no great extent of overburden, which presumably means that some of the joints and other fissures are open to the surface. All this and heavy usage by visitors means that organic debris from outside finds its way to many parts of the cave which, in consequence, has what is by British standards, a rich fauna. The same can be said of Ogof y Ci, another stream cave with several entrances and little depth of overburden. In the larger systems, too, it is noticeable that animals, particularly terrestrial forms, tend to be more numerous in parts of the caves near the surface. This is particularly obvious in the case of the fly *Heleomyza serrata* (Fig. 5.6).

Having said this, it is nevertheless true that considerable populations of troglobitic crustaceans can be found in parts of caves which are well away from the surface, are not subject to flooding, are devoid of bat or other guano and where there are no overt signs of any particulate organic material coming from outside. This is the case in parts of Ogof Ffynnon Ddu I where estimates of the population density of *Proasellus cavaticus* in the film of flowing water on a number of stalagmite slopes

Fig. 5.6. The Dipteran fly (*Heleomyza serrata*) is common in many South Wales caves, usually where the rock cover is thin (photo by P. R. J. Chapman).

yielded results varying from 0 to $80 \, \text{m}^{-2}$ (Jefferson, 1969). These animals appear to be feeding on a floccular material present on the stalagmite surfaces, which seems to consist largely of the filamentous chlamydo-bacteria recorded by Mason-Williams (1967). These bacteria, being heterotrophic, could not be responsible for the initial build-up from inorganic constituents of the organic content of the floccular material, and its origin is far from clear. Various possibilities which have been discussed (Jefferson, 1976) include an exogenous source in the form of dissolved organics entering the cave in seepage water, an endogenous production of organic matter by chemoautotrophic bacteria which derive their energy from inorganic chemical reactions and another endogenous source in the traces of fossil organic material present in limestones.

Whatever its origin there are certainly appreciable quantities of organic matter present in caves, even in situations which seem to be inaccessible to plant or animal debris carried in from the surface. Silts from two pools in Ogof Ffynnon Ddu, fed only by seepage water, were found to have total organic contents of 2.3% and 3.3% dry weight (Jefferson, 1969). These pools contained *Niphargus* which are omnivorous, but are known to ingest a certain amount of silty material, and may well utilize any organic matter contained in it. The substrate on the floor of relatively dry parts of caves also contains organic matter, but in widely varying quantities. Determinations carried out on sediments from Ogof Ffynnon Ddu and Dan yr Ogof gave dry weight total organic contents ranging from 0.1% to 3.9% with an average of 1.5% (Jefferson, 1976). Although these figures indicate that on the whole rather less organic matter is present than in rich surface soils, clearly some material, of whatever origin, is available for utilization by animals.

Discussion

The cave fauna of South Wales, like that of Britain as a whole, is limited in comparison with those of many other parts of the world. This applies particularly to troglobites and, above all, to the ancient forms, the palaeotroglobites, although in this respect South Wales fares better than most of the other cave areas in Britain. Restricted as our troglobite fauna is, the distribution within the British Isles of the few species that there are, poses some interesting questions.

It is axiomatic in any discussion of the distribution of troglobites that, in view of the habitat to which they are confined, their ability to extend their range is restricted, and the rate at which they can spread is very much lower than in the case of surface forms. Among aquatic species the superficial interstitial forms have more widespread habitats available, and more chance of surface transport, than the deeper phreatic species, and so are able to spread rather more rapidly than these. The terrain between one karst area and another must represent an almost insurmountable barrier to deep phreatic forms, even though within any one such area they may be able to move with relative ease provided the limestone is suitably fissured.

The neotroglobite faunas of the various British cave areas are all rather similar. There are differences, some no doubt real, others merely the result of the inevitable incompleteness of records, but the general picture is of relative uniformity. This is to be expected if these forms have achieved their status only recently when, at some stage in post-glacial time, the disappearance of the widespread interconnecting surface populations converted those surviving underground from troglophiles to troglobites. The situation with the aquatic palaeotroglobites is in marked contrast to this. There are striking differences between some of the faunas, and a strong presumption of the influence of geological events at a time before the origin of the neotroglobites. In this respect South Wales shows considerable similarity to Mendip, but these two areas contrast strongly with Devon on the one hand, and more northerly areas on the other. South Wales and Mendip stand out as having the association of two typically cave-dwelling deep phreatic forms, *Proasellus cavaticus* and *Niphargus fontanus*. Devon has its endemic *Niphargellus glenniei*, an interstitial form but well represented in caves, while the more northerly areas have nothing comparable except for *Bathynella* in Yorkshire.

It seems likely that these aquatic troglobites are survivors of a British underground fauna which was in existence by late Tertiary times, and it has been suggested (Jefferson, 1974) that its differentiation into several distinct faunas might have been the result of isolation on separate islands in the past. There is a considerable amount of geological evidence that at the end of the Pliocene and in early Pleistocene times, sea level was nearly 200 m higher than at present, and was responsible for the so-called '600-foot wave-cut platform' which is particularly clear in some parts of Wales. Such a high sea level would have broken the British Isles into a number of separate islands: Devon, Mendip, Wales, the chalk downlands of southern England and others further north. The fauna of these islands would be expected to differentiate under the influence of isolation, and when the islands eventually rejoined, the surface forms would spread and intermingle, but the underground faunas would retain a measure of distinctness owing to their

exceedingly low rate of dispersal. There were, of course, fluctuations in sea level as the glaciations came and went, but these seem to have been superimposed on this eustatic fall of nearly 200 m taking place over a time span of two or more million years.

The distribution of aquatic troglobites in Britain is notably southern, and they do not extend any further north elsewhere in Europe. This is usually explained as being the result of the destruction of fauna in the areas covered by ice during the Pleistocene glaciations. When the ice retreated, surface forms could recolonize northwards, but, for the reasons already discussed, underground forms could not, or could only spread exceedingly slowly, except for the most superficial interstitial animals such as *Bathynella* and *Niphargus aquilex* which have made some progress in that direction.

There seems to be no satisfactory alternative to this explanation in general terms, but in detail there are problems, and these relate especially to South Wales. This is an area which has been repeatedly glaciated, including during the Devensian glaciation, and yet troglobitic crustacea are probably more common here than in any other part of Britain. There are two possible explanations; recolonization since the last retreat of the glaciers; or survival beneath the ice cover. The idea of recolonization from Mendip has much to commend it. The two areas are not far apart and it has been pointed out (Jefferson, 1969, 1974) that in immediately post-glacial times they were not separated by sea water as the upper part of the Bristol Channel was then dry land. There is also the similarity between the two faunas. There are, however, difficulties. The troglobitic crustacea of South Wales are most widespread along the North Crop, which is the part furthest from Mendip, and it is difficult to think of convincing explanations for their scarcity in some of the other limestone areas of South Wales. Then, the similarity of the two faunas, although very marked, is not complete. There seem to be some differences between the *Proasellus cavaticus* of the two areas, and although these might have diverged since the recolonization, a longer period of isolation seems more likely. There is also the curious fact that *Niphargus kochianus* has not been found in Wales. It seems odd that if *P. cavaticus* and *N. fontanus* have been able to spread to the North Crop, the distinctly more superficial *N. kochianus* has not been able to reach the area at all.

The alternative explanation of survival beneath the ice cover is not devoid of problems either. Glennie (1967) has put forward reasons for believing that animals such as *Niphargus* might be able to survive beneath an ice-cap which would insulate the underlying rock so that at only moderate depths the earth's heat would produce temperatures higher than the mean at the surface. This could well be true, but the survival of animals in such situations for very long periods seems more problematical. The food for such animals could hardly come from the surface during the time of ice cover, and an endogenous source would have to be invoked, either one of those discussed as possibilities earlier, or residues or surface organic matter – a sort of sub-fossil energy – carried underground during pre- or inter-glacial times and having to last, in the case of the longer glaciations, for many thousands of years.

There is also the problem of explaining why, if these animals could survive beneath the ice in South Wales, they apparently did not do so in any of the more northern regions. It can hardly be a simple matter of present-day temperatures being too low for them, since cave temperatures in some northern areas are comparable with those of the North Crop in South Wales. It may be that survival was possible only in places, like South Wales, near the edge of the ice-sheets. Alternatively, as Hazelton & Glennie (1962) have suggested, areas such as Yorkshire may not have had sufficient depth of fissured limestone beneath the ice to achieve the required temperature. In the present state of knowledge neither the possibility of survival beneath the ice, nor of post-glacial recolonization, can be ruled out. Indeed, it has even been suggested (Henry, 1977) that *Proasellus cavaticus* only reached Britain in early postglacial times when sea level was low, and that it spread, through superficial deposits, up the Thames Valley as an interstitial animal. However, in all this uncertainty, two things do seem certain: some ancient troglobitic animals are present in South Wales; and the question of how they came to be there will be with us for some time to come.

Appendix 5.1. Cave biology in North Wales

Relatively little cave biology other than work on bats, particularly the lesser horseshoe bat, has been undertaken in North Wales, but fauna records are available from 11 caves or mines in the limestone areas of Clwyd and the immediate border region. There are also a few fauna records from some non-limestone mines in other parts of the area and very few flora records (of entomogenous fungi) from the Tremeirchion caves.

None of the caves or mines has yielded any of the troglobitic aquatic crustaceans such as *Proasellus cavaticus* and *Niphargus fontanus* found further south. This may not mean that they are absent, but may merely be a reflection of the limited amount of collecting done in the area. On the other hand, their absence would be in keeping with the markedly southern distribution of these relatively deep phreatic forms and may well be real. *N. aquilex* has been recorded from a surface locality in Gwynedd, in river gravel, but this is a superficial interstitial species and has a wider distribution than the more truly troglobitic forms. The most interesting aquatic animals so far reported from underground in the area are two species of water beetle, *Hydroporus ferrugineus* and *H. obsoletus*. These dytiscid beetles are not common in any situation and it seems likely that they are troglophilic.

As might be expected, terrestrial forms are better represented in North Wales caves, and some of them are undoubted troglophiles. This is the case, for instance, with the millipede *Polymicrodon polydesmoides* and the dipteran *Speolepta leptogaster*. It probably also applies to several of the other flies which have been recorded. Many of the dozen or so species of springtail (Collembola) which have been found are certainly troglophiles and at least one, *Onychiurus schoetti*, can provisionally be considered a troglobite. Further work might well reveal other troglobitic collembolans and perhaps mites – a group which is at present particularly poorly represented in the records from the area.

Appendix 5.2. List of Welsh caves and mines from which invertebrates have been recorded

Note: These are the localities for the fauna listed in Appendix 5.3. Mines have only been included if situated in limestone. The localities are grouped according to their Watsonian vice-counties; those at the end having the prefix N refer to North Wales.

		National Grid Reference
Brecon (Vice-county No. 42)		
1.	Agen Allwedd	SO 187 159
2.	Bridge Cave	SN 912 140
3.	Cwmdwr Quarry Cave I	SN 857 156
4.	Dan yr Ogof	SN 838 160
5.	Downey's Cave	SN 848 154
6.	Eglwys Faen	SO 193 157
7.	Hospital Cave	SN 839 152
8.	Ogof Clogwyn	SO 213 124
9.	Ogof Craig a Ffynnon	SO 219 128
10.	Ogof Cynnes	SO 141 154
11.	Ogof Ffynnon Ddu I	SN 848 152
12.	Ogof Ffynnon Ddu II	SN 863 159
13.	Ogof Gam	SO 188 158
14.	Ogof Nedd Fechan	SN 912 142
15.	Ogof Pant Canol	SN 849 153
16.	Ogof Pen Eryr	SO 207 152
17.	Ogof Rhyd Sych	SO 041 102
18.	Ogof Triachwech	SN 901 160
19.	Ogof y Ci	SO 040 105
20.	Ogof (y) Daren Cilau	SO 205 153
21.	Ogof yr Esgyrn	SN 837 161
22.	Pant Mawr Pothole	SN 891 162
23.	Penderyn Quarry Cave	SN 955 091
24.	Porth yr Ogof	SN 928 124
25.	Powell's Cave	SN 850 154
26.	Shakespeare's Cave	SO 217 125
27.	Step Cave	SN 837 161
28.	Town Drain	SN 911 137
29.	Tunnel Cave (Cathedral Cave)	SN 839 161
30.	Twll Gwynt Oer	SN 859 159
31.	Weighbridge Cave	SN 866 164
32.	White Lady Cave	SN 911 137
33.	Will's Hole	SN 915 080
Carmarthen (Vice-county No. 44)		
34.	Easter Cave	SN 737 189
35.	Llygad Llwchwr	SN 669 178
36.	Ogof (Cil) yr Ychen	SN 613 166
37.	Ogof Foel Fawr	SN 734 188
38.	Pal y Cwrt II	SN 674 182
39.	Pwll Swnd	SN 762 184
Glamorgan (Vice-county No. 41)		
40.	Bacon Hole	SS 561 868
41.	Barland Quarry Sink	SS 576 896
42.	Draethen Mine (Pot)	ST 214 876
43.	Draethen Roman Mine	ST 217 877
44.	Guzzle Hole	SS 574 886
45.	Lesser Garth Cave	ST 125 821
46.	Llethrid Cave	SS 531 912
47.	Minchin Hole	SS 556 868
48.	Ravenscliff Cave	SS 546 873
49.	Red Chamber	SS 426 867
50.	Tooth Cave	SS 532 911
Monmouth (Vice-county No. 35)		
51.	Digger's Hole	SO 545 147
52.	Lady Park Wood Cave	SO 546 146
53.	Otter Hole	ST 526 961
Pembroke (Vice-county No. 45)		
54.	Hoyle's Mouth	SN 109 002
West Gloucester (Vice-county No. 34)		
55.	Bowler Combe Cave C4 (see also 60)	SO 561 154
56.	Clearwell Mines	SO 578 085
57.	Roman Hole	SO 651 199
58.	St Briavels Cave	SO 570 052
59.	Seymour Swallet	SO 591 139
60.	Symond's Yat East Caves (see also 55)	SO 561 154

Denbigh (Vice-county No. 50)

N1.	Cat Hole Mine	SJ 196 629
N2.	Dinorben Farm Sink	SH 967 747
N3.	Maeshafn Cave	SJ 198 605
N4.	Ogof Dydd Byraf/Minera	SJ 255 518
N5.	Waen Las Mine	SJ 193 608

Flint (Vice-county No. 51)

N6.	Leet Caves	SJ 189 638
N7.	Ogof Hesp Alyn	SJ 192 655
N8.	Ogof Nadolig	SJ 192 656
N9.	Pant y Mwyn Mine	SJ 188 646
N10.	Tremeirchion Caves	SJ 086 723

Salop (Vice-county No. 40)

N11.	Lower Ceiriog Cave	SJ 267 374

Appendix 5.3. Classified list of invertebrates recorded in Welsh caves and mines

Note: Only records at the level of genus or species have been included. Except in a few special cases records not going beyond the genus have not been included for localities where one or more species belonging to that genus have been recorded. Localities are indicated by the code numbers given in Appendix 5.2.

Platyhelminthes

Turbellaria
Tricladida
  Planariidae
    Polycelis felina (Dalyell), 1
    Phagocata vitta (Dugès), 1, 4, 12, 14, 19, 24
    Crenobia alpina (Dana), 18, 24
  Dendrocoelidae
    Dendrocoelum lacteum (Müller), 4

Annelida

Chaetopoda
Oligochaeta
  Aeolosomatidae
    Aeolosoma hemprichi Ehrenberg, 12
  Naididae
    Stylaria lacustris (L.), 8
  Enchytraeidae
    Henlea nasuta (Eisen), 46
    Fridericia sp., 56
  Lumbriculidae
    Stylodrilus heringianus Claparède, 24
  Lumbricidae
    Allolobophora chlorotica (Savigny), 53
    Allolobophora sp., 1
    Dendrobaena rubida (Savigny) f. *tenuis* (Eisen), 1
    Dendrobaena veneta (Rosa), 53
    Eiseniella tetraedra (Savigny) f. *typica*, 1, 4, 24
    Octolasion cyaneum (Savigny), 53

Mollusca

Gastropoda
Mesogastropoda
  Hydrobiidae
    Potamopyrgus jenkinsi (Smith), 50
Stylommatophora
  Cochlicopidae
    Azeca goodalli (Férussac), 19
    Cochlicopa lubrica (Müller), 19, N4
  Endodontidae
    Discus rotundatus (Müller), 19, 45
  Zonitidae
    Oxychilus alliarius (J. S. Miller), 19
    Oxychilus cellarius (Müller), 12, 19, 24, 47, 53
    Retinella nitidula (Draparnaud), 14

  Helicidae
    Helix hortensis Müller, 19

Arthropoda

Crustacea
Cladocera
  Daphniidae
    Daphnia obtusa Kurz, 44
Ostracoda
  Cypridae
    Candona lucens Baird, 17
    Cypridopsis subterranea J. P. Wolf, 11
    Potamocypris sp., 24
Copepoda
  Cyclopidae
    Macrocyclops albidus (Jurine), 8
    Eucyclops agilis (Koch), 8
    Paracyclops fimbriatus (Fischer), 11
    Acanthocyclops vernalis (Fischer), 4, 11, 46
    Acanthocyclops viridis (Jurine), 4, 8, 11, 46
Isopoda
  Asellidae
    Proasellus cavaticus (Leydig), 1, 3, 4, 8, 9, 10, 11, 12, 14, 15, 17, 18, 19, 24, 29, 41, 45, 53
    Proasellus cf. *P. meridianus* (Racovitza), 44
  Trichoniscidae
    Trichoniscus pusillus Brandt, 11
    Trichoniscus pygmaeus Sars, 56
    Androniscus dentiger Verhoeff, 9, 12, 17, 19, 24, 36, 45
  Oniscidae
    Oniscus asellus L., 19, 28, 31, N3, N6
Amphipoda
  Gammaridae
    Crangonyx subterraneus Bate, 15
    Gammarus pulex (L.), 4, 6, 11, 14, 19, 24, 28, 46, 53, N5
    Niphargus aquilex Schiödte, 4?, 24?, 49
    Niphargus fontanus Bate, 1, 3, 4, 7, 8, 9, 10, 11, 12, 15, 17, 18, 19, 20, 21, 24, 27, 29, 30, 31, 35, 37, 53, 56
Mysidacea
  Mysidae
    Praunus flexuosus (Müller), 53
Myriapoda
Symphyla
  Scutigerellidae
    Scutigerella causeyae Michelbacher, 19
  Scolopendrellidae
    Scolopendrella notacantha Gervais, 52
    Symphylella isabellae (Grassi), 44, 57
    Symphylella jacksoni (Bagnall), 35
    Symphylella vulgaris (Hansen), 3
Diplopoda
  Glomeridae
    Glomeris marginata (Villers), 19, 24, 53
  Brachychaeteumidae
    Brachychaeteuma melanops Brade-Birks, 53
  Craspedosomidae
    Polymicrodon polydesmoides (Leach), 3, 11, 15, 24, 53? (juv.), N3, N6, N8
  Polydesmidae
    Brachydesmus superus Latzel, 14, 15, 24, 56, 58
    Polydesmus angustus Latzel, 44, 45, 53
    Polydesmus denticulatus C. L. Koch, 19
    Polydesmus gallicus Latzel, 19
  Blaniulidae
    Blaniulus guttulatus (Bosc), 5, 19, 35, 51, 53
    Proteroiulus fuscus (Am Stein), 56
  Iulidae
    Cylindroiulus punctatus (Leach), 59
    Tachypodoiulus niger (Leach), 29

Culicidae
   Culex pipiens L., 4, 8, 11, 25, 26, 31, 40, 45, 47, 48, 53, 54, N1, N3, N4, N6, N9, N10
   Culiseta annulata (Schrank), 1, 53, N1, N3
Chironomidae
   Anatopynia notata (Meigen), 8
   Xenopelopia falcigera (Kieffer), 17
   Prodiamesa olivacea (Meigen), 24
   Chaetocladius perennis (Meigen), 44
   Limnophyes sp., 53
   Metriocnemus sp., 14, 19
   Chironomus sp., 8
   ?Stictochironomus sp., 19
   Rheotanytarsus sp., 14 (larval cases)
   ?Spaniotoma sp., 11
Simuliidae
   Simulium ornatum Meigen, 34
Mycetophilidae
   Bolitophila cinerea Meigen, 4
   Macrocera stigma Curtis, 55
   Speolepta leptogaster (Winnertz), 4, 9, 10, 11, 12, 15, 17?, 18, 20, 24, 29, 30, 31, 33, 35, 36, 39, 42, 45, 53, 54, 60, N3, N6
   Rymosia fasciata (Meigen), 29
   Rymosia sp., N1, N3, N6, N9
   Tarnania dziedzickii (Edwards), 60
   Tarnania fenestralis (Meigen), N10
   Exechia sp., N9
   Exechiopsis subulata (Winnertz), 42
   Mycetophila ocellus Walker, 11
Sciaridae
   Sciara forficulata Bezzi, 11?
   Sciara sp., 3, 12, 24, 29, N4
   Lycoriella sp., N5
   Bradysia sp., 1, 3, 20, 29, 53, 56, N3, N4
Rhagionidae
   Rhagio sp., 24
Empididae
   Tachydromia sp., 46
   Wiedemannia rhynchops (Nowicki) s. *insularis* Collin, 24
Phoridae
   Megaselia rufipes (Meigen), 12
   Megaselia sp., 59
   Diplonevra concinna (Meigen), 59
   Triphleba antricola (Schmitz), 22
Lauxaniidae
   Sapromyza sordida Haliday, N10
Heleomyzidae
   Tephrochlamys rufiventris (Meigen) v. *canescens* (Meigen), 5, N10
   Scoliocentra amplicornis Czerny, N9
   Scoliocentra caesia (Meigen), 29, 35
   Scoliocentra villosa (Meigen), N3, N8, N10
   Heleomyza serrata (L.), 1, 9, 11, 12, 17, 18, 19, 25, 31, 38, 45, 54, 56, 60, N1, N3, N4, N5, N6, N8, N9, N10
Sphaeroceridae
   Copromyza nigra (Meigen), 20, 53, 60, N4, N10
   Leptocera silvatica (Meigen), 53, N5
Fanniidae
   Fannia canicularis (L.), N10
Siphonaptera
  Hystrichopsyllidae
   Hystrichopsylla talpae (Curtis), 59
Coleoptera
  Carabidae
   Leistus spinibarbis (F.), 11
   Nebria brevicollis (F.), 11, 35
   Trechus (Trechoblemus) micros (Herbst), 12, 53
   Trechus obtusus Erichson, 24
   Bembidion decorum (Panzer), 24

   Bembidion tibiale (Duftschmid), 24
   Pterostichus aethiops (Panzer), 11
   Abax parallelepipedus (Piller & Mitterpacher) (syn. *A. ater*), 28
   Calathus fuscipes (Goeze) (syn. *C. cisteloides*), 11
   Agonum albipes (F.) (syn. *A. ruficorne*), 14
   Amara aulica (Panzer), 24
  Dytiscidae
   Hydroporus ferrugineus Stephens, 24, N11
   Hydroporus nigrita (F.), 17
   Hydroporus obsoletus Aubé, N5
   Oreodytes sanmarki (Sahlberg) (syn. *O. rivalis*), 17
   Platambus maculatus (L.), 24
   Agabus guttatus (Paykull), 1, 24
  Hydrophilidae
   Helophorus aquaticus (L.), 14
   Helophorus brevipalpis Bedel, 14, 46
   Helophorus laticollis Thomas, 46
  Hydraenidae
   Hydraena gracilis Germar, 24, 46
  Leiodidae
   Choleva agilis (Illiger), 1, 2, 11, 19
   Choleva cf. *C. agilis* (Illiger) et *C. angustata* (F.), 1
   Choleva spadicea (Sturm), 24, 45, 53
   Catops nigricans (Spence), 11
   Catops sp., 1, 24
  Staphylinidae
   Lesteva hanseni Lohse, 26
   Lesteva pubescens Mannerheim, 1, 3, 4, 11, 12, 19, 20, 24, 53, 59
   Ochthephilus aureus (Fauvel) (syn. *Ancyrophorus aureus*), 1, 4, 11, 15, 19, 53
   Dianous coerulescens (Gyllenhal), 24, 35
   Lathrobium fulvipenne (Gravenhorst), 5
   Xantholinus sp., N6
   Staphylinus morio (Gravenhorst sensu auct. Brit.), 24
   Quedius mesomelinus (Marsham), 1, 5, 11, 24, 56, 59
   Aloconota sp., 53
   Atheta sp., 46
   Chiloporata longitarsis (Erichson), 11
  Pselaphidae
   Euplectus sp., 56
  Scarabaeidae
   Aphodius fimetarius (L.), 14
  Elmidae
   Limnius volckmari (Panzer), 24
  Cantharidae
   Rhagonycha femoralis (Brullé) v. *limbata* Thomson, 42
  Cryptophagidae
   Cryptophagus distinguendus Sturm, 60
  Cerambycidae
   Rhagium bifasciatum F., 42
   Rhagium mordax (Degeer), 42
Arachnida
Opiliones
  Nemastomatidae
   Nemastoma bimaculatum (F.), 24
Araneae
  Theridiidae
   Theridion sp., 56
  Nesticidae
   Nesticus cellulanus (Clerck), 9, 53
   Nesticus sp., 19
  Tetragnathidae
   Meta menardi (Latreille), 4, 11, 18, 19, 24, 25, 29, 35, 42, 45, 47, 53, 54, 56, N1, N2, N6
   Meta merianae (Scopoli), 2, 11, 13, 24, 29, 32, 35, 45, 53, N9
  Linyphiidae
   Gonatium rubellum (Blackwall), 42
   Diplocephalus protuberans (O.P.–Cambridge), 17
   Araeoncus crassiceps (Westring), 24

Porrhomma convexum (Westring), 19, 24, 53
Porrhomma egeria Simon, 10
Porrhomma pygmaeum (Blackwall), 24
Porrhomma rosenhaueri (L. Koch), 19, 45
Porrhomma sp., 22
Lepthyphantes pallidus (O.P.–Cambridge), 1, 19
Lepthyphantes zimmermanni Bertkau, 11, 24
Lepthyphantes sp., 42?
Acari
  Veigaiidae
    Veigaia agilis (Berlese), 53
    Veigaia nemorensis (C. L. Koch), 12
    Veigaia transisalae (Oudemans), 53
    Veigaia sp., 8, 22
  Parasitidae
    Parasitus sp., 46
    Eugamasus berlesei Willmann, 53
    Eugamasus cornutus (G. et R. Canestrini), 8
    Eugamasus loricatus (Wankel), 6, 17
    Eugamasus magnus (Kramer), 1, 24
    Parasitellus ferox (Trägårdh), N11
    Pergamasus crassipes (L.), 46
    Pergamasus integer Bhattacharyya, 1
  Ixodidae
    Ixodes hexagonus Leach, N11
  Pachygnathidae
    Pachygnathus rostratus (Trägårdh), 1, 4

Nanorchestidae
  Speleorchestes poduroides Hirst, 1
Rhagidiidae
  Rhagidia gigas (Canestrini), 1
  Rhagidia longipes Trägårdh, 1?
  Rhagidia odontochela Turk, 10
  Rhagidia punkva Zacharda, 11, 12
  Rhagidia spelaea (Wankel), 1, 3, 8, 11, 12, 15, 29, 52, 56
  Rhagidia terricola (Koch), 1
  Rhagidia sp. nov.?, 1
  Rhagidia sp., 9, 53
  Coccorhagidia clavifrons (Canestrini), 1
Alicorhagiidae
  Alicorhagia clavipilus (Thor), 12
Eupodidae
  Linopodes motatorius (L.), 1
Ereynetidae
  Ereynetes sp. nov.?, 24
Glycyphagidae
  Glycyphagus domesticus (de Geer), 1
Anoetidae
  Myianoetus diadematus Willmann, 25
Belbidae
  Damaeus crispatus Kulczynski, 15

References

Bouvet, Y. (1975). Les trichoptères du group de *Stenophylax*: conditions de vie et réactions aux variations des facteurs due milieu. *Annales de Spéléologie*, **30**, 207–29.

Bouvet, Y. (1976). Ecologie et reproduction chez les trichoptères cavernicoles du groupe de *Stenophylax* (Limnephilidae, Stenophylacini). *Proceedings of the 1st International Symposium on Trichoptera (1974)*, pp. 105–9. The Hague: Junk.

Chapman, P. (1979). The biology of Otter Hole, near Chepstow. *Transactions of the British Cave Research Association*, **6**, 159–67.

Cubbon, B. D. (1970). Flora records of the Cave Research Group of Great Britain from 1939 to June 1969. *Transactions of the Cave Research Group, GB*, **12**, 57–74.

Cubbon, B. D. (1972). Flora records for 1969–70. *Transactions of the Cave Research Group, GB*, **14**, 201–3.

Dixon, J. M. (1974). Biospeleology in the North West of England. In *Limestones and Caves of North-West England*, ed. A. Waltham, pp. 149–81. Newton Abbot: David & Charles.

Dobat, K. (1973). Beobachtungen an markierten und unmarkierten Köcherfliegen (Trichoptera) in der Bärenhöhle im Lonetal (Schwäbischer Jura). *International Journal of Speleology*, **5**, 63–85.

Dobat, K. (1975). Die Höhlenfauna der Schwäbischen Alb. *Jahrbuch Gesellschaft Naturkunde Württemburg*, **130**, 260–381.

Dresco-Derouet, L. (1960). Étude biologique comparée de quelques espèces d'Araignées lucicoles et troglophiles. *Archives de Zoologie Expérimentale et Generale*, **98**, 271–354.

Edington, A. (1977). Biospeleology of Dan yr Ogof. *Transactions of the British Cave Research Association*, **4**, 331–41.

Glennie, E. A. (1967). The distribution of the hypogean Amphipoda in Britain. *Transactions of the Cave Research Group, GB*, **9**, 132–6.

Hazelton, M. (1955 *et seq.*). Biological records of the Cave Research Group of Great Britain. *Biological Supplement of the Cave Research Group, GB*, Parts 1–5, *Biological Records Cave Research Group, GB*, Parts 6–8, *Transactions of the Cave Research Group, GB*, **7(3)**, 10–19; **9**, 162–241; **10**, 143–65; **12**, 3–26; **13**, 167–97; **14**, 205–30; *Transactions of the British Cave Research Association*, **5**, 164–98.

Hazelton, M. (1971, 1972). Vice-county records of fauna collected from the hypogean and related zones. *Transactions of the Cave Research Group, GB*, **13**, 198–223; **14**, 231–72.

Hazelton, M. (1974). Biological Recorder's Report. *British Cave Research Association Bulletin*, **6**, 4–6.

Hazelton, M. (1975). The biology of the Mendip caves. In *Limestones and Caves of the Mendip Hills*, ed. D. I. Smith, pp. 313–51. Newton Abbot: David & Charles.

Hazelton, M. & Glennie, E. A. (1962). Cave fauna and flora. In *British Caving*, 2nd edn, ed. C. H. D. Cullingford, pp. 347–95. London: Routledge & Kegan Paul.

Henry, J. P. (1977). Origine et ancienneté de *Proasellus cavaticus* (Leydig) (Crustacea, Isopoda, Asellota des eaux souterraines). *Proceedings of the 7th International Speleological Congress*, pp. 243–6. Sheffield.

Jeannel, R. (1959). Situation géographique et peuplement des cavernes. *Annales de Spéléologie*, **14**, 333–8.

Jefferson, G. T. (1969). British cave faunas and the problem of their food supply. *Proceedings of the 4th International Congress of Speleology (1965)*, **4–5**, 129–33. Yugoslavia.

Jefferson, G. T. (1974). The distribution of hypogean animals in the British Isles. *Abhandlungen 5th International Kongress Speläologie (1969)*, **4 B8**, 1–6. Stuttgart.

Jefferson, G. T. (1976). Cave faunas. In *The Science of Speleology*, ed. T. D. Ford & C. H. D. Cullingford, pp. 359–421. London: Academic Press.

Jefferson, G. T. (1981). Diptera in British Caves. *Proceedings of the 8th International Congress of Speleology*, **1**, 106–7. Bowling Green, Kentucky.

Jefferson, G. T. (1982). Life in the twilight zone. *Caves and Caving*, **18**, 10–13.

Jefferson, G. T. (1983). The threshold fauna, a neglected area of British cave biology. *Studies in Speleology*, **4**, 53–8.

Jones, J. G. & Edington, A. (1968). An ecological survey of hydrocarbon-oxidising micro-organisms. *Journal of General Microbiology*, **52**, 381–90.

Lloyd, O. C. (1977). Ferns in cave entrances. *Proceedings of the 7th International Speleological Congress*, p. 288. Sheffield.

Locket, G. H. & Millidge, A. F. (1953). *British Spiders*, vol. 2. London: Ray Society.

Mason-Williams, A. (1959). The formation and deposition of moonmilk. *Transactions of the Cave Research Group, GB*, **5**, 135–8.

Mason-Williams, A. (1961). Biological aspects of calcite deposition. *Symposium Internazionale di Speleologie* (Memoria V della Rassegna Speleologica Italiana), pp. 1–4. Como.

Mason-Williams, A. (1966). Micro-organisms in relation to food and energy sources in caves. *Proceedings of the British Speleological Association*, **4**, 69–74.

Mason-Williams, A. (1967). Further investigations into bacterial and algal populations of caves in South Wales. *International Journal of Speleology*, **2**, 389–95.

Mason-Williams, A. (1969a). Comments on the bacterial populations of small pools in caves. *Proceedings of the 4th International Congress of Speleology (1965)*, **4–5**, 161–6. Yugoslavia.

Mason-Williams, A. (1969b). A note on the effects of tracer dyes on microbial populations of streams. *Proceedings of the 4th International Congress of Speleology (1965)*, **4–5**, 167–72. Yugoslavia.

Mason-Williams, A. & Benson-Evans, K. (1958). A preliminary investigation into the bacterial and botanical flora of caves in South Wales. *Cave Research Group, GB*, Numbered Publication 8.

Mason-Williams, A. & Benson-Evans, K. (1967). Summary of results obtained during a preliminary investigation into the bacterial and botanical flora of caves in South Wales. *International Journal of Speleology*, **2**, 397–402.

Mason-Williams, A. & Holland, L. (1967). Investigations into the 'wall-fungus' found in caves. *Transactions of the Cave Research Group, GB*, **9**, 137–9.

Matile, L. (1962). Morphologie et biologie d'un diptère cavernicole *Speolepta leptogaster* Winnertz (Mycetophilidae). *Memoir Musée Nationale de Histoire Naturelle. Series A. Zoologie*, **20**, 219–42.

Proudlove, G. (1982). Cave fish. *Caves and Caving*, **15**, 6–7.

Racovitza, E. G. (1907). Essai sur les problèmes biospéologiques. *Biospeologica*, **1**, 371–488.

Railton, C. L. (1969). Glennie's caving in South Wales. *Transactions of the Cave Research Group, GB*, **11**, 97–100.

Schiner, J. R. (1854). Fauna der Adelsberger-, Lueger- und Magdalenen Grotte. In *Die Grotten und Höhlen von Adelsberg, Lueg, Planina und Laas*, ed. A. Schmidl, pp. 231–72. Vienna: Braumüller.

Schmitz, H. (1913). Biologisch–anatomische Untersuchungen an einer hölenbewohnenden Mycetophilidenlarve *Polylepta leptogaster* Winn. *Natuurhist. Genootschap in Limburg*, **Jahrboek 1912**, 65–96.

Thinès, G. & Tercafs. R. (1972). *Atlas de la vie souterraine. Les animaux cavernicoles*. Brussels: de Visscher.

Turk, F. A. (1972). Biological notes on Acari recently recorded from British caves and mines with descriptions of three new species. *Transactions of the Cave Research Group, GB*, **14**, 187–94.

Turquin, M.-J. (1972). La faune de la Grotte de Hautecourt (Ain). *Sciences*, **3(2)**, 145–54.

Turquin, M.-J. (1973). Une biocoenose cavernicole originale pour le Bugey; le Puits de Rappe. *Sciences*, **3**, 235–56.

6

The Stone Age cave archaeology of South Wales

H. STEPHEN GREEN

The primary archaeological importance of the caves of South Wales lies in the Palaeolithic period, where they provide almost the total body of available information concerning human settlement on the fringe of the then known world. The Palaeolithic occupation of South Wales seems to have been almost totally confined to the last (Devensian) glaciation when sea level was low and Britain was no more than an isthmus of Europe (Fig. 6.1). During the later prehistoric and historic periods caves served a variety of functions as shelters or places of temporary settlement, as refuges and as burial repositories. The present account is confined to the Stone Age *sensu lato*, that is to say until the period, around 1000 BC when the regular and skilled use of flint seems to undergo a marked decline

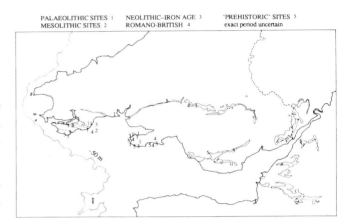

PALAEOLITHIC SITES 1   NEOLITHIC–IRON AGE 3   'PREHISTORIC' SITES 5
MESOLITHIC SITES 2   ROMANO-BRITISH 4   exact period uncertain

Fig. 6.2. Distribution of archaeological sites in South Wales caves in relation to the limestone outcrops.

(Green, 1980). The convention 'bc' is used below to represent radiocarbon years.

The distribution of sites (Fig. 6.2) shows their restrictions to the limestone outcrops but, within this, we may detect a bias among Palaeolithic caves to the present coastal areas, where the grazing grounds would have been most extensive. It is only in the later periods that serious exploitation of the (now) inland caves took place. No discussion of sea levels will be attempted here but on maps, other than Figure 6.1, the −50 m contour is shown to indicate the approximate Upper Palaeolithic coastline, except during the glacial maximum when it may have plunged to around −100 m (Bowen, 1978).

The hunting and gathering periods

The Palaeolithic settlement of South Wales, as represented by caves, is confined to the period of the last glaciation, with the possible exception of Bacon Hole, Gower, where Ipswichian occupation has been claimed (Stringer, 1977).

Lower Palaeolithic occupation in South Wales is represented by single handaxes from Penylan, Cardiff (Lacaille, 1954), and from Rhossili, Gower (Green, 1981a). But the Rhossili handaxe, if truly comparable with Roe's Wolvercote type, may belong to the Ipswichian–early Devensian period (Roe, 1976),

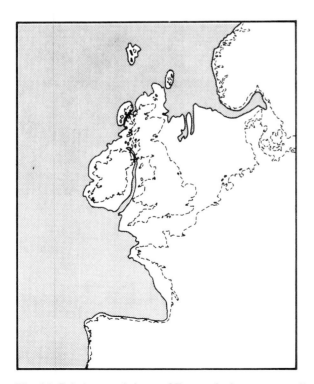

Fig. 6.1. Britain as an isthmus of Europe; both present coastline and the −100 m contours are shown.

leaving the Penylan handaxe alone as the only possible relic of an earlier occupation of South Wales. In reality, typology cannot be used to proclaim with confidence any particular age for the Penylan handaxe, especially in view of the latter's manufacture on a quartzite pebble; one can only say that it might be of Lower or Middle Palaeolithic age. The argument for or against widespread Lower Palaeolithic settlement of the whole of Britain cannot be reviewed here in detail. However, Wymer's (1977) suggestion that the evidence for such settlement has been destroyed by the scouring action of the Wolstonian and Devensian glaciations requires qualifications since the unglaciated upland area of southwest England is similarly free of large concentrations of handaxes. A study of the European distribution of Acheulian sites (Collins, 1976) has shown that concentrations are restricted to the major river valleys of northwestern France and southeast England. It would be extremely surprising, therefore, if the Lower–Middle Palaeolithic settlement both of Wales and the remainder of highland Britain were ever other than distinctly marginal. In North Wales and the Marches, such marginal occupation is represented by the rich site of Pontnewydd Cave, Clwyd, now in course of excavation (Green, 1981b, 1984; Green et al., 1981) and by the three handaxes now known from Herefordshire (Pye, 1972, 1978).

Middle Palaeolithic

There are four sites to consider here: Long Hole, Paviland, Bacon Hole (all Gower) and Coygan, near Carmarthen (Fig. 6.3). The evidence for Middle Palaeolithic settlement at Long Hole is based largely upon stratigraphic evidence and the only artefacts which may be of this period are a possible limestone flake and a bone tool (Campbell, 1977). Mousterian occupation has long been claimed at Paviland (Sollas, 1913) but the argument is purely typological, and, in the absence of any stratigraphic support, cannot be sustained. The so-called 'proto-Levalloisian' chert flake published by Grimes (1935, 1951) need not be diagnostic of any particular period. Little can be said at this stage of the possible, last interglacial occupation at Bacon Hole (Stringer, 1977). A series of possible bone tools has been discovered and study of these is still in progress. The absence of stone artefacts or any other corroborative evidence for the presence of man on the site must render suspect the preliminary identification of these bones as tools.

The only undoubted Middle Palaeolithic site in Wales is that of Coygan Cave (Grimes, 1935; Clegg, 1969). Even here human occupation of the cave was very slight and the principal finds comprise only two triangular (Fig. 6.4) and one other crudely shaped handaxe. Such triangular handaxes (of so-called 'bout-coupé' type) have been considered (cf. Bordes, 1954; Mellars, 1969; Tuffreau, 1971; Shackley, 1977) to belong to a chronological horizon in the Ipswichian or early Devensian but the evidence for this needs critical re-examination. It may be that the occupation is not far removed in time from the apparently too-young radiocarbon determination of 31 250 ± 310 bc (GrN-4400) from the Lower Calcrete layer underlying the Mousterian industry. A further date of 36 734 + 2713/−2024 bc (BM-499) on reindeer antler 'associ-

Fig. 6.3. Coygan Cave with Herbert Eccles' excavations in progress in 1913 (photo by permission of Carmarthen County Museum).

ated with Mousterian artefacts, including a characteristic handaxe' seems more acceptable (Burleigh et al., 1976).

In Britain, Collins (1969), has defined a 'Paxton Stage' characterized by such handaxes, but the chronology of most British and Belgian examples is less than clear (Mellars, 1974; Ulrix-Closset, 1975). Collins (1969) has claimed that the bout coupé handaxes from Little Paxton itself 'are well dated to early in the last glaciation'. They derive from the First Terrace of the Great Ouse and are associated with a fauna characteristic of the last glacial. However, what is known of the dating of this terrace, although not necessarily, of course, of its contained artefacts, would favour a late Devensian age (Horton, 1970; Horton, Shepherd-Thorn & Thurrell, 1974) and therefore, offers merely a terminus ante quem for the artefacts.

Early Upper Palaeolithic (EUP)

Reconsideration by Jacobi (1980) of the Upper Palaeolithic of Britain in its continental context suggests that it is divided by the Devensian glacial maximum into two periods of Upper Palaeolithic occupation of 36 000–25 000 bc (EUP) and 10 000–8000 bc (LUP), with a hiatus during the period of the glacial maximum when Britain was almost certainly uninhabited (Fig. 6.5). The date of the inception of the EUP phase is uncertain but an age of around 36 000 bc is generally accepted on the basis both of a radiocarbon date of 36 320 + 1470/−1240 bc (GrN-6324) from Kent's Cavern, Devon (Campbell & Sampson, 1971; Campbell, 1977), and of

Fig. 6.4 Welsh hand-axes; top left – Penylan; top right – Rhossili;
bottom – Coygan Cave.

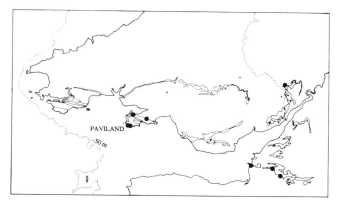

Fig. 6.5. Distribution of early Upper Palaeolithic sites.

knowledge of continental chronology which suggests that the British EUP is a blend of elements typical of France and central Europe arriving via the Low Countries. Because of the complex processes involved in cave sedimentation, and in the absence of informative stratigraphies from modern scientific excavation, it is not possible to determine with confidence whether genuine cultural mixing has resulted in a distinctively British cultural assemblage or whether, more probably, what we have is a cultural palimpsest whose different elements can no longer be retrieved other than typologically. However, three chronological horizons have been defined by Jacobi (1980). The first is marked by the appearance of leaf points, radiocarbon-dated around 36000 bc at Kent's Cavern (Campbell, 1977) and from early contexts in Europe, most particularly in Belgium where, at the sites of Spy and Couvin, leaf points appear at the interface of the Middle and Upper Palaeolithic sequence (Otte, 1976). That the leaf points do represent a genuinely earlier phase in Britain is suggested by their abraded condition, which contrasts with the undamaged Aurignacian and Font Robert material (Jacobi, 1980). The second horizon is marked by numerous 'burins busqués' of Aurignacian II age, around 30000 BP (Waterbolk, 1971; Mellars, 1974). Rare 'Font Robert' points (*sensu lato*), dating *c.* 26000 bc on the basis of the Belgian evidence from Maisières (Otte, 1977), typify the third phase.

The South Wales sites: Paviland Cave (Figs 6.5, 6.6, 6.7) is the only major site of this period in Wales, and indeed in Britain, being 10-times richer in artefacts than Kent's Cavern, the only other numerically important EUP site in the British Isles. Both sites were large caves sited for the hunting of herbivores grazing on the Bristol and English Channel plains, respectively. The bulk of the EUP material from Paviland is probably of Aurignacian II age in French terms, a range of

Fig. 6.6. Paviland Cave in 1912; excavation by Professor W. J. Sollas in progress.

Fig. 6.7. Paviland Cave in 1979, showing the setting of the cave overlooking the Bristol Channel plain.

perhaps 31 000–28 000 bc. The radiocarbon date of 25 650 ± 1300 bc (BM-1367) on a *Bos* bone from Paviland (Molleson & Burleigh, 1978) is, even on the basis of the range of two standard deviations, scarcely acceptable within that time span and is likely to refer to the later 'Font Robert' occupation. However, the Paviland date is not based on stratified material nor, indeed, even on material, such as humanly worked bone, which had any necessary connection with the occupation of the site. The archaeological significance of the date is therefore uncertain.

The principal discoveries at Paviland made by Buckland (1823) and Sollas (1913), are well known and have recently been discussed by Jacobi (1980). The main elements are:

1 leaf points, indicative possibly of settlement as early as 36 000 bc.
2 a rich 'Aurignacian' industry of around 30 000 bc.
3 the famous 'red lady' burial, possibly of similar antiquity to the Aurignacian industry (Mellars, 1974), but believed by Molleson (1976), on the basis of the radio-carbon determination of 16 510 ± 340 bc (BM-374) on a sample of bone from the skeleton, to represent a much later burial.
4 one tanged 'Font Robert' point (Sollas, 1913) of around 26 000 bc.
5 a late Upper Palaeolithic (LUP) industry of late glacial date *c.* 10 000–8000 bc.

The 'red lady' skeleton, believed by Buckland to be that of a woman, is now known to represent a male, aged around 25

years and about 1.70 m in height. The objects accompanying the burial, in addition to the enveloping ochre which has stained the bones red, comprise two ivory bracelets and some two dozen ivory wands, conventionally interpreted as blanks for beads, and sea shells perforated for suspension or for use in fringing leather clothing. (I am indebted to Dr Graham Oliver of the National Museum of Wales, for confirming that the perforations are consistent with perforation for suspension rather than for the removal of the flesh as Buckland suggested.) The question of the age of the burial has been reviewed by Jacobi (1980), who dated the burial to the period of Aurignacian occupation, on the basis, in part, of specifically Aurignacian parallels to the ivory wands and bracelets from sites in Belgium and, in part, on the basis of the total absence of contemporary human settlement as far as 600 km south of Paviland around 16 000 bc – at that time the nearest glacier ice may have lain only an hour's walk north of the cave (Bowen, 1970). Molleson (1976) regarded a mammoth's skull, complete with tusks, found near the burial as probably associated with it on the grounds of its proximity. Jacobi (1980), on the other hand, identified the skull as raw material for the Aurignacian ivory workshop which produced the bracelets and wands. The association of the mammoth skull, however indirectly, with the burial is wholly plausible given the well-known date of 16 050 + 1400/−1200 bc (Birm-146) from Cae Gwyn Cave, Clwyd; a doubtful late Devensian occurrence at Brean Down (Stuart, 1974); and the recent dating of mammoth as late as 10 700 bc at the ARC gravel pit at Condover, Salop. The picture sketched by Molleson is one of great charm and power, since it would involve the transport of a corpse over a tremendous distance for burial in a venerated site, but, given the possibility of contemporary settlement at Little Hoyle Cave in Dyfed (Green, 1986), it would be unwise to dismiss the 16 050 bc determination for the burial.

The regional pre-eminence of the Paviland Cave site over contemporary settlements is well illustrated by consideration of the numbers of artefacts recorded from each site (Campbell, 1977). Whereas Paviland has produced over 5000, all other South Wales sites, certain and probable, muster <50 between them. Of the possible EUP sites listed by Campbell, we may dismiss the finds from Deborah's Hole, Coygan and Ogof yr Ychen as not being closely dateable and those from Longbury Bank (Little Hoyle) as LUP or later in date. Sites with certain occupation, other than Paviland, comprise only Long Hole and Nottle Tor in Gower, where the scanty finds provide certain evidence of settlement during the leaf point phase. Further west, it is now clear that there was Aurignacian settlement at Hoyle's Mouth but the evidence for this remains to be published. We should note also the distinct probability of 'Font Robert' occupation at Cathole (Delporte, 1976).

The late Upper Palaeolithic (LUP)

The period of conditions favourable to the reoccupation of the British Isles by herds of large herbivores and, in their wake, by man, is now dated to the period of around 10 000–8000 bc (Jacobi, 1980). The LUP in Wales sees the development of a new pattern of settlement. Paviland was no longer the pre-eminent site in the Bristol Channel Plain region but

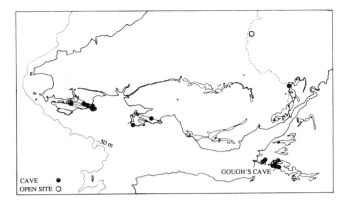

Fig. 6.8. Distribution of late Upper Palaeolithic sites.

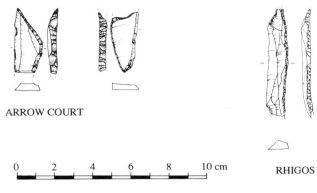

Fig. 6.9. Late Upper Palaeolithic finds from Arrow Court (after Campbell), Rhigos and Gwernvale. Scale 1:1.

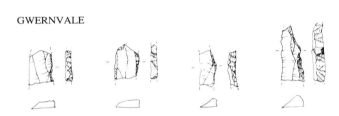

was replaced by Gough's Cave, at the foot of the Mendip Hills (Collcutt, 1979), which may have become an all-season base camp, with open and cave sites in the southwest and South Wales representing outlying sites used as temporary shelters on hunting expeditions. The number of LUP cave sites with certain evidence of occupation numbers eight (compared with only four sites of EUP age) and the mean number of artefacts found at sites in Wales (excluding the EUP finds at Paviland) is about 50/site. The sites include Priory Farm Cave, the Hoyle, Little Hoyle (= Longbury Bank), Potter's Cave, Nanna's Cave, Paviland and Cathole (Fig. 6.8). Indeed, the number of LUP sites is, in reality, larger since one should include the open sites of Gwernvale, Powys (Healy & Green, 1984), possibly Rhigos, Mid Glamorgan, and certainly Arrow Court, Herefordshire, just over the English border (Fig. 6.9). The LUP pattern of settlement seems to represent more widespread and intensive exploitation of the late glacial environment. The increased interest evidenced in the lesser river valleys and uplands is suggestive, perhaps, of greater competition for the available game. Such intensified human competition may have been a factor in the local extinction, observed also on a worldwide scale, of a number of Pleistocene land mammals (Stuart, 1974). However, a recent consideration of this subject (Kowalski, 1980) would rule out an anthropogenic factor and suggests instead that the extinction of species of the Pleistocene megafauna in Eurasia is related to the disappearance of the steppe–tundra ecosystem and to the appearance of a continuous taiga belt, developments unparalleled in earlier interglacials.

Material equipment differs in detail from that of the EUP (Fig. 6.10). The bonework includes an eyed needle at Cathole, used in the manufacture of garments or, perhaps, of snowshoes as among the sub-arctic Indians of Canada (Rogers, 1970). The predominant change among the stone tools is the appearance of a large number of backed tools, designed for insertion into slotted hafts and presumably used in a variety of ways as components of spears (one example was apparently found deeply embedded in a jaw of *Megaloceros* (Bohmers, 1956)), or of knives. The different backed tools of the Mesolithic period (termed 'microliths') show a progressive diminution in size. This may have been a function of different game animals hunted in the changing climatic conditions of the LUP and Mesolithic, as Campbell (1977) has suggested. However, cur-

Fig. 6.10. Upper Palaeolithic finds from South Wales. Upper row; leaf-point, Aurignacian busked burin and scrapers, tang of Font Robert point (all Paviland). Lower row; backed tools, burin (Priory Farm Cave), shank of bone needle (Cathole).

rent models suggest a heavy dependence on large herbivores throughout the period and a more likely explanation is increasingly economical use of flint, under the pressures of growing population (and hence competition for resources) and rising sea level, accompanied by the spread of forest and ground cover in the Mesolithic period, rendering raw materials less accessible.

The Upper Palaeolithic site of Ogof Daren Cilau, Llangattwg, Brecknock, published by Campbell (1977) definitely seems to be a site-that-never-was. The circumstances which led to the error are recorded in manuscripts housed in the University Museum, Oxford.

Cave-use

From the beginning of the Flandrian interglacial, archaeological finds from caves are of greatly diminished importance for the interpretation of the past. This is well illustrated by Figure 6.11 which contrasts open-air and cave settlement in Wales during the Mesolithic period. There are various explanations of this. In the first place, caves, with their natural thermal advantages, could have presented an attraction for human occupation during the colder phases of the glacial Pleistocene. This is not to suggest, however, that Palaeolithic man was dependent upon natural shelter, for his ability to construct heated houses is well known, particularly in European Russia (Klein, 1973). Indeed, assessment by Binford (1978) of patterns of cave-use among 240 modern and historically attested hunter–gatherer communities yielded no examples of cave-use for residential purposes in temperate or polar regions. Among the Nunamiut Esquimaux of Alaska, Binford found two situations in which cave-use was practised. The first was as natural shelters when travelling on hunting or trapping expeditions and the second was as thawing-out sites. Such sites were used by trappers in winter to thaw the frozen bodies of animals taken in traps and should be recognizable archaeologically because of their characteristics of large

hearths (used to thaw the bodies for skinning), debris from tool and trap repair, remains of trappers' meals and the nearly complete and articulated skeletons of the skinned animals. Clearly the archaeological evidence for cave-use must be assessed on its own merits but the evidence for modern cave-use in cold climate conditions does not suggest a possible base-camp role for caves.

Of some relevance here is the destruction, by glacial action, of open-air settlements. This is illustrated by the distribution of EUP sites in Britain (Campbell, 1977), where open sites are preserved only outside the limit of the Devensian ice. During the LUP, however, open sites are distributed more widely, albeit accompanied by increasing use of caves in areas where these were available. This leads to our second point that caves have frequently been accorded an importance out of proportion to their true significance because of the fact that they are easily recognized and all too easily excavated (or, more often, merely cleared out) and temporary open-air hunter-encampments are, in general, unlikely to be discovered except by chance. The Mesolithic period in Wales is, therefore, of some importance in so far as it suggests a model of interglacial cave-use, by hunter–gatherers, as transit sites of one kind or another. None of the caves has produced more than a few hundred artefacts, a small total when compared with figures from base-camps running into thousands.

The farming periods

The Neolithic and Bronze ages

The *floruit* of the use of caves as tombs for the dead is undeniably Neolithic. Analysis of 18 cave burial sites in northern England by Gilks (1973) has yielded the following results:

Early Neolithic	2
Late Neolithic	12
Early Bronze Age	3
Structural features	3
Grave goods	17
Contemporary settlement	6

Caves were used as natural chamber tombs in the Neolithic period and it is only to be expected that their use should diminish with the appearance of single-grave burial as the normal fashion, at the beginning of the second millennium bc. The chronology of cave burial in Britain, such as it is, would seem frequently to be later in date within the Neolithic than the majority of chamber tomb constructions and may be evidence of the continuation of a method of disposal of the dead (but now in a natural rock vault) in a chronological or cultural context where the building of a megalithic tomb would have performed no socially reinforcing role (Renfrew, 1973).

In Wales the best example of the use of a cave for later Neolithic burial is the well-known site of Gop, near Prestatyn, in the north (Dawkins, 1901). In South Wales, the Tooth Cave in Gower (Harvey *et al.*, 1967) has produced burials which are likely to belong to the first half of the second millennium bc. In all eight inhumations were found in two chambers of the cave, but there does seem to be good evidence for the use of the cave as a dumping area for domestic rubbish, prior to its

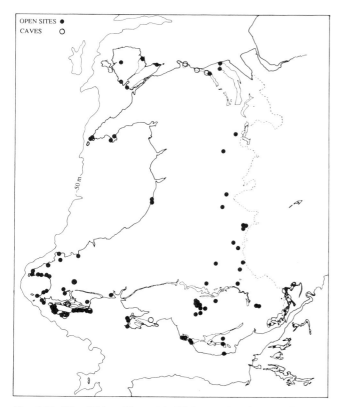

OPEN SITES ●
CAVES ○

−50 m

Fig. 6.11. Mesolithic settlement in Wales, contrasting use of caves and open sites.

use as an ossuary. For instance, in Chamber I plentiful charcoal was found but there was no sign of burning *in situ* and it seems most likely that fires were actually lit only in the entrance. (The keeping-clean of such occupation floors by the original occupants is a problem rarely considered by archaeologists (Movius, 1969), but is full of implications for the interpretation of occupation deposits in caves and elsewhere.) One cannot be sure which, if any, of the items found were grave-goods and which discarded artefacts. The composition of the finds is of some interest, however, for it includes a bone spatula and other items most probably used for leather-working. In addition there were four flint knives, including one of the discoidal polished variety; a bone bead; a perforated bone point; and pottery which, but for the apparent presence of a shoulder on one sherd, might have been classified as Rinyo–Clacton ware. One may recognize in this grouping elements of Rinyo–Clacton and late Beaker affinity which are known sometimes to occur together in settlement contexts (Green, 1980). The potential leather-working tool kit (the knives for cutting, the scraper for scraping hides, the spatula for burnishing and the perforated point for drilling) find a close parallel in a Beaker-period tool kit, again from a cave, at Wetton Mill in Staffordshire (Green, 1976). This association of leather-workers–pastoralist equipment in caves recurs in Middle Bronze Age contexts, discussed below.

No other examples exist of burials certainly associated with artefacts, but, if English parallels are valid, the bulk of unassociated burials (Fig. 6.12) may be of later Neolithic date, although there is the possibility that some are Mesolithic (Clark, 1980). Such examples are known from Aveline's Hole and Gough's (New) Cave near by in Somerset. Certainly, none seem likely to be Iron Age, as Boon (1980) has noted, although more recent examples occur; for instance the Roman burials at Ogof yr Esgyrn (Mason, 1968). Cave settlement does not seem to have taken place on any scale during the earlier Neolithic and, in this, the Welsh sites mirror the situation within Britain as a whole (Bradley, 1978). One possible explanation for this is the small population size coupled with concentration on cereal cultivation at the expense of pastoral farming, to which the heavily wooded earlier Neolithic landscape was less well suited. But by the later Neolithic there is ample evidence for widespread pastoralism and, in this, the shelter afforded by natural caves would have been advantageous (Green, 1980). However, factors other than the purely practical govern cave-use and must be taken into account. For example, ethnographic evidence suggests the possibility of a complete taboo on cave-use among some groups (Jennings, 1979).

Finds, mostly of later Neolithic date, attest interest in caves during this period. Sites with evidence of settlement include Ogof Gofan, Pembrokeshire (Davies, 1969), where Peterborough ware was found in apparent association with a hearth near the cave entrance; Ogof Morfran, also Pembrokeshire mainland, where the finds include a late Neolithic sherd decorated with holes below the rim pierced before firing and paralleled at Dyffryn Ardudwy chamber tomb, Merioneth (Davies, 1975; Powell, 1973); Nanna's Cave, Potter's Cave and Daylight Rock Fissure, all on Caldey Island (Lacaille & Grimes, 1961). It is perhaps worth commenting here that the

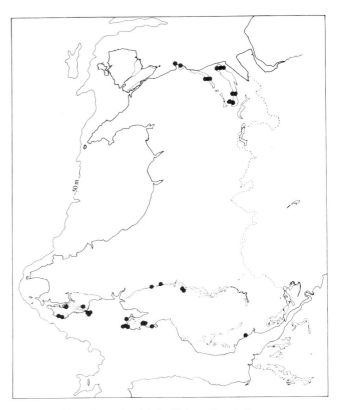

Fig. 6.12. Map of cave burials in Wales, all periods.

Neolithic pot from Nanna's Cave, whose reconstruction puzzled Lacaille & Grimes (1961), has in fact broken along a coil-line, in the manner of some sherds from Coygan (Wainwright, 1967).

Around the end of the second millennium bc, there is evidence of actual settlement at Culver Hole, Gower, Ogof yr Esgyrn, Brecknock, and at Lesser Garth, Glamorgan. The finds from the first site apparently comprise pottery alone but Ogof yr Esgyrn has produced an interesting range of artefacts (Fig. 6.13) including a biconical gold bead, a bronze rapier, a bronze razor, two bronze awls, bone awls, a bone weaving-comb, a bone spatula, a flint knife and fragments of pottery vessels (Mason, 1968). It is possible to regard the artefacts, except for the gold bead which is an early Bronze Age type, as

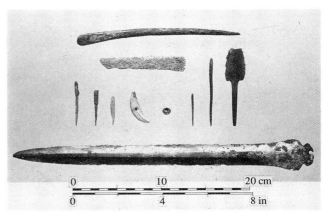

Fig. 6.13. Middle Bronze Age finds from Ogof yr Esgyrn.

a contemporaneous group of the last, 'Penard', phase of the middle Bronze Age (1050–900 bc). The everyday objects represented suggest actual occupation but the rapier, razor and the ?ancestral gold bead suggest that a family, or families, of some consequence may have been involved. The presence of the rapier is suggestive of unsettled conditions and we should note that the Penard phase is marked in Wales by the appearance both of a new weapon of war, the sword, and of the first hill-forts (Burgess, 1980; Savory, 1980). It may be, there-fore, that caves were re-used in this period as temporary refuges.

The broadly contemporaneous occupation of the Lesser Garth Cave may be a similar example (Hussey, 1966). The finds here include sherds of a number of Middle Bronze Age pottery vessels; two bone weaving-combs of precisely the type found at Ogof yr Esgyrn; a bone needle; bone awls; a bone handle for an awl, possibly of bronze; a flint scraper and a broken flint pick. The objects, as a group, clearly attest activities connected predominantly with pastoral farming and, aided by the animal bones discovered, we can reconstruct a picture of the exploitation of sheep for their wool and of cattle, pig and, perhaps, horse for meat. Hussey (1966) makes reference to the presence of grinding stones in the cave which might suggest the grinding of grain there, obtained, perhaps, by barter.

There is also a small Middle Bronze Age personal hoard of objects from Priory Farm Cave, Pembroke (Laws, 1908). The objects, all of bronze, comprise a saw-blade, a tanged chisel, a mock torsion ring and the blade of a 'transitional' palstave. These finds probably represent the property of an individual buried for safekeeping in a time of temporary danger.

It is unwarranted at this stage to extend this survey into an interpretative account, for in all periods caves must be seen as no more than a form of natural shelter, which might be utilized on convenient occasions for a variety of purposes. The effect of these temporary occupations is to present a palimpsest in which the nature of the successive occupations may be irretrievable. The functions of caves must further be seen in the context of contemporary settlement patterns and of sub-sistence and economic strategies as a whole, as indeed Bradley (1978) has tried to do. Our understanding of cave-use in the Holocene offers an important key to understanding life in the Pleistocene. In the same way, study of the historical evidence for cave use in Australia (Jennings, 1979) and elsewhere is vital for our understanding of the part caves played in daily life. We will end with a quote from Bowdler (1975) made with reference to caves in Australia: 'We are left with a paradox; the type of site most sought after by the archaeologist was probably of least importance to the people who sporadically occupied it.'

Acknowledgements

I am indebted to Elizabeth Healy for Figure 6.5 and to both Elizabeth Healy and Yolanda Stanton for reading the chapter in draft form; to Stephen Hartgroves, Paul Hughes, Eric Broadbent and John Thomas for the illustrations, and to Monica Cox for typing the manuscript.

Postscript

This chapter was originally written about ten years ago. Since then, I have excavated at Little Hoyle and Hoyle's Mouth caves in Dyfed, and in the Elwy Valley of Clwyd. For many reasons, it has only been possible to incorporate a few of these results in the present text.

References

Binford, L. R. (1978). *Nunamiut Ethnoarchaeology*. London: Academic Press.

Bohmers, A. (1956). Statistics and graphs in the study of flint assem-blages. *Palaeohistoria*, **5**, 1–25.

Boon, G. C. (1980). Two Iron Age glass beads in the National Museum. *Bulletin of the Board of Celtic Studies*, **28(iv)**, 745–6.

Bordes, F. (1954). Les limons Quaternaires du Bassin de la Seine. *Archives de l'Institut de Paléontologie Humaine, Paris*, Mémoire 26.

Bowdler, S. (1975). Caves and aboriginal man. *Australian Natural History*, **18**, 216–19.

Bowen, D. Q. (1970). The palaeoenvironment of the 'Red Lady' of Paviland. *Antiquity*, **44**, 134–6.

Bowen, D. Q. (1978). *Quaternary Geology*. Oxford: Pergamon Press.

Bradley, R. (1978). *The Prehistoric Settlement of Britain*. London: Routledge & Kegan Paul.

Buckland, W. (1823). *Reliquiae Diluvianae*. London: John Murray.

Burgess, C. (1980). The Bronze Age in Wales. In *Culture and Environment in Prehistoric Wales*, ed. J. A. Taylor, pp. 243–86. Oxford: British Archaeological Reports, No. 76.

Burleigh, R., Hewson, A. & Meeks, N. (1976). British Museum natural radiocarbon measurements VIII. *Radiocarbon*, **18**, 16–42.

Campbell, J. B. (1977). *The Upper Palaeolithic of Britain*. Oxford: Clarendon Press.

Campbell, J. B. & Sampson, C. G. (1971). A new analysis of Kent's Cavern, Devonshire, England. *University of Oregon Archaeological Papers*, **3**, 1–40.

Clark, G. (1980). *Mesolithic Prelude*. Edinburgh: Edinburgh Univer-sity Press.

Clegg, J. (1969). Excavations at Coygan Cave, near Laugharne. *Carmarthen Antiquary*, **5 (1964–69)**, 13–20.

Collcutt, S. N. (1979). Notes sur le 'LUP' de la Grande-Bretagne. In *La Fin des Temps Glaciaires en Europe*, ed. D. de Sonneville-Bordes. Paris: C.N.R.S. Colloque No. 271.

Collins, D. (1976). The geography of the European Lower Palaeolithic. In *Colloque X, L'Evolution de l'Acheuléen en Europe. Union Internationale des Sciences Préhistoriques et Protohistoriques. IXe Congrès*, ed. J. Combier, pp. 156–65.

Collins, D. & Collins, A. (1969). Cultural evidence from Oldbury. *Bulletin of the London Institute of Archaeology*, **8–9**, 151–76.

Davies, M. (1969). *Archaeology in Wales*, **9**, 13–14. Council for British Archaeology.

Davies, M. (1975). *Archaeology in Wales*, **15**, 38. Council for British Archaeology.

Dawkins, B. (1901). On the Cairn and Sepulchral Cave at Gop, near Prestatyn. *Archaeological Journal*, **8**, 322–41.

Delporte, H. (1976). L'Organization du Périgordien supérieur en France. *Union Internationale des Sciences Préhistoriques et Proto-historiques. IXe Congrès; Colloque* **XV**, 7–51.

Gilks, J. A. (1973). The Neolithic and Early Bronze Age pottery from Elbolton Cave, Wharfedale. *Yorkshire Archaeological Journal*, **45**, 41–54.

Green, H. S. (1976). Neolithic and Bronze Age flint, stone and bonework. In *The Excavation of Wetton Mill Rock Shelter*, ed. J. Kelly, pp. 66–74. City of Stoke on Trent Museum Archaeological Society. Report No. 9.

Green, H. S. (1980). *The Flint Arrowheads of the British Isles*. British Archaeological Reports, No. 75. Oxford.

Green, H. S. (1981a). A Palaeolithic flint handaxe from Rhosili, Gower. *Bulletin of the Board of Celtic Studies*, **29**, 337–9.

Green, H. S. (1981b). The first Welshman: excavations at Pontnewydd. *Antiquity*, **55**, 184–95.

Green, H. S. (1984). *Pontnewydd Cave: A Lower Palaeolithic Hominid Site in Wales: The First Report*. Cardiff: National Museum of Wales.

Green, H. S. (1986). The Palaeolithic settlement of Wales research project: a review of progress 1978–1985. In *The Palaeolithic of Britain and its Nearest Neighbours: Recent Trends*, ed. S. N. Collcutt, pp. 36–42. Cardiff: National Museum of Wales.

Green, H. S., Stringer, C. B., Collcutt, S. N., Currant, A. P., Huxtable, J., Schwarcz, H. P., Debenham, N., Embleton, C., Bull, P., Molleson, T. I. & Bevins, R. E. (1981). Pontnewydd Cave in Wales – a new Middle Pleistocene hominid site. *Nature*, **294**, 707–13.

Grimes, W. F. (1935). Coygan Cave, Llansadyrnin, Carmarthenshire. *Archaeologia Cambrensis*, **90**, 95–111.

Grimes, W. F. (1951). *The Prehistory of Wales*. Cardiff: National Museum of Wales.

Harvey, J. C., Morgan, R. & Webly, D. P. (1967). Tooth Cave, Ilston, Gower. *Bulletin of the Board of Celtic Studies*, **22(ii)**, 277–90.

Healy, E. & Green, H. S. (1984). The Lithic industries. In *The Gwernvale Long Cairn, Crickhowell, Brecknock*, ed. W. J. Britnell, and in *Gwernvale and Penywyrlod*, ed. W. J. Britnell & H. N. Savory, pp. 113–35. Cardiff: Cambrian Archaeological Monographs.

Horton, A. (1970). *The Drift Sequence of Sub-Glacial Topography in Parts of the Ouse and Nene Basin*. Institute of Geological Sciences, Report 70/9. London.

Horton, A., Shepherd-Thorn, E. R. & Thurrell, R. G. (1974). *The Geology of the New Town of Milton Keynes*. Institute of Geological Sciences, Report 74/16. London.

Hussey, M. J. (1966). Final excavations at the Lesser Garth Cave, Pentyrch. *Transactions of the Cardiff Naturalists Society*, **93 (1964–66)**, 18–36.

Jacobi, R. M. (1980). The Upper Palaeolithic of Britain with special reference to Wales. In *Culture and Environment in Prehistoric Wales*, ed. J. A. Taylor, pp. 15–100. Oxford: British Archaeological Reports, No. 76.

Jennings, J. N. (1979). Man and other animals in Australian caves. *Transactions of the British Cave Research Association*, **6**, 93–130.

Klein, R. (1973). *Ice Age Hunters of the Ukraine*. Chicago: Chicago University Press.

Kowalski, K. (1980). Origin of mammals of the Arctic tundra. *Folia Quaternaria, Krakow*, **51**, 3–16.

Lacaille, A. D. (1954). A hand-axe from Pen-y-lan, Cardiff. *Antiquaries Journal*, **34**, 64–7.

Lacaille, A. D. & Grimes, W. F. (1961). The prehistory of Caldey: Part 2. *Archaeologia Cambrensis*, **110**, 30–70.

Laws, E. (1908). Bronze implements from the shores of Milford Haven. *Archaeologia Cambrensis*, **8**, 114–15.

Mason, E. J. (1968). Ogof-yr-Esgyrn, Dan-yr-Ogof Caves, Brecknock. *Archaeologia Cambrensis*, **117**, 18–71.

Mellars, P. (1969). The chronology of Mousterian industries in the Périgord region. *Proceedings of the Prehistoric Society*, **35**, 134–71.

Mellars, P. A. (1974). The Palaeolithic and Mesolithic. In *British Prehistory: A New Outline*, ed. C. Renfrew, pp. 41–99. London: Duckworth.

Molleson, T. (1976). Remains of Pleistocene man in Paviland and Pontnewydd Caves, Wales. *Transactions of the British Cave Research Association*, **3(2)**, 112–16.

Molleson, T. & Burleigh, R. (1978). A new date for Goat's Hole Cave. *Antiquity*, **52**, 143–5.

Movius, H. L. (1969). The Chatelperonnian in French archaeology. *Antiquity*, **43**, 111–23.

Otte, M. (1976). L'Aurignacien en Belgique. In *Colloque XVI: L'Aurignacien en Europe. Union Internationale des Sciences Préhistoriques et Protohistoriques. IXe Congrès*, ed. J. Kozlowski, pp. 144–63.

Otte, M. (1977). Données générales sur le Paléolithique Supérieur Ancien de Belgique. *L'Anthropologie*, **81**, 235–72.

Powell, T. G. E. (1973). Excavation of the Megalithic tomb at Dyffryn Ardudwy. *Archaeologia*, **104**, 1–49.

Pye, W. R. (1972). *Archaeology in Wales*, No. 12, p. 16. Council for British Archaeology, Group 2.

Pye, W. R. (1978). *Archaeology in Wales*, No. 18, p. 37. Council for British Archaeology, Group 2.

Renfrew, C. (1973). Monuments, mobilization and social organization in Neolithic Wessex. In *The Explanation of Culture Change*, ed. C. Renfrew, pp. 539–58. London: Duckworth.

Roe, D. A. (1976). The evolution of the Acheulian in Britain. In *Colloque X, L'Evolution de L'Acheuléen en Europe. Union Internationale des Sciences Préhistoriques et Protohistoriques. IXe Congrès*, ed. J. Combier, pp. 31–46.

Rogers, E. S. (1970). *Indians of the Subarctic*. Toronto: Royal Ontario Museum.

Savory, H. N. (1980). *Guide Catalogue of the Bronze Age Collections*. Cardiff: National Museum of Wales.

Shackley, M. L. (1977). The bout coupé handaxe as a typological marker for British Mousterian industries. In *Stone Tools as Cultural Markers*, ed. R. V. S. Wright, pp. 332–9. Australian Institute of Aboriginal Studies, Canberra. Prehistory and Material Culture Series, No. 12. New Jersey: Humanities Press.

Sollas, W. J. (1913). Paviland Cave: an Aurignacian station in Wales. *Journal of the Royal Anthropological Institute*, **43**, 325–74.

Stringer, C. (1977). Evidence of climatic change and human occupation during the last interglacial at Bacon Hole Cave, Gower. *Gower*, **28**, 36–44.

Stuart, A. J. (1974). Pleistocene history of the British vertebrate fauna, *Cambridge Philosophical Society Biological Reviews*, **49**, 225–66.

Tuffreau, A. (1971). *Quelques Aspects du Paléolithique Ancien et Moyen dans le Nord de la France*. Amiens: Numéro Spéc. de la Soc. de Préhistoire du Nord.

Ulrix-Closset, M. (1975). *Le Paléolithique Moyen dans le Bassin Mosan en Belgique*. Liège: Université de Liège.

Wainwright, G. J. (1967). *Coygan Camp*. Cardiff: Cambrian Archaeological Association.

Waterbolk, H. T. (1971). Working with radiocarbon dates. *Proceedings of the Prehistoric Society*, **37**, 15–33.

Wymer, J. J. (1977). Handaxes from beneath glacial till at Welton-le-Wold, Lincolnshire, and the distribution of palaeoliths in Britain. *Proceedings of the Prehistoric Society*, **43**, 355–60.

7

Recent advances in cave archaeology in southwest Wales

MEL DAVIES

In South Wales the Carboniferous Limestone outcrops across the Gower Peninsula from Mumbles to Worms Head. It then continues under Carmarthen Bay and into south Pembrokeshire including the northern half of Caldey Island (Fig. 7.1). There are caves of archaeological interest from Bacon Hole in the east to Ogof Garreg Hir in the west – most of them coastal at the present time; 32 will be described here including the outlying site, Coygan Cave, which is only 2 km from the sea, near the village of Laugharne.

It is now some years since two major publications appeared dealing with the excavations in these caves. The second edition of *British Caving* (Jackson, 1962) described 12 of the caves and Campbell (1977) assessed eight of these in addition to another five. Much work has been carried out since Campbell, assisted in some small way by the author, completed his survey in 1975, with new bone caves discovered in Gower and very full excavations pursued in many of the caves on Caldey Island. The National Museum of Wales has started a programme of cave excavations which includes detailed ecological and chronological work using a variety of new techniques (Green, 1984, 1986), and workers from the University College of Wales, Aberystwyth, and the University College of Swansea have, through their geological investigations, further narrowed the artificial gap between that science and archaeology (Bowen, 1980, 1986; Campbell, Andrews & Shakesby, 1982). Large quantities of virgin sediments remain to be excavated from the well-dug caves of Minchin Hole and Bacon Hole, and many sites on the Castlemartin coast in south Pembrokeshire will probably remain untouched because of the inaccessibility of the sheer 50 m cliffs which they are contained in, or until they finally collapse into the sea.

Although the area to be covered here stretches for 60 km east to west it must be remembered that throughout much of the Upper Palaeolithic period the whole strip was dry land since seawater was locked up in the enlarged polar ice-caps. The hunter who arrived in Gower from the east searching for shelter in caves, albeit only temporarily, would soon detect the prominent limestones of Mumbles Head. From there Pwlldu Head is visible, then Oxwich Point and Porteynon Point, both of limestone. Excavation shows that the traverse westwards

from there to Worms Head is littered with occupied caves (including six newly discovered ones). From the summit of Worms Head, just 30 m above another cave with an occupation site, the limestones of Caldey Island to the west, and the limestone promontory of Coygan Cave 23 km to the northwest are clearly visible in good weather. Furthermore, they would have been as identifiable to the hunter of yesteryear as they are to the geologist's eye today. In the hinterland north of this coastal area, there is a Carboniferous Limestone outcrop following the edge of the Coalfield, but while caves are fairly plentiful in it, bone caves are rare. It appears, therefore, that prehistoric hunters were forced to hug what is now the coast in order to be near their prey – the deer, bison and mammoth grazing in the tundra (and later the scrub woodland) of what is now Carmarthen Bay.

While he did not excavate personally in the caves of southwest Wales, Jacobi (1980a) has analysed the available evidence based on artefacts and fauna for the Upper Palaeolithic and the Mesolithic periods. Bowen (1980) sketched what he called 'a conservative outline of the Pleistocene sequence of environmental change', pulling together data collected not only by himself, including participation in the Minchin Hole excavations of Sutcliffe (1981), but also by his students and colleagues at Aberystwyth. In summary, at the time he visualized a pre-125 kilo-annees (ka) glaciation with ice crossing southwest Wales from a northwesterly direction. No till, or boulder clay left by this ice is known from inside any of the caves, but it is thought that the Lower Red Cave Earth stratum in Minchin Hole dates from this cold period. Underlying it is an Inner Beach interpreted as being formed during a pre-Last Interglacial event when sea level was twice some 3 m higher than today, at approximately 194 ka and 216 ka. Overlying it is the 'Patella' beach with another high sea level dated to approximately 120 000 years ago formed during the Last Interglacial. A cold episode followed, called the Last (Devensian) Glacial Stage, represented in Minchin Hole by the Upper Red Cave Earth mixed with limestone breccias (termed *thermoclastic scree* by Campbell, 1977). The mammal fauna in these various layers is described below, but at this stage it is necessary to stress the chronostratigraphic outline, mainly

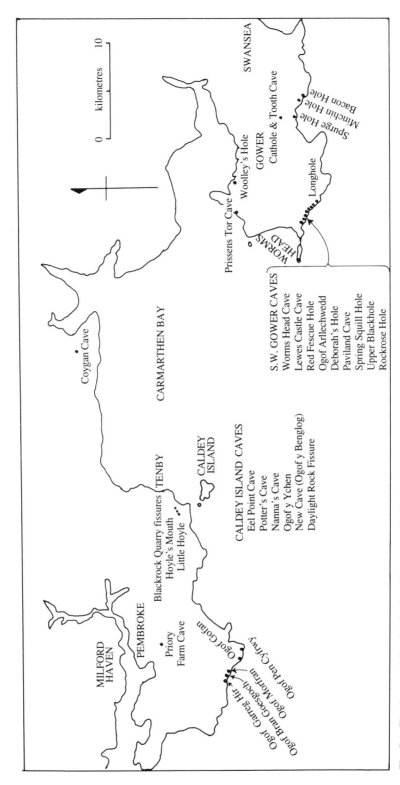

Fig. 7.1. Sketch map of the distribution of bone caves in South Pembrokeshire and the Gower.

based on the Minchin Hole type-site which will be applied at the other caves to be described. Bowen later went on to expand this sequence (Bowen & Henry, 1984) and completed a drilling programme into glacial sediments in Gower which has yet to be published in full (Bowen, 1986). When this is done the sequences in Bacon Hole as well as Minchin Hole can be tied in with Bowen's view of the Quaternary history of southwest Wales.

South Pembrokeshire

Ogof Garreg Hir (SR 9428 9387)

Starting in the west the first cave is Ogof Garreg Hir, discovered in August 1972 (Davies, 1972, 1976, 1977). The entrance, which is only 1.2 m high by 0.5 m wide, is situated about 21 m down a sheer cliff 46 m high, and access is only possible via a rope descent in an open-sided chimney and a 20 m long traverse along a ledge which narrows to 0.5 m, as the sea never leaves the foot of the cliff. Excavations were carried out in December 1972, April 1974 and March 1977 for a distance of 2.5 m into the cave, leaving a further 2 m undisturbed. The matrix was a moist, reddish-brown cave earth containing sub-angular stones, but any stratification that might have existed had been destroyed by storm–wave action, a layer of stalagmite that once covered most of the outer cave having been completely eroded away except for fragments clinging to the east wall. The cave earth contained a bone awl and patinated flint blade (Fig. 7.2), also a flint microlith more accurately described as a narrow, backed bladelet, finely retouched along most of one edge. With these were two human bones from the left side of a small adult, and animal remains comprising red deer, roe deer, wolf, pig, hare, fox and several species of bird. Jacobi (1980b) considered the blade and microlith to be from the later Mesolithic period (later than 8500 BP). The bone assemblage is unusual in that so many different species were represented by so few bones. This, and the two human bones, suggests an occupation in a passage or chamber which was once situated seaward of the present small tunnel, and which has disappeared into the sea due to cliff erosion. The visitor who cowers on the tiny ledge outside this cave during a winter storm will need no convincing of the

Fig. 7.2. Implements found in Ogof Garreg Hir. (a) Flint blade, 8.7 cm long. (b) Bone awl, 9.8 cm long.

violently destructive force of the waves! The absence of marine molluscs such as limpets and mussels from what is a kitchen midden deposit suggests that the sea was a long way off from the cave during the occupation, and this confirms an early date not later than the Mesolithic.

As will be shown below, caves which today are perched inaccessibly in vertical sea-cliffs would have had easy access in post-glacial times via scree slopes formed by periglacial freeze–thaw processes during the Last Glaciation. A eustatic recovery of sea level with ameliorating climate followed by wave erosion has removed the lower parts of these screes, but fragments remain in the gullies of the Great Orme coast of North Wales, in the 'Slades' of Gower, and running into the larger cave entrances such as Minchin Hole in Gower.

Ogof Bran Goesgoch (SR 9432 9386)

First noted in 1969 by R. A. Kennedy and the author, being no. 12 in a cave census being undertaken on the Castlemartin cliffs. It seemed to have archaeological potential but it was not until a subsequent visit in 1977 that three bones thrown up by burrowing rabbits were noticed on its floor. These were human, from the left hand and right foot, respectively. No further disturbance was noted two years later, but in 1982 a wire fence was erected across the cave to ensure protection from sheep and rabbits. Excavation has not been carried out into what is probably an inhumation site dating to prehistoric times (Davies, 1976, 1977). In common with other caves on the Castlemartin cliffs, it is safeguarded by Army authorities, and access is not permitted when NATO forces are using the firing range for tank exercises. Climbing as well as cave exploration is controlled, and because of the nesting sites used by auks and choughs, advice on access is given by a Range Conservation Group. Most of the coastal area is also notified as a Site of Special Scientific Interest.

Ogof Morfran (SR 9471 9377)

Discovered in 1969 and excavated the same year and in three successive years up to 1974 by R. A. Kennedy, P. Wilkins and the author (Davies & Kennedy, 1969; Davies, 1975), the occupied part of the cave is a pleasant, south-facing rock shelter about half-way down the 42 m high, sheer cliff. The eastern end becomes a passage which crosses over a pothole before opening out in another tunnel in a cliff facing east, but access is by rope descent over the cliff onto a wide ledge in front of the shelter. Although only 8% of the cave floor has been excavated, mainly in a single trench 5 m long taken to bedrock, there was a wide range of finds. Bones of ox were hacked and occasionally blackened by fire, and an antler of red deer in which 31 cm of the beam had survived was seen to have its brow tine smoothed as if it had been shaped and cut for use as a digging tool. As well as flint, charcoal, limpets, mussels and a sandstone saddle-quern attested to occupation, and dating was provided by sherds of a hand-made, bucket-shaped vessel of probably late Bronze Age provenance (Savory, 1971). Sherds from a deeper level seemed to be Neolithic pottery, but lightly cemented to the wall of the cave by stalagmite was a sherd of Samian ware of the form known as Dragendorff 37. Loose in the cave earth were two Roman coins and part of a bronze pin from the period AD 268–74. There were no human bones but some might yet be found in the 92% of the shelter which remains unexcavated. It was clear that storm-waves occasionally break into the cave, and timbering fixed inside the trench in 1974 was found collapsed and strewn about during an inspection some years later. Herring gulls also build nests in the cave and cause a surprising amount of damage, including breaking, by repeated pecking, the antler which was left *in situ* for a few weeks to dry out.

Ogof Gofan (SR 9581 9301)

An entrance to this cave on the west side of Saddle Head was first noticed in 1966 but it could not be reached as it was under an overhang about half-way down a sea-cliff 41 m high (Fig. 7.3). A second, slightly more accessible entrance was discovered two years later but it still required a rope descent. Bones were uncovered in an inner chamber 37 m from the accessible entrance, so far in fact, that it was concluded that they must have been brought in via the 'overhang' entrance which was only 9 m away. As both entrances now open in sheer, wave-lapped cliffs, the conditions of deposition must reflect a coastal configuration quite different from what it is today. A limited excavation by R. A. Kennedy revealed charcoal and flint waste. Some bones were collected from the surface of the passage connecting the entrances which are about 30 m apart. The full bone list (Bateman, 1969) includes cat, pig, ox, roe deer, sheep or goat and hare. Also in this passage, but lying loose in a small hollow with fragments of roe deer bones, was a sherd of late Neolithic pottery representing a late stage in the development of Peterborough ware (Savory, 1968). A section cut by storm-waves in the deposits just inside the overhang entrance shows about 0.5 m of stony red clay

Fig. 7.3. The difficult descent to Ogof Gofan.

resting on bedrock, topped by a thick stalagmite floor. Some of the stalagmites in the innermost chambers are 5 m high but there is no evidence that prehistoric man penetrated this far. It is believed, however, that he traversed the passage between the two entrances because several stalagmites in this area had been deliberately broken or had their tops lopped off. The pieces lying about no longer fit neatly on their stumps because of further growth, but they are certainly not growing under today's climatic regime as this has been checked in both summer and winter. It is concluded that ancient man was either something of a vandal or deliberately tried to clear obstructions away to make the passage route more convenient (Davies, 1969). Further excavations in this cave would encounter severe problems with the thickness of the stalagmite deposits.

Ogof Pen Cyfrwy (SR 9586 9287)

This cave is situated about 120 m further out on Saddle Head than Ogof Gofan in an exposed position and at the same altitude. As a result most of its contents have long been washed out by the sea. Furthermore, a pothole inside the cave falls to sea level and there is evidence that water erupts through this during storms. The cave was discovered in 1969 and a piece of stalagmited clay removed from the wall was claimed to contain a jaw fragment of a juvenile hyena. A brief excavation in March 1978 revealed only rodent bones cemented in fragments of breccia. Slope patterns in stalagmite remains 2 m thick indicate two separate periods of growth, but there has been no further excavation as this cave is again only accessible by an exposed rope climb.

Other archaeological caves are known in the Castlemartin area; several entrances are known and some have been excavated briefly without finding anything; other holes have been detected from the sea by a boat traverse and not yet reached. The coast can only be properly explored by boat backed up by a cliff-top team equipped with ropes whose members can descend the cliff as directed by searchers in the boat. This was attempted by members of Cwmbran Caving Club in 1969 using a sailing dinghy and again in 1971 in a motor boat, assisted by the local Coastguard. Both trips were successful and not all the holes spotted then have yet been reached or even accurately relocated. Some of the caves are used by greater horseshoe bats so great care must always be exercised during exploration. It should be added that, although modern potholers claim these caves as 'discoveries', the egg collectors of the last century and earlier had reached them all. These eggs were collected for human consumption and one elderly local man in 1969 was able to confirm how widespread the practice had been.

The Tenby area

A group of four caves is situated near Tenby and, being easily accessible, have been more or less thoroughly explored and full reports published.

Priory Farm Cave (SM 976 019)

Using the original records and finds made by Style and Dixon in 1906–7, Grimes (1933) was able to describe a Pleistocene excavation in a laminated clay in which mammoth, hyena, reindeer, horse and cave bear remains had been found. Patinated backed blades of Creswellian type and blades and microliths were found in a gravel layer, but it was not possible to work out the stratigraphical relationship between the gravel and the bone-bearing clay. Campbell (1977) terms one of the flint tools an awl and agrees with a later Upper Palaeolithic dating.

Hoyle's Mouth (SN 112 003)

This cave, near Tenby, was reopened by a team led by Dr H. N. Savory (1973) who dug a trench into the cave and along the entrance platform outside it. However, it proved impossible to avoid earlier disturbances, and the flint implements found can only be ascribed to a Creswellian phase on typological grounds. Fortunately, the forms corresponded very well with those of specimens preserved in Tenby Museum from earlier excavations, 60 of which were traced by Campbell (1977), with 18 of them being figured and described. He termed the cave a 'base camp' from which forays could be made by hunters who might also use smaller caves temporarily within a radius of about 10 km.

Little Hoyle (SN 112 001)

Also known as Longbury Bank Cave, this cave is situated about 400 m from Hoyle's Mouth near Tenby. As the original cave and its interesting collapsed (or solution) doline rear entrance were considered already completely excavated, in 1958 Professor McBurney opened a trench to the west of the lower entrance (McBurney, 1959). Considering the size of the excavation the finds were sparse but he did report bones of reindeer and bear species, with a large Creswellian-type blade. The earlier excavators had also found remains of mammoth, horse and woolly rhinoceros. McBurney seemed fairly confident that he could place an evolved Creswellian industry in a late glacial rather than a post-glacial context.

Excavation in 1986 showed that the stratigraphic level of a dated fauna pre-dated the late glacial occupation. Two undisturbed layers of uncemented limestone breccia were found with a silty matrix containing fauna but no evidence of artefacts, and the fauna showed no marks ascribable to human action. The dated material, comprising a total of four bear and reindeer bones, was found in the lower of the undisturbed layers and gave a mean age, using the three methods radiocarbon-dating, amino acid racemization and Uranium series disequilibrium of 17.85 ± 0.12 ka BP. Thus the animals would seem to be contemporary with the main late Devensian ice advance (Green, 1986; Rae et al., 1987).

Blackrock Quarry Fissures (SN 109 002)

Dawkins (1874) recovered bones of mammoth, hyena, woolly rhinoceros, lion, horse and, supposedly, hippopotamus from a cave revealed by quarrying. The quarry, which is now disused, has been thoroughly examined, but no trace remains of any caves. This mention of hippopotamus, which is an interglacial species, has to be treated with caution, unsupported as it is by similar finds from neighbouring caves.

Caldey Island

Situated 3 km south of Tenby, the island is only 1 km from the nearest part of the mainland, and for much of the Last Glacial Stage was joined to the mainland providing easy passage for large animals, whose bones have been found in the caves. Although the northern half of the island consists of Carboniferous Limestone the bone caves are concentrated in two areas – Eel Point in the northwestern corner, and from High Cliff to a little south of Den Point in the northeastern corner.

Eel Point Caves (SS 1303 9726)

First excavated in about 1840 when abundant remains of mammoth, rhinoceros, hyena, lion, bison, bear, deer and hippopotamus were found, the cave was re-examined in 1950 by Professor W. F. Grimes and Brother James Van Nedervelde. They concluded that quarrying had destroyed the cave except for remnants of stalagmite and a hard breccia (Lacaille & Grimes, 1955). However, these few surviving fragments led Brother James to probe the site again in 1970 when a disappointingly sterile fissure was discovered, and finally, in 1986, ancient, sealed entrances were uncovered which yielded flint of Mesolithic character only 50 cm below the surface, and a range of animal bones and teeth buried much deeper comprising hyena, mammoth, horse and three individual rhinoceroses. There was no evidence that man was a contemporary of these animals, but work here is continuing (Nedervelde & Davies, 1987).

Potter's Cave (SS 1435 9707)

When discovered in 1950 the two north-facing entrances some 6 m apart were blocked by blown sand and fallen rocks, some of which had been dislodged by adjacent quarrying. Excavation over the following five years from the east entrance revealed late Bronze Age–Iron Age pottery, early Bronze Age (Beaker) pottery and Neolithic pottery in mixed upper layers, together with a fauna comprising post-glacial animals. With the fauna were the remains of fish, crustaceans and sea shells testifying to a late date for the occupation when eustatic recovery of sea level was nearly complete. Cemented in stalagmite and also under it human remains and flint were found in the inner cave but no details were provided in the early reports (Lacaille & Grimes, 1955, 1961), and excavation was abandoned due to the hardness of the stalagmite until 1973.

When the 60 cm thick stalagmite layer was removed hyena teeth were discovered, also a patinated, convex, blunted-back blade of Creswellian type deep in a stony, red clay, and on extending the excavation into the west entrance and out onto the sloping platform fronting the caves, a remarkable range of finds was uncovered. Horse bones appeared, well gnawed by hyena, and bones of woolly rhinoceros (Coelodonta antiquitatis) including a scapula fashioned in a way resembling the bone scrapers retained in Manchester Museum from A. L. Armstrong's excavations in Pinhole Cave, Creswell, Derbyshire (Jackson, 1962). If not the work of man then this peculiarly shaped bone must be the product of a hyena biting off the thin, blade-like part of the scapula, but being unable to cope with the thicker caudal margin of the bone. In a chamber connecting the two main passages, but also partly distributed in cavities near by, human burials comprising two individuals were eventually pieced together, and it was perhaps one of these people who owned the magnificent necklace of 49 blue glass beads collected from beneath a boulder on the platform below the west entrance and dating from the first or second century BC. There was certainly no evidence that the burials were as old as the extinct mammal fauna, although the flint proves that humans were hunting in the vicinity at the time that the hyena was dragging its prey into the cave (Nedervelde & Davies, 1975, 1977a; Savory, 1980a,b).

Nanna's Cave (SS 1457 9697)

The history of the exploration of this cave, starting in 1911, is long and complex, but the author has assembled the evidence of previous digs (Lacaille & Grimes, 1961) with that from modern excavations led by Brother James Van Nedervelde (Nedervelde & Davies, 1977b) lasting from 1973 to 1986 (Nedervelde & Davies, 1976–86). Only a summary of the many and varied finds can be given here and to avoid confusion in view of the various interpretations placed on the excavations over the years, this description will deal with the stratigraphy starting from earliest times. The present cave floor, which still seals some undisturbed deposits, lies at, and immediately below 20.81 m O.D., while the steep slope in front of the cave ends on the raised beach platform, dating from the Last Interglacial, at 8.51 m O.D. A small remnant of a cemented breccia with angular limestone fragments survives at the back of this platform, but it does not seem to be archaeological, and its date has not been established. A deep trench has been excavated irregularly in steps from the platform, reaching into the cave. At maximum the trench is 4 m wide and at times 2.5 m deep. As much as 2 m of this was found to be excavation debris from within the cave not all of which had been efficiently sieved. All the human bones recovered in the last decade of excavations have been examined, and compared with fragments stored in Tenby Museum, and it is now quite clear that three adults and at least one juvenile were interred within the cave, although none of the remains have been recovered from an unambiguous, stratified position. Animal remains consist of pig, sheep or goat, *Bos* species, fox and two horse teeth which may have been intrusive. Some of the bones were blackened by fire. This limited faunal range is entirely post-glacial, but Creswellian flint tools have been found, mostly under the platform in a red silt layer, or in a greyish-yellow silt which occurs also within the cave. There are also characteristic Mesolithic flints and bone or antler piercing tools. A range of pottery starts with round-bottomed Neolithic A bowls, Bronze Age pottery, Iron Age and Romano-British pottery. The Roman finds comprise coins, a fibula and a spindle whorl attesting to further occupation in the third or fourth centuries AD. The one peculiarity about Nanna's Cave is the lack of a late glacial fauna which is so common in the neighbouring sites, Potter's Cave and Ogof yr Ychen.

Ogof yr Ychen (SS 1464 9692)

This has been the best documented of the Caldey Island cave excavations of the last two decades. A section is shown in Figure 7.4. When first discovered by Brother James Van Nedervelde in 1970, the cave had already been truncated by nineteenth-century quarrying, and considerable material lost from probable continuations to chambers 3 and 4. The 1970 entrance led to chamber 1 which contained the skeletal remains of a large member of the *Bos* family – either bison or aurochs. This entrance collapsed so a new one was opened to chamber 1, another was widened to gain entry to chamber 2 and further excavation opened entrances to chamber 3, chamber 4 (from both ends) and, in 1984, to the roof of chamber 2. The main bone-bearing stratum was a periglacial scree with yellow, silty clay, which yielded remains of wolf, hyena, woolly rhinoceros and wild boar, the latter being at the top of this particular stratum (Bateman, 1973). In this stratum was a typically Creswellian, convex, blunted-back knife of flint, and a large adinole knife which was 5 m deep in the pit at the mouth of chamber 4. A mere 30 cm below the knife was a scapula of woolly rhinoceros radiocarbon dated to $22\,350 \pm 620\,y$ BP (Birm, 340) (Nedervelde, Davies & John, 1973). In view of the proximity of the knife and the bone, it is considered that prehistoric man was present in the cave some 22 millennia ago and that the climate was bearable. The further excavations yielded the remains of three human adults, one of whom had entered the cave, apparently head-first, through the shaft dropping into chamber 1 such that after disarticulation, part of his mandible became incorporated 6.5 m lower, in the clay of chamber 3.

A child's bones from a much later date were found in the chamber 4 entrance in the Romano-British deposits. The pathology of these skeletons was fascinating in that one of the adult skulls had the marks of an injury on the back which, fortunately, had healed, and in a second two canine teeth had been lost in life, surely an unusual event unless perpetrated deliberately.

Other animal remains, but in an upper cave earth, comprised ox or bison, red deer and roe deer, many of the bones complete and ungnawed, suggesting fatal falls through the various shafts that once opened to the surface. Coarse, black pottery and finely made Mesolithic flints were also securely referred to certain layers (Nedervelde, 1972).

New Cave (Ogof y Benglog) (SS 1470 9688)

Also discovered by Brother James Van Nedervelde. In 1969 excavation revealed a human skull which had entered the cave (which was only the size of a rock shelter) possibly by rolling downslope from an extension lost by quarrying. The mandible was missing, the skull was from a female and it was accompanied by a well-patinated flint (or chert) leaf-shaped arrowhead. At a depth of 1.5 m lower than the 'occupation' layer a backed blade of Creswellian appearance was found. Excavation was stopped by the presence of a hard, cemented breccia and there was no animal fauna. Much probably remains to be discovered in the cave which is only a few metres from Ogof yr Ychen, and there are empty solution hollows along the cliff face connecting them (Nedervelde, 1969).

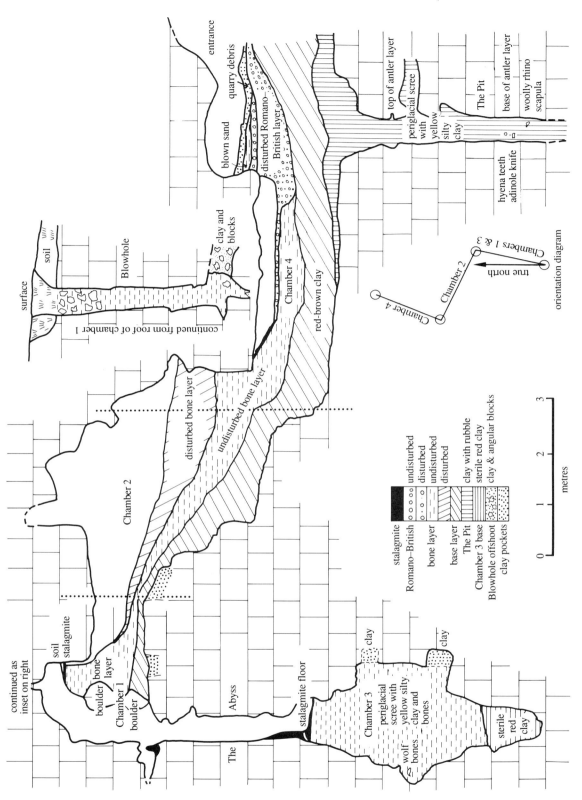

Fig. 7.4. Section of Ogof yr Ychen and its deposits on Caldey Island.

The following labels appear within the figure:

continued as inset on right

soil
stalagmite
boulder — bone layer
Chamber 1
boulder

The Abyss

stalagmite floor

Chamber 2

disturbed bone layer
undisturbed bone layer

surface
soil
Blowhole

continued from roof of chamber 1

clay and blocks

Chamber 4

red-brown clay

entrance
blown sand
quarry debris
disturbed Romano—British layer

top of antler layer
periglacial scree with yellow silty clay

The Pit
base of antler layer
woolly rhino scapula
hyena teeth
adinole knife

orientation diagram
Chambers 1 & 3
true north
Chamber 2
Chamber 4

Chamber 3
periglacial scree with yellow silty clay and bones
wolf bones
clay
clay
sterile red clay

Key:
- stalagmite
- Romano–British — undisturbed / disturbed
- bone layer — undisturbed / disturbed
- base layer
- The Pit — clay with rubble
- Chamber 3 base — sterile red clay
- Blowhole offshoot — clay & angular blocks
- clay pockets

0 1 2 3
metres

Daylight Rock Fissure (SS 1499 9660)

This small cave runs northwestwards from the interior of a short rock tunnel which, penetrating the spine of cliff at this point, led mariners to bestow this name on it. Excavated as early as 1954 the fissure yielded plentiful teeth of woolly rhinoceros, horse, hyena and a milk molar of mammoth (Lacaille & Grimes, 1955), although the site was narrow and only 1.5 m high. Bones identified came from these species but also from lion, bear, reindeer, ox or bison and giant deer (Lacaille & Grimes, 1961). Some of the bones had been gnawed by hyena but there was no evidence for the presence of man in the Pleistocene level. Above it, however, there was a Neolithic occupation with scanty human remains, and Neolithic pottery termed 'Peterborough' ware. Clearly the main habitation floor was on the platform outside, and running into another cave for which there are no excavation details except that Creswellian and Mesolithic tools were found.

Coygan Cave (SN 286 093)

Excavated several times between 1866 and 1963, the cave had the misfortune to be in the vicinity of an active quarry. The author found unexcavated sections of passage in 1969 but blasting caused a roof collapse which sealed it in 1971. The cave was surprisingly de-scheduled as an Ancient Monument, and was completely destroyed soon after despite being a very important middle Palaeolithic site with artefacts and an associated fauna. The McBurney excavations of 1963, completed by Clegg (1969) revealed two handaxes, the larger one being classified as late Acheulian, while the fauna could have fitted into a typical Upper Palaeolithic Welsh context; however, a middle Devensian age is suggested by a radiocarbon date on charcoal of 38.68 + 0.27/−0.20 ka (BM 499) (Green, 1986).

South Gower

The Carboniferous Limestone sea cliffs of the south Gower coast range for 19 km from Worms Head in the west to Pwlldu Head in the east, and there are 13 important caves facing the sea with easy access, and a further two a short distance inland near Parkmill. Another 16 caves are known, one of which has yielded human bones in the past; five have yielded animal bones, and the other 10 contain sediments which are likely to be archaeological. All these sites have been assessed by the Nature Conservancy Council as part of a management programme for the coast, the greater part of which is a Site of Special Scientific Interest, while the National Trust is the major landowner (Davies, 1987). The six minor bone caves mentioned cannot be described here as few excavation details are known (Rutter, 1949), and the 15 to be included will be dealt with only in summary form.

Worms Head Cave (SS 3836 8769)

Situated at the western extremity of the island which can only be reached during the five hours of low tide, and below an exposed climb, the cave is not readily accessible, but W. Riches (1923–4) excavated at the back of the Outer Chamber and at the Inner Chamber finding human bones from three

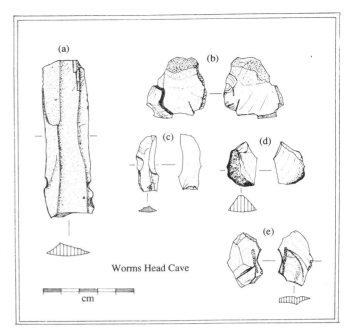

Fig. 7.5. Flint and rhyolite tools found in Worms Head Cave. (*a*) A sharp-edged knife of greenish-grey rhyolite, possibly struck from a glacial erratic, found in wet clay with a few stones 40 cm below the surface. (*b*) A flat flake of white-patinated flint or chert, with signs of blunting on one edge; probably a 'core tablet'. (*c*) Patinated flint flake without secondary working. (*d*) Flint or chert core remnant. (*e*) Pale grey flint or chert flake with blunting on the concave edge; the opposite edge is very sharp and the tool can be held comfortably between two fingers for use as a knife.

individuals, and animal remains comprising bear, wolf, fox, cat, reindeer and bird. Jackson (1962) gave a wider range of Pleistocene bones, which is either an error, or is derived from possible earlier excavations by E. C. Cunnington, who was known to have dug in Gower caves, but who perished in World War I. As the cave is on the South Gower Coast National Nature Reserve, the author has made a careful examination, cleared out much of Riches' spoil and cut a demonstration profile to show to members of the public. In the course of these works it became clear that the human bones lay in an upper stratum separated by sterile, stony clay from the Pleistocene bone layer which was close to bedrock. A number of chert and flint flakes, and a fine rhyolite blade were found, the latter having Upper Palaeolithic affinities (Fig. 7.5). As storm-waves enter the cave much has been lost, but the Inner Chamber is undisturbed except for the pit left by Riches (Davies, 1981, 1986a).

Lewes Castle Cave (SS 4141 8729)

A cave with two entrances open and a third sealed with deposits, its early excavation history, if any, is not known, and the entrance chamber was shown to be of no archaeological interest (Davies, 1982). However, when a passage was dug and extended by the Marisia Caving Club in 1985, a Pleistocene fauna was found comprising hyena, wolf, mammoth, woolly rhinoceros and reindeer. To avoid further disturbance in what is clearly a very important palaeontological cave the Club willingly terminated its excavation.

Red Fescue Hole (SS 4266 8678)

A small archway, only 2 m wide by 1 m high, when cleared of rubbish, was found to contain two human bones, with a flint implement lying on a surface much disturbed by burrowing animals. As the passage continues for at least 2 m before the deposits meet the roof, the cave is clearly of great archaeological potential as the interior is quite undisturbed (Davies, 1986b,c).

Ogof Arllechwedd (SS 4324 8632)

Another small archway, 1.2 m wide by 0.9 m high with a clay floor well traversed by badgers, yielded Bronze Age pottery in 1982. Smoke tests show that this entrance connects with a slightly higher entrance 30 m away behind a spur of rock, but this is almost choked with rubble from the cliff above. The lower passage penetrates for 2 m before becoming too low for progress, but it has not been excavated to avoid damage to what is another potentially rich archaeological site (Davies, 1983c, 1986d).

Deborah's Hole (SS 4338 8629)

A well-known cave with three small entrances, the central and largest one being only hands-and-knees sized. The first chamber is today too small to stand up in, but it is well lit by the low-angle winter sunlight which streams in through the west entrance. The second chamber is loftier but no wider, and when J. G. Rutter (pers. comm.) excavated the cave to a depth of 1 m in the 1940s he found nothing of archaeological interest, yet in 1861 it is claimed that remains of hyena, bear and other animals were found accompanied by flint and worked bone (Jackson, 1962). There is no mound of debris outside the cave and no evidence of any excavation except Rutter's, and recent disturbance by rabbits, but some 25 m to the east there exists a cave 2 m high with a large mound of cast-out debris on the platform outside from which fragments of Bronze Age pottery have been collected after they had been weathered out. This cave, which is unnamed, has obviously been excavated in the past and there is surviving evidence of the edge of the original floor adhering to the cave walls about 1 m off the present floor. Doubt must therefore exist whether the cave bearing the name Deborah's Hole is the genuine Pleistocene site, and only further excavation would dispel it (RCAHM, 1976).

Paviland Cave (SS 4373 8588)

Also known as Goat's Hole, the site is discussed in Chapter 6. It is a major early Upper Palaeolithic site but not necessarily the only one of importance as several other caves, if thoroughly excavated, would probably prove to have been frequently occupied by wandering hunting groups throughout the Upper Palaeolithic period. Paviland has no special features which would have made it more comfortable than, say, Bacon Hole or the many large caves at slightly lower altitude whose contents have now been washed out by the sea. Each of the Paviland excavators discovered that the sea had been at work since his predecessor wielded the spade, but even today it is possible to find wave-washed fragments of flint, which were once tools or at least waste flakes, among the pebbles on the cave floor. Fragments of breccia survive attached to the cave

walls well out of reach of storm-waves, and there is one small passage on the west side choked with pebbles. Sollas (1913) included a section across the cave illustrating a fissure closing in but over 4 m deep below his excavation floor. It is quite possible that archaeological sediments were left undisturbed in this fissure, and they would have survived to this day, protected by fresh sand and pebbles thrown in by the sea. A new excavation here could well prove fruitful as experience has shown that workmen employed for cave excavation rarely pursued deposits into uncomfortable pockets and fissures. The cave has every appearance of having been formed as a sea cave when sea level was 5–8 m higher than at present, but in which interglacial period this occurred it is not possible to say. The chimney leading from the ceiling to another entrance on the cliff face is probably a phreatic feature which would have been useful in leading away smoke from fires lit in the cave.

Spring Squill Hole (SS 4393 8578)

This cave has an entrance 2 m high by 1.2 m wide facing slightly west of south at about 30 m O.D. A flame test suggested a connection with another hole on the cliff facing west, and this was later located some 20 m away. Before attempting to enter the intervening passage by widening the tunnel in order to carry out bat census work, a narrow trial sounding was carried out. Below 20 cm depth hacked bone fragments appeared including a metacarpus of deer, followed by a pig humerus under a slab, and a lunate flint scraper at 28 cm. Clearly the cave was archaeological and, with most of the floor at the entrance and the passage linking it to the second entrance still undisturbed, the bat census was abandoned (Davies, 1985a). Again this is a cave of considerable potential as the flint scraper is at least as old as late Neolithic and the deposits obviously extend to a much greater depth.

Upper Blackhole (SS 4422 8560)

A rubble-floored entrance only 0.5 m high at about 60 m O.D. discovered in 1983 seemed to hold promise as a bat roost. On attempting to enter and while clearing some of the rubble, bones were noticed and two proved to be human. Subsequent examination showed the floor to be a brown, stony cave earth much disturbed by animals to a depth of 20 cm. In this were broken pieces of stalagmite, fragments of anthracite coal (the nearest outcrop being 12 km away), bones of fox, pig, badger and human remains comprising portions of a skull, clavicle, rib, tibia, two metatarsus bones and a premolar. The wear on the premolar suggested the burial was that of a young adult, but there was no dating evidence, and the sounding was backfilled. The cave extends for at least 2 m and must represent an archaeological site of some importance (Davies, 1983e, 1984).

Rockrose Hole (SS 4488 8519)

Pronounced phreatic scallops about half-way up the cliff above the grassy, coastal slope mark the site of this cave which has an impressive arched entrance 2.5 m high and 3.5 m wide, choked with damp stalagmite and scree. There is also a scree slope below the cave and the entrance is situated on the west side of a steep gully which continues uninterruptedly to the foot of the top line of cliffs. An investigation in the rubble on the floor only just inside the drip line revealed two human bones – a mandible fragment and a phalange. Animal bones comprised fox, badger, pig, ox, bird and fish. The presence of limpet shells suggests that the cave was once an inhabited site, the floor of which has suffered upheavals caused by burrowing animals. In the interior a floor can be distinguished which slopes downwards into what must be a large, and probably important, cavern. In his book Campbell (1977) referred to a cave called 'Shallow Shelter' in which a trench failed to uncover archaeological remains, but he gave no positional details or description. In view of the proximity of Rockrose Hole to Longhole (below) which Campbell's team was excavating at the time, and the evidence of a trial trench on the platform, the author regards the two caves as one and the same, and Campbell was merely unlucky not to have taken his excavation further under the cliff overhang (Davies, 1983a,b).

Longhole (SS 4513 8505)

With an excavation history going back to 1861 Longhole has yielded a mixed glacial–interglacial fauna with flint or chert tools which could be considered earlier Upper Palaeolithic (EUP). The cave, which lies at about 55 m O.D., was examined by Campbell who then excavated a trench on the platform 2 m wide by 7 m long, finding an area undisturbed by the earlier workers (Campbell, 1977). His tool forms comprised only struck waste, but he considered there was enough pollen and faunal evidence accompanying them to confirm an EUP attribution. Strangely, he decided to ignore the interior of the cave where there are certainly undisturbed layers visible, protected by thick stalagmite.

Cathole and Tooth Caves (SS 5376 9001, SS 533 907)

These are the only inland caves to be described in Gower so it is convenient to consider them together, although their contents are dissimilar. Cathole has long been known, and was excavated in the nineteenth century, but the first modern investigation was by McBurney (1959), whose results were then re-examined carefully by Campbell (1977) prior to his own excavation in 1968. The occupation was clearly later Upper Palaeolithic and Mesolithic as shown by tool forms illustrated by McBurney and Campbell; Garrod (1926) illustrated a typically Creswellian backed blade from the nineteenth-century excavation. The fauna included lion, cat, hyena, bear, badger, red deer, reindeer, roe deer, hare and smaller mammals, but Campbell commented that the fauna in all the previous excavations was mixed. Much remains to be excavated from within the cave, and the northern tunnel, which is quite separate from the main cave, appears to be untouched.

Tooth Cave is a short distance up-valley from Cathole, a mere 6 m above the valley floor which is known to carry a stream after particularly heavy rain. Opened by digging in 1961, it had two chambers which could be excavated but which once had easier access to the surface (Harvey et al., 1967). The first chamber was roughly a square, 5 m per side, half of which was excavated to a depth ranging from 1.2 to 3 m without reaching bedrock. Scattered human bones were found, some

in discrete piles, and flints suggested a date at the late Neolithic–early Bronze Age horizon. The second chamber was much narrower and yielded more human bones and middle Bronze Age pottery, although the occupation was not considered to be very much later than that in the first chamber. Leather-working was one of the activities suggested and a particularly fine bone needle was found with many bones of sheep, and a few of ox and dog, while cat and pig were present. However, the total quantity of food remains was very limited suggesting that the cave was more of a port-of-call than a residence, and the absence of sea shells (except for a single limpet) in a cave only 3 km from the sea suggests that this was a purely pastoral community. In view of the size of the occupation area it is still possible that earlier remains might be found if excavation were taken under the obstructive stalagmite floor.

Spurge Hole (SS 5468 8730)

A small arch 1.2 m wide was discovered in 1985 in a steep cliff near Southgate, and during clearance of gravelly earth to gain access to the interior of the cave, human bones were uncovered. A subsequent trial sounding by J. G. Rutter and the author revealed an adult burial still partly articulated, with an age at death of 35–45 y, but no dating evidence (Davies, 1985b, 1986e). Many of the bones were eroded, as part of the skeleton was near the surface in deposits disturbed by burrowing animals. It is just possible for a slim person to crawl 10 m into the cave, but the passage does not become wider and the archaeological potential is unknown.

Minchin Hole (SS 5554 8686)

The lower sequence of deposits in this cave has been noted in a previous section and it is noteworthy that there is no evidence for the presence of man in these early millennia. The deposits in the cave are so thick (25 m is a cautious estimate) that no excavator has yet worked his way through them, even in a single trench, and there is no prospect of this being done despite the fact that the cave is generally considered to be a type-site. The access route to the entrance degenerates into a steep scramble for the last few metres so it is difficult to transport equipment into the cave. Dr A. J. Sutcliffe's excavations under the aegis of the British Museum (Natural History) from 1972 to 1984 (Sutcliffe & Currant, 1984) showed that the main bone-bearing stratum was an interglacial one with narrow-nosed rhinoceros (Dicerorhinus hemitoechus), elephant, red and fallow deer (Dama dama), pig, lion, hyena, red fox and three species of vole (Sutcliffe, 1981). Iron Age and Roman pottery were discovered in the period 1914–48 (Rutter, 1949), together with bronze brooches, bone spoons and a spindle-whorl, and it is still possible today to find potsherds lying on the floor between slabs that have collapsed from the roof and walls. The early post-war excavations have not been published in full but it seems that three distinct hearth areas were occupied during the first five centuries AD and finds suggest sporadic use of the cave into medieval times (Savory, 1980a).

Bacon Hole (SS 5605 8683)

With an impressive entrance about 18 m wide (Fig. 7.6), the floor at 11.5 m O.D. carries a complex ascending series of

Fig. 7.6. The great arch of Bacon Hole; bedded screes can be seen high on the right-hand side (photo by C. Howes).

grey sands, orange sands, sandy breccia, cave earth, grey clays, upper sands, upper cave earth and cemented breccias or head (Currant, Stringer & Collcutt, 1984). Excavated in the nineteenth century, the definitive modern excavations were led by Dr C. Stringer (1975, 1977) and a team from the British Museum (Natural History) between 1974 and 1984. He found an interglacial fauna similar to that from Minchin Hole which died out with the first appearance of mammoth in the grey clays, and the dating of this warm fauna was from 130 ka to 80 ka.

The Quaternary sequence of Minchin Hole and Bacon Hole are complementary. Faunal correlations tie the sites together for a period bracketing the interglacial high sea level represented by the Patella Beach in Minchin Hole and the sandy breccia in Bacon Hole. While Bacon Hole has produced good biostratigraphical information for this period and later, Minchin Hole extends the lithostratigraphic and amino-acid records back to an earlier interglacial; however, full publication of these correlations is awaited.

Prissens Tor Cave (SS 4262 9370)

Also known as Spritsail Tor Cave, it has two entrances 4 m apart in a north Gower sea cliff linked by a low tunnel, the whole now floored by blown sand, although the cave is 20 m above the beach. It is not clear how much of the cave was destroyed by nineteenth-century quarrying, but excavations at that time, and further evaluation since, have identified a Pleistocene fauna, and later occupation, one phase at least being dated to the third–fourth century AD (RCAHM, 1976). During heavy rain, water is channelled naturally over the cliff face to fall in front of the cave, and it was noted in 1982 that when the blown sand was washed away a cave platform consisting of stony, red clay lay revealed. This platform must have lain inside the cave before the overhang was quarried away (Davies, 1983d). The edge of this platform, which is traversed by a footpath, is crumbling away, its contents being scattered on the sandy slope ending virtually at beach level. Over the years remains of hyena, fox, pig, woolly rhinoceros, reindeer, *Bos* species, and antler fragments have been retrieved from this debris, together with patinated flint waste

flakes. It is not possible to claim an association between the flint and the fauna because a Bronze Age midden was identified by Penniman on the summit of the Tor (Rutter, 1949), and the artefacts from it could have slipped down. They have the appearance of Mesolithic debitage which might point to the existence of an open working floor on the Tor.

Another event occurred during August 1986 when a tremendous rainstorm caused clay and sand to be washed out from a fissure alongside the west entrance. An array of bones and teeth were revealed, among which were canines of hyena and bear, with teeth of woolly rhinoceros, mammoth and horse. Bones from at least three of these species were also present, and some bore the marks of gnawing by hyena while others were complete. This site has now been re-covered by blown sand.

Woolley's Hole (SS 4607 9359)

This cave has a well-hidden entrance in a quarried, steep, dip-slope 10 m from the top of the cliff in north Gower. Discovered in 1954 the passage had to be cleared of earth for the first 15 m or so, and ended in a narrow shaft (Taylor, 1957). This was subsequently widened to allow entry to a single chamber 8 m below. The cave then became 'lost' for many years until rediscovered after much searching by the author. A wet clay mound in the chamber at the foot of the shaft was found to contain remains of a large badger and further investigation among the boulders on the floor of the chamber revealed scattered human remains (Davies, 1985c). It seems unlikely that the skeleton entered via the shaft, but ancient passages from the chamber are sealed by stalagmite, so the mystery must remain unsolved.

Conclusion

This review has attempted to concentrate on the basics of what has been carefully excavated during the last two decades, while much space has been allocated to a large number of trial excavations in new caves in Castlemartin and Gower. That the pace of discovery has accelerated, particularly in Gower, will be realized when it is seen that the Royal Commission on Ancient Monuments Inventory published in 1976 describes 17 caves in the Gower peninsula. Eight of these have received attention or further publication since that date, and in addition a further eight new cave discoveries with archaeological remains have been described. Rather than going over old excavations and seeking undug portions of well-known caves like Longhole, Hoyle's Mouth and Cathole, the time has come to start fresh excavations in new caves, armed with the full panoply of investigatory tools available to the present-day worker. It should be possible, in Gower at least, to select a cave and excavate it from the upper Roman levels to at least Last Interglacial levels. Instead of tackling an unmanageably vast cave like Bacon Hole, a medium-sized entrance should be selected like Rockrose Hole, large enough for habitation but beyond the reach of enormous scree run-ins. Castlemartin caves are generally difficult of access, but even here caves such as Ogof Bran Goesgoch can be reached by an easy footpath. It is, of course, unwise to extrapolate findings from one cave and apply them to others in a different lithological zone of limestone or to another county where similarities of climate today might not have prevailed many millennia ago. This is not to advocate widespread cave archaeological excavation, but there are a sufficient number of caves among those already discovered to keep investigative teams busy for many years to come and, no doubt, many unsuspected caves await discovery.

References

Bateman, J. (1969). Ogof Gofan Bone List (priv. comm., 2–4–69).

Bateman, J. (1973). Faunal remains from Ogof-yr-Ychen, Caldey Island. *Nature*, **245**, 454–5.

Bowen, D. Q. (1980). The Pleistocene scenario of Palaeolithic Wales. In *Culture and Environment in Prehistoric Wales*, ed. J. A. Taylor, pp. 1–14. British Archaeological Reports, British Series, 76.

Bowen, D. Q. (1986). Aspects of the geomorphology and Quaternary history of Gower. *Nature Conservancy Council Gower Excursion*, July 1986.

Bowen, D. Q. & Henry, A., eds. (1984). *Field Guide: Wales, Gower, Preseli, Fforest Fawr*. Cambridge: Quaternary Research Association.

Campbell, J. B. (1977). *The Upper Palaeolithic of Britain. A Study of Man and Nature in the Late Ice Age*, 2 vols. Oxford: Clarendon Press.

Campbell, S., Andrews, J. T. & Shakesby, R. A. (1982). Amino acid evidence for Devensian ice, west Gower, South Wales. *Nature*, **300**, 249–51.

Clegg, J. (1969). Excavations at Coygan Cave, near Laugharne. *The Carmarthen Antiquary*, **5** (1964–69), 13–20.

Cullingford, C. H. D. (1962). *British Caving: An Introduction to Speleology*, 2nd edn. London: Routledge & Kegan Paul.

Currant, A. P., Stringer, C. B. & Collcutt, S. N. (1984). Bacon Hole Cave. In *Field Guide: Wales, Gower, Preseli, Fforest Fawr*, ed. D. Q. Bowen & A. Henry, pp. 39–44. Cambridge: Quaternary Research Association.

Davies, M. (1969). Ogof Gofan. *Cwmbrân Caving Club Journal*, **4**, 9.

Davies, M. (1972). Ogof Garreg Hir. *Archaeology in Wales*, **12**, 18–19.

Davies, M. (1975). Ogof Morfran. *Archaeology in Wales*, **15**, 43.

Davies, M. (1976). Ogof Brân Goesgoch. *Archaeology in Wales*, **16**, 47.

Davies, M. (1977). Ogof Brân Goesgoch – a new archaeological cave in south Pembrokeshire. *Cambrian Caving Council Journal*, **4**, 45–6.

Davies, M. (1981). Worms Head Cave. *Archaeology in Wales*, **21**, 22.

Davies, M. (1982). Archaeological caves of the south Gower coast. *Wm. Pengelly Cave Studies Trust Newsletter*, **40**, 1–8.

Davies, M. (1983a). Rockrose Hole, Overton. *Archaeology in Wales*, **23**, 40.

Davies, M. (1983b). Archaeological caves in Gower – old and new. *Wm. Pengelly Cave Studies Trust Newsletter*, **43**, 1–5.

Davies, M. (1983c). Ogof Arllechwedd, Middleton. *Archaeology in Wales*, **23**, 38.

Davies, M. (1983d). Prissens Tor Cave, Llanmadoc. *Archaeology in Wales*, **23**, 39.

Davies, M. (1983e). Upper Blackhole, Overton. *Archaeology in Wales*, **23**, 41.

Davies, M. (1984). Upper Blackhole. *Wm. Pengelly Cave Studies Trust Newsletter*, **44**, 7–8.

Davies, M. (1985a). Spring Squill Hole. *Wm. Pengelly Cave Studies Trust Newsletter*, **46**, 5–9.

Davies, M. (1985b). Spurge Hole. *Wm. Pengelly Cave Studies Trust Newsletter*, **46**, 1–4.

Davies, M. (1985c). Woolley's Hole rediscovered. *South Wales Caving Club Newsletter*, **100**, 28.

Davies, M. (1986a). Worms Head Cave, Rhosili. *Archaeology in Wales*, **26**, 34.

Davies, M. (1986b). Red Fescue Hole, Rhosili. *Archaeology in Wales*, **26**, 34.

Davies, M. (1986c). Red Fescue Hole, Gower. *Wm. Pengelly Cave Studies Trust Newsletter*, **48**, 1–5.

Davies, M. (1986d). Ogof Arllechwedd, Middleton. *Archaeology in Wales*, **26**, 34.

Davies, M. (1986e). Spurge Hole, Southgate. *Archaeology in Wales*, **26**, 34.

Davies, M. (1987). *The Caves of South Gower Coast: Worms Head to Porteynon Point*. Oxwich: Nature Conservancy Council.

Davies, M. & Kennedy, R. A. (1969). Cave sites (Ogof Gofan, Ogof Pen Cyfrwy, Ogof Morfran). *Archaeology in Wales*, **9**, 13–14.

Dawkins, W. B. (1874). *Cave Hunting*. London.

Garrod, D. A. E. (1926). *The Upper Palaeolithic Age in Britain*. Oxford: Clarendon Press.

Green, H. S. (1984). *Pontnewydd Cave: A Lower Palaeolithic Site in Wales. The First Report*. Cardiff: National Museum of Wales.

Green, H. S. (1986). The Palaeolithic settlement of Wales research project: a review of progress 1978–85. In *The Palaeolithic of Britain and its Nearest Neighbours, Recent Trends*, ed. S. N. Collcutt, pp. 36–42. Sheffield: University of Sheffield.

Grimes, W. F. (1933). Priory Farm Cave, Monkton, Pembrokeshire. *Archaeologia Cambrensis*, **88**, 88–100.

Harvey, J. C., Morgan, R. & Webley, D. P. (1967). Tooth Cave, Ilston, Gower: an early Bronze Age occupation. *Bulletin of the Board of Celtic Studies, Cardiff*, **22**, 277–85.

Jackson, J. W. (1962). In *British Caving*, ed. C. H. D. Cullingford, pp. 267–8, 332. London: Routledge & Kegan Paul.

Jacobi, R. M. (1980a). The Upper Palaeolithic of Britain with special reference to Wales. In *Culture and Environment in Prehistoric Wales*, ed. J. A. Taylor, pp. 15–99. British Archaeological Reports, British Series, 76.

Jacobi, R. M. (1980b). The early Holocene settlement of Wales. In *Culture and Environment in Prehistoric Wales*, ed. J. A. Taylor, pp. 131–206. British Archaeological Reports, British Series, 76.

Lacaille, A. D. & Grimes, W. F. (1955). The prehistory of Caldey. *Archaeologia Cambrensis*, **104**, 85–165.

Lacaille, A. D. & Grimes, W. F. (1961). The prehistory of Caldey, Part 2. *Archaeologia Cambrensis*, **110**, 30–70.

McBurney, C. B. M. (1959). Report on the first season's fieldwork on British Upper Palaeolithic cave deposits. *Proceedings of the Prehistoric Society*, **25**, 260–9.

Nedervelde, Br. J. Van (1969). Caldey Island (New Cave). *Archaeology in Wales*, **9**, 14.

Nedervelde, Br. J. Van (1972). Ogof-yr-Ychen, Caldey Island. *Archaeology in Wales*, **12**, 19–20.

Nedervelde, Br. J. Van & Davies, M. (1975). Potter's Cave, Caldey Island. *Archaeology in Wales*, **15**, 45.

Nedervelde, Br. J. Van & Davies, M. (1976). *Caldey Island Caves – Nanna's Cave Excavations During 1976* (privately printed).

Nedervelde, Br. J. Van & Davies, M. (1977a). Potter's Cave, Caldey Island. *Archaeology in Wales*, **17**, 48.

Nedervelde, Br. J. Van & Davies, M. (1977b). Nanna's Cave, Caldey Island. *Archaeology in Wales*, **17**, 46.

Nedervelde, Br. J. Van & Davies, M. (1977c). *Excavations in Potter's Cave and Nanna's Cave, Caldey Island* (privately printed).

Nedervelde, Br. J. Van & Davies, M. (1978). *Nanna's Cave, Caldey Island, Excavations During 1978* (privately printed).

Nedervelde, Br. J. Van & Davies, M. (1979–80). (Nanna's Cave, Potter's Cave, Ogof-yr-Ychen), *Caldey Island* (privately printed).

Nedervelde, Br. J. Van & Davies, M. (1981–2). (Nanna's Cave, Ogof-yr-Ychen, Daylight Rock), *Caldey Island* (privately printed).

Nedervelde, Br. J. Van & Davies, M. (1983–5). *Caldey Island Cave Archaeology* (Nanna's Cave, Ogof-yr-Ychen) (privately printed).

Nedervelde, Br. J. Van & Davies, M. (1986). *Nanna's Cave, Eel Point Caves, Caldey Island* (privately printed).

Nedervelde, Br. J. Van & Davies, M. (1987). *Caldey Island Annual Excavation Report 1987* (privately printed).

Nedervelde, Br. J. Van, Davies, M. & John, B. S. (1973). Radiocarbon dating from Ogof-yr-Ychen, a new Pleistocene site in west Wales. *Nature*, **245**, 453–4.

Rae, A. M., Ivanovich, M., Green, S. H., Head, M. J. & Kimber, R. W. L. (1987). A comparative dating study of bones from Little Hoyle Cave, South Wales, UK. *Journal of Archaeological Science*, **14**, 243–50.

RCAHM (1976). *An Inventory of the Ancient Monuments in Glamorgan*, vol. 1, part 1 (19 caves described).

Riches, W. (1923–4). Worms Head Cave. *Annual Report of the Royal Institution of South Wales*, pp. 20–5. Swansea.

Rutter, J. G. (1949). *Prehistoric Gower*. Swansea: Welsh Guides.

Savory, H. N. (1968). Ogof Gofan Pottery (priv. comm., 3–10–68).

Savory, H. N. (1971). Ogof Morfran (priv. comm., 7–9–71).

Savory, H. N. (1973). Excavations at the Hoyle, Tenby, in 1968. *Archaeologia Cambrensis*, **122**, 18–34.

Savory, H. N. (1980a). The Neolithic in Wales. In *Culture and Environment in Prehistoric Wales*, ed. J. A. Taylor, pp. 207–32. British Archaeological Reports, British Series, 76.

Savory, H. N. (1980b). *Guide Catalogue of the Bronze Age Collections*. Cardiff: National Museum of Wales.

Sollas, W. J. (1913). Paviland Cave: an Aurignacian station in Wales. *Journal of the Royal Anthropological Institute*, **43**, 325–74.

Stringer, C. B. (1975). A preliminary report on new excavations at Bacon Hole Cave. *Gower*, **26**, 32–7.

Stringer, C. B. (1977). Evidence of climatic change and human occupation during the Last Interglacial at Bacon Hole Cave, Gower. *Gower*, **28**, 36–44.

Sutcliffe, A. J. (1981). Progress report on excavations in Minchin Hole, Gower. *Quaternary Newsletter*, **33**, 1–18.

Sutcliffe, A. J. & Currant, A. P. (1984). Minchin Hole Cave. In *Field Guide: Wales, Gower, Preseli, Fforest Fawr*, ed. D. Q. Bowen & A. Henry, pp. 33–7. Cambridge: Quaternary Research Association.

Taylor, M. C. (1957). Woolley's Hole. *British Caver*, **28**, 55–9.

8

Cave archaeology in North Wales

MEL DAVIES

It is the purpose of this account to review published works on cave excavations in North Wales, and to provide details of nine recent excavations with interpretations based on the most up-to-date palaeoclimatic evidence. Dr J. W. Jackson's chapter in *British Caving* (Cullingford, 1962) included an admirable survey of excavations up to 1959, and since that date various caves have been dug by Dr H. S. Green, J. D. Blore and the author, assisted by members of the North Wales Caving Club. A summary of the earlier work was included in reviews by Valdemar & Jones (1970) and Valdemar (1970). Valdemar went on to arrange for a check to be made by the National Museum of Wales on nineteenth-century bone identifications

relating to Cefn Caves and Pontnewydd Cave prior to publishing a valuable bibliography in 1972. The positions of the sites are noted on Figure 8.1.

Before considering the details of excavations it is necessary to set the scene in relation to ice-age events in North Wales as these are significantly different from those prevailing in South Wales. The outcrop of the Carboniferous Limestone is situated mainly in Clwyd and there is a small but important outcrop around Llandudno in Gwynedd. The Anglesey Limestone is not known to contain any archaeological caves, although they have been diligently sought. The earliest period from which cave deposits have been found in North Wales is

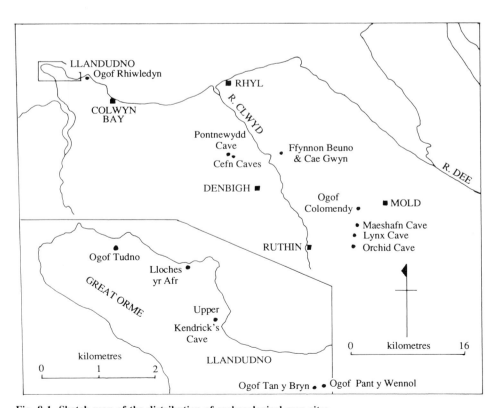

Fig. 8.1. Sketch map of the distribution of archaeological cave sites in North Wales.

Table 8.1 *Correlation of climatic periods and archaeological remains* (after Taylor, 1975).

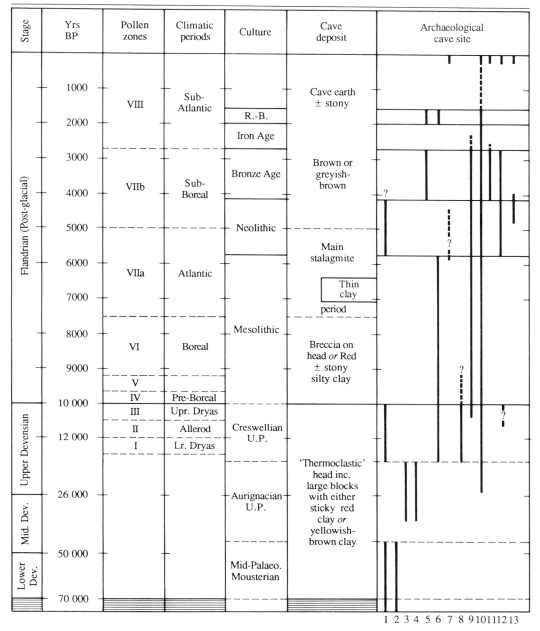

Key to cave sites

1. Cefn Caves	2. Pontnewydd Cave	3. Ffynnon Beuno
4. Cae Gwyn Cave	5. Maeshafn Cave	6. Lynx Cave
7. Ogof Tudno	8. Ogof Tan y Bryn	9. Lloches yr Afr
10. Ogof Pant y Wennol	11. Ogof Rhiwledyn	12. Upper Kendrick's Cave
13. Ogof Colomendy	Ipswichian Interglacial	

the later Middle Pleistocene, but this is only from Pontnewydd Cave. Evidence for later periods is summarized in Table 8.1. The warm, temperate period of the Ipswichian Interglacial lasted tens of thousands of years, and ended about 70 000 years ago. This is when the details of our hard-rock coastlines were largely acquired and erosion at the modern sea level is merely

modifying them (Bowen, 1977). The mammals then present in North Wales included hippopotamus (*H. amphibius*), straight-tusked elephant (*Palaeoloxodon antiquus*) and soft-nosed rhinoceros (*Dicerorhinus hemitoechus*), and remains of all three have reportedly been excavated from Pontnewydd Cave and Cefn Caves. The fact that these early remains are only known from two caves reinforces the importance of the sites.

During the Ipswichian Interglacial a higher sea level gave rise to raised beaches which are still discernible in part around the coasts of Wales. A good example is known on limestone bedrock at Red Wharf Bay, Anglesey (SH 532 816) (Bowen, 1974), but it has proved impossible to relate the red till and head deposits there at 3 m O.D. to sterile clays visible in small caves a little higher above sea level in a quarried cliff near by on Castell Mawr. With the onset of colder conditions heralding the start of the Devensian Glaciation, the three warmth-loving animal species retreated southwards, two of them finally becoming extinct while hippopotamus survives in Africa. Glaciers formed in Snowdonia and ice from that centre reached the North Wales coast (Bowen, 1974), only to be overcome by the stronger ice sheet flowing from the north, termed the Irish Sea Ice. This flow penetrated the Vale of Clwyd, sealing the cave entrances at Cae Gwyn (SJ 085 724) with stratified silts and clays. Strangely there is no clear evidence of archaeological material sealed in any other North Wales cave except that glacially derived debris flows were identified inside Pontnewydd Cave by S. N. Collcutt (in Green, 1984). The thicknesses of the Welsh and Irish Sea ice sheets are controversial but certainly the archaeological caves discussed here were all at an altitude low enough to be covered by glacial drift. Excavation often shows, however, that the severe periglacial climate that presaged the ice accumulation resulted in widespread cavern breakdown. The resulting block litter in cave entrances may have restricted penetration of the passages by glacial drift. It is unlikely that the ice sheets covered North Wales for the whole of the Devensian period. Bowen's opinion (1974) was that until 50 000 y BP the area was cold and treeless, and experienced episodes of interstadial warmth. There were occasional periods of polar desert and deposition, forming river terraces. On the coastline, cliffs had been abandoned by the falling sea level as more water became locked up in the icecaps, and these cliffs gradually became buried by debris weathered from their upper slopes. None of these events left any evidence in North Wales caves except possibly the large angular blocks found at the deeper levels in some of the excavations described below.

Between 50 000 and 26 000 y BP there was a slight climatic amelioration, and the cold fauna excavated from Ffynnon Beuno (SJ 085 724) probably dates from this period (Hicks, 1886). Between 26 000 and 18 000 y BP conditions worsened; there was a swift accumulation of ice, the greatest thickness being present 18 000 years ago. North Wales became free of ice by 14 500 years ago (Shotton, 1977). One of the fullest summaries of the effects of Pleistocene climatic change on cave deposits is given by J. G. Evans (1975) in discussing Tornewton Cave, Devon.

Having discussed ice-age events in North Wales, what of man? A flint tool in the shape of a completely patinated plano-convex spearhead (McBurney, 1965) was discovered in Ffynnon Beuno during Hicks' careful excavation, but it is not possible to place it stratigraphically in relation to the glacial maximum of 18 000 y BP. However, it was in 'close proximity to a mammoth tooth' (Hicks, 1886), and worked bones were

also found. Rowlands (1971) obtained a radiocarbon date of 18 000 y BP for a mammoth carpal from either Ffynnon Beuno or Cae Gwyn Cave. The bone was from a collection held locally since it was excavated about 1886 (Hicks, 1888). However, examination of the collection suggests that the date is not stratigraphically significant as it cannot be tied in with the published stratigraphy in the cave, except that the bone seems to have been in deposits subsequently sealed by the till of the Irish Sea Ice (Peake, 1973). In fact, it is not known which of the two caves contained the bone. In common with many of the smaller North Wales archaeological caves, Ffynnon Beuno was not completely cleared out by the nineteenth-century excavators. When examined in 1972, it was found to contain pebbly sediments containing erratic rock types, bone fragments and a broken rhinoceros tooth. Davies (1975) recommended that a modern excavation could profitably be carried out in the cave. About 6 m above Ffynnon Beuno main entrance lies the present entrance to Cae Gwyn Cave. Another entrance was sealed by the glacial silts and clays in fields to the north until revealed by a collapse following a wet winter. In a water-worn tunnel roughly half-way between the two entrances Hicks found a partly patinated flint end-scraper (Oakley, 1972), and plentiful bones of a cold fauna. A year later he discovered (Hicks, 1888) a patinated flint flake with hyena and reindeer remains just inside the collapsed entrance noted above. Here some 7 m of glacial drift was revealed by excavation, and for the first time the presence of man in North Wales before what later became accepted as the glacial maximum was proved. The flint flake was situated in 'reddish, sandy clay with here and there fragments of stalactites, stalagmites and angular blocks of limestone'. There is reason now to question whether the flint flake was in its original position. The main period of stalagmite growth during Devensian and Flandrian times, as proved by modern excavations, was postglacial. The argument depends on the quantity of stalagmite found, as small accretions may have occurred before the period of glacial maximum.

Modern excavations started with G. E. Hesketh's dig in *Maeshafn Cave*, Clwyd (SJ 198 605) (Hesketh, 1954–5). A more complete list of finds than was available to Dr J. W. Jackson (in Cullingford, 1962) can now be given, and it is worth recording in detail because in only three other caves to be discussed was Romano-British material found. With bones of small ox, human remains comprising five adults and a child were found, together with: a Bronze Age arrowhead; a bronze penannular brooch, possibly first century AD; a Roman zoomorphic plate brooch, possibly second century AD; a bronze ring possibly for a cloak, finger or toe, undated; and a bronze Roman fibula, first century AD. The cave floor slopes very steeply downwards from the entrance, and it has been disturbed by miners so it is not possible to place much reliance on the stratification. Also the end of the cave is subject to flooding. Due to the presence of stalagmite on some of the bones, it is possible that they are considerably older than the Romano-British material found.

Lynx Cave, Llanferres, Clwyd (SJ 194 593), was excavated during 1962–4 by J. D. Blore (1965) who found extensive disturbance, at least of the top soil, 'making dating by stratification almost impossible'. The major find was the mandible of an almost-adult lynx, but 15 cm below it a Romano-British brooch, AD 100–150 was found. The lynx is a rare species in Britain, the latest record of it being in a late Devensian context from Aveline's Hole, Mendip (Stuart, 1974, 1977), so such a reversal of dating might be thought to throw suspicion on the identification. However, Dr D. Bramwell of the Peakland Archaeological Society confirmed the identification of the bone and also noted Arctic lemming (*Lemmus lemmus*), another late Devensian Aveline's Hole animal, and reindeer (*Rangifer tarandus*) which is believed to have disappeared from most of Britain at the end of the Devensian (Stuart, 1974). Blore added two flints to his finds, one of which is recognizably a Creswellian blade with blunting along its single shoulder. The Creswellian culture in Wales is taken to have evolved by 14 000 y BP and to have survived, at least in some areas, until about 9 000 y BP (Tratman, 1975). Human bones were found scattered about, and on the evidence of the fibulae, at least six individuals were present. There is no means of knowing whether the bones date from the Creswellian period or from the Romano-British period.

In recent years a number of caves, archaeological to a greater or lesser degree, have been discovered in the Llandudno area (Fig. 8.1). One of the minor sites is *Ogof Tudno* (SH 764 842), sometimes known as Badger Cave, situated high on the magnificent limestone peninsula of the Great Orme. In 1976 T. A. Stone of Llandudno excavated a trench under an overhang just to the east of the cave's main entrance (see Appendix 8.1). It revealed a connection with the main entrance and signs of limited occupation sealed by stalagmite. These comprised a hearth containing bones of sheep or goat, pig, hare, fox, fish and others, with numerous limpet and mussel shells. The remains were disturbed by badger burrowing, and it is clear from the coal and bottle-glass fragments that miners had cleared out the main entrance to provide access for their wheelbarrows during a search for copper ores. The system has a complex of interconnecting crawl-tunnels where further relics may yet be found.

Ogof Tan y Bryn is situated in a private garden and a preliminary examination by D. James of Conwy and the author showed that about 2 m of deposits including clay, earth and stalagmite had been removed sometime in the past, possibly as a source of garden soil. This disturbance extends for 5.5 m into the cave and beyond this point there is a rising floor of red clay running 4 m into a chamber which is about 4 m high. At the apex of the roof there is a blockage of earth and tree roots indicating the position of a blocked shaft which, when open, would have formed a trap for animals and debris. A brief 1 m² trial sounding 20 cm deep in the entrance of the cave yielded eight patinated flints with hacked bone fragments from large animals, contained in an undisturbed stony, yellowish-brown, silty clay. Four of the flints are shown in Figure 8.2, and two of these are clearly Creswellian tools – a blunted-back blade and a leaf-shaped blade. Thus with Lynx Cave and Cefn Caves where Valdemar & Jones (1970) recognized a Creswellian industry among the flints excavated in the nineteenth century, the find at Tan y Bryn pushes the spread of late glacial man even further into the heart of North Wales. The actual

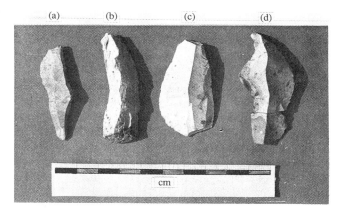

Fig. 8.2. Flint implements from Ogof Tan y Bryn, Llandudno. (*a*) Patinated blunt-backed blade. (*b*) Thick flake used as a blade. (*c*) Patinated leaf-shaped blade. (*d*) Patinated flake used as a blade.

implements show no more affinity with those from Creswell Crags caves (Campbell, 1969) 160 km to the east, than with those from Aveline's Hole and Gough's Cave on Mendip, and Nanna's Cave on Caldey Island far to the south (Lacaille & Grimes, 1955, 1961).

Lloches yr Afr, a rock shelter alongside the Marine Drive on the Great Orme (SH 779 838), was excavated by the author in 1973–4. The dig showed that the site was an open shelter as a result of the blasting away of the outer wall during the construction of the Drive, an operation completed in 1877. As the contents were seriously deteriorating under the effects of weathering, the bulk of the sediments were cleared out. The deposits were found to be laid against a faulted face of limestone containing copper ore, solutions of which had been absorbed by most of the bones thus rendering them green. They were excavated to a maximum depth of 1.78 m over a 3 m length along the face, and a profile is shown in Figure 8.3. Use of the shelter started just at the end of the Devensian cold period when a hyena left behind the remains of the bones it had been crunching. The bones were mixed with the thermoclastic scree, or head, of layer 6, a layer of congelifracts or 'eboulis' produced by freeze–thaw fracturing of the cave roof. The angular stones were found to increase in size with depth in this layer suggesting that the climate had been even more severe before the hyena arrived on the scene. Table 8.2 gives 69% as the ratio of stones greater than 2 mm in size, but the sample analysed had already had its large stones removed before testing. Hyena is claimed to have disappeared from Britain by late Devensian times (Stuart, 1977), and here in Lloches yr Afr its activities are indeed recorded at the top of the cold-climate deposit. With an amelioration of climate, soft calcareous deposits termed yellowish-grey silt (layer 5) were laid down suggesting a warm, damp climate, and this layer, and the next one (layer 4), are considered to be the equivalent of stalagmite in other caves. The formation of stalagmite in caves is a subject often misunderstood by archaeological writers who lack experience of the subterranean environment. It grows in a warm, wet climate with consistently high temperatures, i.e., with non-continentality of climate. In Wales today there is very little active stalagmite growth. Even during winter rains, when some caves drip furiously, there is little evidence of stalagmite

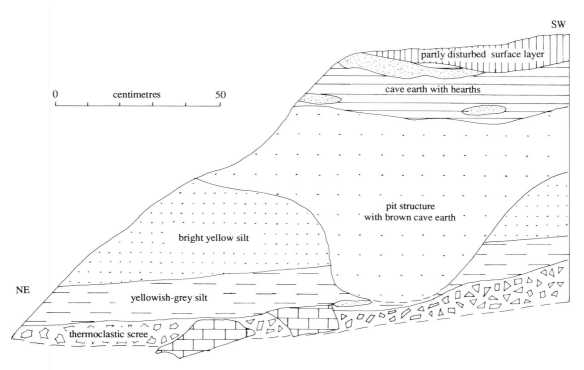

Fig. 8.3. Profile of deposits in Lloches yr Afr, Great Orme's Head, Llandudno.

Table 8.2 *Pedological analyses of materials from Lloches yr Afr*

Sample	Stones % > 2 mm	Untreated Sample		< 2 mm Fraction (Dried and Ground)						
		Colour	pH	Mechanical Analysis				Loss on Ignition %	CaCo₃ %
				Sand	Silt	Clay	Texture		
Layer 2	57	10 yr 3/3 dark brown	8.2	30	47	23	Clay loam	5.3	69
Pea-sized gravel	75	10 yr 5/4 yellowish-brown	8.6	58	28	14	Sandy loam	1.9	68
Bright yellow silt	31	10 yr 7/4 very pale brown	8.2	16	48	36	Silty clay	3.2	96
Yellowish-grey silt	46	10 yr 6/4 light yellowish-brown	8.5	22	44	34	Clay loam	2.4	80
Base layer– Thermo-clastic scree	69	7.5 yr 5/4 yellowish-brown	8.6	39	36	25	Clay loam	0.8	71

deposition. Exceptions are found in caves situated under limekiln spoil heaps. To find a favourable climate one has to go back to the 'Atlantic' period which started about 8000 y BP (although some authors put it 500 years later), when rainfall was up to 11% on average above present-day levels, and short-term variations could have included some years with rainfall at 25% or even 50% above the present average (Taylor, 1975). Preceding the Atlantic period were the pre-Boreal and Boreal periods, times of pronounced dryness with long severe winters and short continental summers. The two periods do not seem to be differentiated in the cave record and cave deposits for the time 10000 to about 8000 y BP are generally red, silty clays, becoming stony due to congelifracts with depth. Sometimes the clay is almost completely absent and the stalagmite rests directly on angular boulders which are then cemented into a breccia. Above the Atlantic-period stalagmite, cave deposits termed sub-Boreal from about 5500 y BP vary from dry, brown cave earth to wet clay or even to water-washed sand. Temperatures were 1.0–1.5 °C higher than the present, but it must have been just as dry, until the marked deterioration into the sub-Atlantic about 2650 y BP with, at worst, unsettled raw winters and cool, cloudy summers. There may have been a reversion to slow, minor stalagmite growth during this period, as shown by thin stalactites or straws. In Lloches yr Afr layers 4 and 5 contained a flint microlith of a type fabricated during the Mesolithic period. With the change of climate to something similar to the present day, human occupation intensified and the cave deposits (layers 3 and 2) consist of successive charcoal layers, some interspersed with wood ashes representing hearths, and illustrated by high loss on ignition value (5.3%) (Table 8.2). Near the bottom of layer 2 the single undecorated potsherd found is tentatively ascribed to the early Neolithic period. The hearths contained abundant food remains comprising ox, sheep or goat, deer and pig. The molluscan fauna has been identified by Dr J. G. Evans as:

Layer 2 {
Littorina littorea (Linne), winkle
Patella vulgata (Linne), limpet
Mytilus edulis (Linne), mussel
Nucella lapillus, dog whelk
}

Layer 3 {
Winkle and mussel
Pomatias elegans (Muller), snail
Crab fragments
}

All except *P. elegans* are marine and probably represent human food debris. Layer 3 consists of a pit structure of unknown purpose and alongside it there was a post-hole remnant. The roughly cut post had been hammered quite firmly into the cave floor, and presumably served to carry a cooking pot over a fire. Pit and post-hole remains have not previously been recorded from British caves, but both have been noted in France (Bordes, 1972).

Ogof Pant y Wennol is another in the Llandudno group of caves, and was excavated by the author from 1974 to 1977. The cave is 5.2 m wide by 8.6 m deep and has yielded undisturbed deposits indicating occupation in Mesolithic and Neolithic times. Within the cave there were about 34 m² available for excavation. The main trench occupied 2 m², and modern or mixed modern–prehistoric superficial deposits have been cleared from a further 10 m² while excavation to Neolithic levels only were made in a further 8 m². Thus 14 m² remains undisturbed within the cave together with a further 25 m² approximately on the entrance platform. There were also indications of three subsidiary passages all filled with promising deposits. The long and detailed excavation enabled a clear picture of the use of the cave to be drawn. Animals used the cave long before man put in an appearance. Woolly rhinoceros (*Coelodonta antiquitatis*), horse, reindeer and hyena remains were found, and the hyena had also left scores of shattered bone fragments representing their meal debris. Even strong horse teeth had been crunched and in one case five fragments could be assembled to reconstruct a left maxillary molar of horse with pressure indentations on either side still discernible. The matrix of these early remains was a sticky red clay, but this had been mixed with a loose, stony, yellowish-brown clay by a natural process akin to solifluction. Both are regarded as late Devensian cold-climate deposits. Beneath them were stone-free clay, silt or sand deposits of which about 1 m thickness was excavated and found to be quite sterile. Grey sand had been accumulating in natural hollows in this ancient clay; there were no indications as to its age but it was assumed to be pre-Devensian in view of its stone-free nature. This suggests deposition during an earlier interglacial or while the cave was still sealed off from the outside.

Above the stony late Devensian layer was a red silty clay layer containing some stones. The stones were only sub-angular in contrast with the shattered ones beneath and the layer was regarded as a Boreal deposit. Man arrived and made use of the cave as indicated by thin charcoal layers. He was still a hunter–fisherman but had grown skilful in the manufacture of flint tools. Figure 8.4(*a*) shows a slightly patinated microlith of a type used in pairs for arrow-tips, e.g., at the Mesolithic site of Star Carr (Clark, 1971). The microlith in Figure 8.4(*b*) was perhaps used as a cutting tool and had delicate, deliberate blunting down the right-hand edge. The third microlith (Fig. 8.4(*c*)) was made from black chert and is neatly blunted along the left-hand edge. All three microliths could have been used as projectile points. In the same layer was Figure 8.4(*d*), also of black chert and probably used as a cutting implement.

Resting on the red silty clay layer was a stalagmite floor, tufaceous in some areas but rock-hard in others. It was not dated precisely but by analogy with other caves it was held to be laid down during the Atlantic climatic period. Over it was a greyish-brown cave earth, much affected by water in that it had been transported from elsewhere and contained human as well as animal bones, flint implements and waste flakes, and the Neolithic pottery known as the Ebbsfleet type in the Peterborough series. Some of the flints were characteristic Neolithic tools: Figure 8.4(*e*) shows a leaf-shaped arrowhead and 8.4(*f*) a convex scraper. Both are quite unpatinated, a feature which is almost universal in Welsh cave flint of Neolithic age or younger. Outside the main trench there were two areas of deposits containing human bones. Both were fissures or spaces between large slabs and were probably originally burial sites. The more complete one had evidence that the burial cavity had been filled with rubble, which included thin sheets of broken

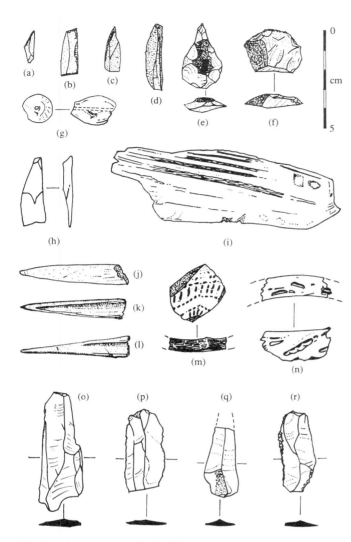

Fig. 8.4. Artefacts from North Wales caves:
Ogof Pant y Wennol:
(*a*) **Mesolithic flint microlith;** (*b*) **Mesolithic flint microlith;**
(*c*) **Mesolithic chert microlith;** (*d*) **Mesolithic chert blade;**
(*e*) **Neolithic flint leaf-shaped arrowhead;** (*f*) **Neolithic flint scraper.**
Ogof Rhiwledyn:
(*g*) **Bronze Age amber bead;** (*h*) **Slate point;** (*i*) **Roe deer antler pick.**
Upper Kendrick's Cave:
(*j*) **Awl made from antler fragment;** (*k*) **Bone awl;** (*l*) **Bone awl;**
(*m*) **Beaker pottery (Bronze Age);** (*n*) **Rim-sherd of Neolithic pottery;** (*o–r*) **Neolithic unpatinated flint knives.**

stalagmite, and covered with a slab of rock. In the other the bones had decomposed in air spaces or become cemented in stalagmite. In all at least four adults and two children were found to have been buried in the excavated part of the cave – both of the children, as shown by X-rays, being babies. The animal bones comprised the usual range of post-glacial fauna including ox, red deer, pig, sheep or goat, possibly roe deer, fox, hare, rodents, bird, frog, fish and crab. Surprisingly crude stone 'choppers' were used until a late date in which a heavy pebble was smashed to produce a crude edge, but small flat pebbles were also carefully 'waisted' so that a thong could be tied round them. While rare in Britain such net-sinkers are recognized at Swiss Neolithic lake villages (Milisauskas, 1978).

Fig. 8.5. Ogof Rhiwledyn, Little Orme, Llandudno.

Ogof Pant y Wennol with at least $39\,m^2$ still available for excavation is now fully protected by being fenced off on private property at SH 808 816.

Another rich site near Llandudno has been excavated by J. D. Blore (1977) (see Appendix 8.1). This is *Ogof Rhiwledyn*. It is a difficult cave to access on the north face of the Little Orme headland (SH 815 827) (Fig. 8.5). Access must have been easier in prehistoric times for Blore found the remains of three children in the cave aged 12–14, 8–9 and 4 years. There were no adults but excavation was stopped by the solidity of a stalagmite floor which was up to 1 m thick in places. The small size of the cave's interior, and the paucity of charcoal coupled with the total lack of sea-shell food suggested that the actual inhabitation occurred on the entrance platform, and the cave itself used for burial only. An interesting find was an amber bead (Fig. 8.4(g)), ascribed to the Beaker people named after a type of pottery which first appeared about 4500 y BP, although none of this was found in the cave. One of two slate points discovered is shown in Figure 8.4(h), and it is suggested that these were leather-workers' tools. Tools had also been made from bone and antler – a pick fashioned from a roe deer tine is illustrated in Figure 8.4(i). It has a double V-notch cut around mid-length and extensive hammer marks at the point. The food remains comprised bones of sheep, a small ox and a larger ox, but there were also bones of pig and roe deer which had no signs of butchery, and remains of fox, stoat, cat, hare, rabbit, rodents, frog and 11 species of bird. Many of these animals clearly only entered the cave in fairly modern times and there is evidence of movement and redeposition of the sediments within the cave, for example, fragments of a single human fibula were found as far apart as 3.5 m. Blore (1977) concluded that the occupation of the cave was 'during the transitional period between the two cultures Neolithic and Bronze Age around 1800 yrs BC' and he placed this towards the end of the sub-Boreal climatic period. However, on the basis that the stalagmite formed mostly during the Atlantic climatic period, some of the human bones must have been interred then, or at least early in sub-Boreal times.

The final cave in the Llandudno series is *Upper Kendrick's Cave* (SH 780 828) situated on a south-facing slope of the Great Orme. There are in fact two caves, both used by Thomas

Fig. 8.6. Stone disc and flint flakes from Upper Kendrick's Cave, Llandudno.

Kendrick in 1879, the lower one now faced by a Victorian summerhouse, and the upper cave 8 m above, accessible via two entrances reached by a cliff-path. It is not absolutely clear in which cave Kendrick made his finds because he was a retired copper miner accustomed to digging but not recording. He found the remains of three human adults and one child together with bones of bison and other animals (Humphreys, 1907–8). Some of these remains may now be seen in Llandudno Public Library while others have been acquired by the British Museum. The Llandudno collection has been examined by the author and comprises bones of at least three human adults, horse, deer, pig, ox and five metacarpals from either roe deer or sheep bearing an incised series of lines drawn to some obscure pattern. The British Museum collection has been described by Sieveking (1971) (Fig. 8.6) and comprises a small polished stone axe-head of typical British Neolithic type, a stone spindle-whorl, a small flint axe-head or knife of black stone roughly flaked around three sides and polished at the blade, a bone weaving-comb of early Iron Age type, a collection of nine bovid and deer teeth decorated and perforated, and finally a decorated horse mandible. Sieveking compared these artefacts with similar finds from the continent and others bearing some similarities from sites in Britain. He concluded that the decorated horse mandible at least was likely to be of Upper Palaeolithic date, and this was later confirmed by a radiocarbon dating of 10 000 y BP (Gillespie, 1985; Appendix 8.1). Following the discovery of extant photographs of Thomas Kendrick, who died in 1897, the author excavated five trenches in the upper cave between 1977 and 1979 because Kendrick had obviously used that cave as a workshop after extending it (Davies, 1983; Appendix 8.1). Bone finds were sparse considering the extent of the excavation but included one human adult, ox, sheep or goat, roe deer, wild boar, fox, hare, several bird species, frog, crab and at least eight species of mollusca with limpets predominating. Flint and pottery indicated a Neolithic occupation accompanied by hearths lying under a full nineteenth-century floor with one door, a window with glass and a slate slab. Local people could just recall a family living in the cave but its connection with Kendrick, if any, was unknown. The prehistoric artefacts found included two awls made of antler and one is shown in Figure 8.4(*j*), two awls of bone (Fig. 8.4(*k,l*)), Bronze Age Beaker pottery with a chevron pattern (Fig. 8.4(*m*)), and decorated rim sherds of Neolithic pottery (Fig. 8.4(*n*)), as well as other fragments from this period and from the Romano-British period. Fairly typical

unpatinated flint knives (Fig. 8.4(*o–r*)) of Neolithic provenance were found generally in loose rubble immediately below and sealed by a brecciated layer. Most interesting was a pierced and decorated wolf canine from about 2.7 m below cave datum, the incised pattern being similar to that on the bovid and deer teeth held by the British Museum. Only one-half of the upper cave floor has been excavated and in this a shaft was opened which connected with the bottom cave, some 8 m below. Only archaeological rubble seemed to separate them; so much remains to be excavated.

Another modern cave excavation in North Wales was at *Ogof Colomendy* (SJ 202 628), started by North Wales Caving Club members in pursuit of cave extensions. When a root-infested cave earth began to reveal bones, some of which were clearly human (Carr, 1975; Appendix 8.1), an organized excavation was carried out briefly both in 1976 and 1977 (Davies, 1977–8; Appendix 8.1). The results were quite striking for such a small dig. The deposits consisted entirely of a brown loose cave earth containing sub-angular limestone fragments and archaeological material, but was thoroughly disturbed by tree roots. Excavation was confined to opening a route into the inner cave, and the potentially rich entrance platform sealed under collapsed blocks has been left untouched. The only dating evidence was a patinated, broken leaf-shaped flint arrowhead of Neolithic provenance which had been in a fire, but this find does not necessarily date the burials due to the absence of stratification. The human remains represented a minimum of three individuals and evidence of peculiar fracturing of the long bones and teeth were found. One mandible had two of its molars fractured, one of them so badly that half the crown and part of the root was missing. This was not due to any disease but rather to some form of impact. Also every long bone was fractured, something unusual in human remains from caves. From an analysis of the position of fracture and its location in the four tibiae and two femora it was suggested that some post-mortem ritual fracturing was carried out before the dead were finally interred. With the maltreated human remains were bones of three oxen, at least three sheep or goats, two dogs, three pigs and a single bone of red deer. All were hacked or smashed but not into sizes small enough to render them suitable for boiling or for extraction of marrow. Either animal food was plentiful at this location or the broken bones were not meal remains at all but offerings placed uneaten with the dead. It also appears that the number of bones and animal individuals is disproportionately large for such a small cave. There are earlier suggestions of ritual concerned with cave burials, for example D. P. Webley noted a human skull of probable early Bronze Age date in Tooth Cave, Gower, which had its top deliberately broken and the pieces placed in a neat pile (Harvey, Morgan & Webley, 1967). A. King considered that 'the extremely rich (cave) deposits . . . can no longer be accepted as occupation debris', and 'the finds must be associated with votive deposits' (King, 1974).

Quite different from anything that has been discussed so far in this chapter is *Pontnewydd Cave* in Clwyd (SJ 015 710), thoroughly excavated by Dr H. S. Green and his team since 1978 (Green, 1984). Although previously excavated in the nineteenth century and used as a store (Molleson, 1976), with consequent damage during the war, Dr Green located otherwise undisturbed debris flows and fluvial deposits, with *in situ* stalagmite floors which could be dated by the Uranium-series technique, with burnt flint being dated by the thermoluminescence technique (Green, 1986). His scenario had the River Elwy outside the cave re-excavating its glacial drift-choked valley after each glacial stage, and thus fluidizing the cave entrance deposits which then infilled the cave in the form of debris flows. Except for marine deposits washed into coastal caves, and ingress of congelifracts at some caves during the Devensian Glaciation, this form of cave fill had not previously been noted in Wales, and at Pontnewydd it gave rise to some problems in interpretation. Nevertheless, the human remains found, consisting of teeth and mandibular fragments from three individuals, and classed by Wymer (1985) as '*Homo sapiens* with some traits we associate with Neanderthal man' were stated by Green to represent human occupation that took place during the period 250 000–230 000 years ago. The stone tools had the following typology: pointed handaxes, Levallois points and flakes, side-scrapers, cleavers, choppers and disc cores, but no tools from the later Upper Palaeolithic period. The mammal fauna showed changes from one layer to the next due to climatic variations and included wolf, bear, hyena, a leopard-like felid, horse, narrow-nosed rhinoceros, Merck's rhinoceros (*Dicerorhinus kirchbergensis*), red deer, roe deer, beaver (*Castor canadensis*) and smaller mammals (Currant, in Green, 1984).

Located <1 km from Pontnewydd Cave are the *Cefn Caves* (SJ 021 705), also excavated by Green (1986) to reveal a sequence which was similar to that in Pontnewydd but which in a previous excavation had also yielded a later Upper Palaeolithic Creswellian point (Campbell, 1977) but no hominid remains. The mammal fauna has not yet been published in full and excavation is continuing into what promises to be an equally interesting cave.

Elsewhere in North Wales the sporting activities of potholers sometimes turn up bones in caves. These are easily spotted and are drawn to the attention of the responsible authorities. Displays of bones and artefacts are produced for the annual conferences of the caving bodies and permanent exhibitions are maintained both underground and in a museum by the William Pengelly Cave Studies Trust. For some years the Nature Conservancy Council has operated a cave palaeontological display at its visitor centre in Gower while the North Wales Caving Club put on a display in the town of Mold, Clwyd. Not to be outdone the Cwmbran Caving Club mounted a display for some weeks at an exhibition in Pontypool, Gwent. In 1988 the Dan yr Ogof Cave Company set up a comprehensive cave geological and archaeological display in a new building in the Swansea Valley. The end result of all this publicity can only be to the good of the conservation of caves and their irreplaceable contents. These are finite resources whose investigation and evaluation must be carried out responsibly and with the greatest care (Davies, 1977).

Appendix 8.1 References for modern excavations

Abbreviations

CCC   Cambrian Caving Council Annual Journal (The Red Dragon)

NWCC   North Wales Caving Club Newsletter produced monthly
WPCST   William Pengelly Cave Studies Trust Ltd
SWCC   South Wales Caving Club Newsletter produced quarterly–annually
AW   Archaeology in Wales – the annual publication of the Council for British Archaeology, Group 2 (Wales)

Ffynnon Beuno
Davies, M., *SWCC*, **72**, February 1973, Rhinoceros remains in a Flintshire cave.

Lloches yr Afr
Davies, M., *AW*, **13**, 1973; *AW*, **14**, 1974; *NWCC*, **12**, November 1973; *NWCC*, **13**, December 1973; *NWCC*, **19**, July 1974; *SWCC*, **78**, December 1974; *WPCST*, **24**, January 1975 (with photograph).

Maeshafn Cave
Hesketh, G. E. *Transactions of the Flintshire Historical Society*, 1954–5.

'Mammoth' Cave
Davies, M., *WPCST*, **29**, July 1977.

Ogof Colomendy
Carr, E., *NWCC*, **32**, August 1975.
Davies, M., *NWCC*, **32**, August 1975; *AW*, **16**, 1976; *NWCC*, **41**, May 1976; *NWCC*, **49**, January 1977; *WPCST*, **28**, January 1977; *NWCC*, **54**, June 1977; *WPCST*, **29**, July 1977; *CCC*, **4**, 1977–8, 18–23, Ogof Colomendy – further animal remains and a third human skeleton; *AW*, **17**, 1977.

Ogof 'Corkscrew' Cave
David, G. C., *AW*, **19**, 1979.

Ogof Pant y Wennol
Davies, M., *AW*, **14**, 1974; *NWCC*, **19**, July 1974; *NWCC*, **21**, September 1974; *NWCC*, **24**, December 1974; *SWCC*, **77**, September 1974; *SWCC*, **78**, December 1974; *AW*, **15**, 1975; *NWCC*, **32**, August 1975; *NWCC*, **35**, November 1975; *SWCC*, **81**, December 1975; *WPCST*, **25**, September 1975 (with plan); *AW*, **16**, 1976; *WPCST*, **28**, January 1977; *WPCST*, **29**, July 1977.
Stone, T. A. & Smith B., *AW*, **19**, 1979.

Ogof Rhiwledyn
Davies, M., *AW*, **13**, 1973; *NWCC*, **15**, February–March 1974.
Blore, J. D., 1977 (published privately as North Face Cave, Little Orme, 1962–76).

Ogof Tan y Bryn
Davies, M., *AW*, **15**, 1975; *NWCC*, **30**, June 1975; *SWCC*, **81**, December 1975.

Ogof Tudno
Stone, T. A., *AW*, **15**, 1975.
Davies, M., *NWCC*, **32**, August 1975; *NWCC*, **35**, November 1975.
Stone, T. A., *AW*, **16**, 1976.

Orchid Cave
Guilbert, G., *AW*, **22**, 1982.

Upper Kendrick's Cave
Davies, M., *AW*, **15**, 1975; *NWCC*, **60**, December 1977.
Davies, M. & Stone, T. A., *WPCST*, **31**, March 1978 (with figures).
Davies, M., *AW*, **18**, 1978; *AW*, **19**, 1979; *WPCST*, **32**, February 1979, pp. 7–9; *Studies in Speleology*, 1983, **IV**, 45–52, The excavation of Upper Kendrick's Cave, Llandudno; (in press), *Studies in Speleology*, 1988.
Stone, T. A. & Davies, M., *AW*, **17**, 1977.
Gillespie, R. *et al.* (1985). Radiocarbon dates from the Oxford AMS system. *Archaeometry Date list 2*, **27** part 2, 237–46.

Lynx Cave (unpublished)
Blore, J. D., 1965. *Lynx Cave, Denbighshire, Preliminary Report*, 1962–4.

References

Bordes, F. (1972). *A Tale of Two Caves*. London: Harper & Row.

Bowen, D. Q. (1974). The Quaternary of Wales. In *The Upper Palaeozoic and Post-Palaeozoic Rocks of Wales*, ed. T. R. Owen, pp. 373–426. Cardiff: University of Wales Press.

Bowen, D. Q. (1977). Hot and cold climates in prehistoric Britain. *Geographical Magazine*, **August**, 685–98.

Campbell, J. B. (1969). Excavations at Creswell Crags: preliminary report. *Derbyshire Archaeological Journal*, **89**, 47–58.

Campbell, J. B. (1977). *The Upper Palaeolithic of Britain*. Oxford: Clarendon Press.

Clark, J. G. D. (1971). *Excavations at Starr Carr*. Cambridge: Cambridge University Press.

Cullingford, C. H. D., ed. (1962). *British Caving: An Introduction to Spelaeology*. London: Routledge & Kegan Paul. 592 pp.

Davies, M. (1975). Our Welsh cave archaeological heritage. *Cambrian Caving Council Journal*, **2**, 41–2.

Davies, M. (1977). Towards the conservation of archaeological caves in Wales, *Cambrian Caving Council Journal*, **4**, 17–18; also in *South Wales Caving Club Newsletter*, **87**, 16–17.

Evans, J. G. (1975). *The Environment of Early Man*. London: Elek.

Green, H. S. (1984). *Pontnewydd Cave: a Lower Palaeolithic Site in Wales. The First Report*. Cardiff: National Museum of Wales.

Green, H. S. (1986). The Palaeolithic settlement of Wales research project: a review of progress 1978–85. In *The Palaeolithic of Britain and its Nearest Neighbours, Recent Trends*, ed. S. N. Collcutt. Sheffield: University of Sheffield.

Harvey, J. C., Morgan, R. & Webley, D. P. (1967). Tooth Cave, Ilston, Gower: an early Bronze Age occupation. *Bulletin of the Board of Celtic Studies*, **22(3)**, 277–85.

Hesketh, G. E. (1954–5). An account of excavations in the cave Big Covert, Maeshafn, Llanferres. *Journal of the Flintshire Historical Society*, **15**, 141–8.

Hicks, H. (1886). Results of recent researches in some bone caves in North Wales (Ffynnon Beuno and Cae Gwyn). *Quarterly Journal of the Geological Society*, **42**, 3–19.

Hicks, H. (1888). On the Cae Gwyn Cave, North Wales. *Quarterly Journal of the Geological Society*, **44**, 561–77.

Humphreys, G. A. (1907–8). Prehistoric remains, Kendrick's Cave, Great Orme's Head, Llandudno. *Proceedings of the Llandudno Field Club*, **2**, 49–57.

King, A. (1974). A review of archaeological work in the caves of northwest England. In *Limestones and Caves of Northwest England*, ed. A. C. Waltham, pp. 182–200. Newton Abbot: David & Charles.

Lacaille, A. D. & Grimes, W. F. (1955). The prehistory of Caldey, part 1. *Archaeologia Cambrensis*, **104**, 85–165.

Lacaille, A. D. & Grimes, W. F. (1961). The prehistory of Caldey, part 2. *Archaeologia Cambrensis*, **110**, 30–70.

McBurney, C. B. M. (1965). The Old Stone Age in Wales. Chapter 2 in *Prehistory and Early Wales*, ed. I. Ll. Foster & G. Daniel. London: Routledge & Kegan Paul.

Milisauskas, S. (1978). *European Prehistory*. London: Academic Press.

Molleson, T. (1976). Remains of Pleistocene man in Paviland and Pontnewydd caves, Wales. *Transactions of the British Cave Research Association*, **3(2)**, 112–16.

Oakley, K. P. (1972). *Man the tool-maker*. London: British Museum.

Peake, D. S. (1973). *Correlation of Quaternary Deposits in the British Isles*. Geological Society of London Special Report No. 4, 59–67.

Rowlands, B. M. (1971). Radiocarbon evidence of the age of an Irish Sea glaciation in the Vale of Clwyd. *Nature Physical Science*, **230(191)**, 9–11.

Shotton, F. W. (1977). Chronology, climate and marine record: the Devensian stage; its development, limits and substages. *Philosophical Transactions of the Royal Society*, **B280**, 107–18.

Sieveking, G. de G. (1971). The Kendrick's Cave mandible. *British Museum Quarterly*, **35**, 230–50.

Stuart, A. J. (1974). Pleistocene history of the British vertebrate fauna. *Biological Reviews*, **49**, 225–66.

Stuart, A. J. (1977). The vertebrates of the last cold stage in Britain and Ireland. *Philosophical Transactions of the Royal Society*, **B280**, 295–312.

Taylor, J. A. (1975). The role of climatic factors. In *The Effect of Man on the Landscape – The Highland Zone*. Council for British Archaeology Research Report 11.

Tratman, E. K. (1975). The cave archaeology and palaeontology of Mendip. In *Limestones and Caves of the Mendip Hills*, ed. D. Ingle Smith & D. P. Drew, pp. 352–9. Newton Abbot: David & Charles.

Valdemar, A. E. (1970). A preliminary report on the archaeological and palaeontological caves and rock shelters of Wales. *Transactions of the Cave Research Group, GB*, **12(2)**, 113–26.

Valdemar, A. E. (1972). Bibliography of Welsh cave archaeology and palaeontology. *Transactions of the Cave Research Group, GB*, **14(1)**, 33–42.

Valdemar, A. E. & Jones, R. D. (1970). An initial report on the archaeological and palaeontological caves and rock shelters in North Wales. *Transactions of the Cave Research Group, GB*, **12(2)**, 99–107.

Wymer, J. J. (1985). Early man in Britain – time and change. *Modern Geology*, **9**, 264–72.

9

Cave diving in South Wales

MARTYN FARR

Cave diving is the most specialized and potentially the most dangerous technique used in underground exploration. As such its adoption has been far from widespread and participants are generally very determined individuals. South Wales has witnessed considerable activity in this field and cave divers have been particularly fortunate with regard to their explorations recorded in the Welsh Sump Index prepared by members of the Cave Diving Group (1986).

The adoption of breathing apparatus to explore submerged passages in South Wales dates from 1946. For example, using home-made apparatus one group attempted to supply a diver with surface-fed, compressed air. Dives of up to 15 m depth were made into the sumps of White Lady Cave (SN 910 136) and Cwm Pwll y Rhyd (SN 912 137) while Pwll Ddu (SN 913 120) was also examined to a depth of about 6 m. Such techniques were generally impractical, and from the early 1940s moves were made to develop self-contained apparatus. Graham Balcombe and Jack Sheppard, who pioneered cave diving in the Mendip Hills, Somerset, were key figures at this time, and using closed-circuit, oxygen-rebreathing equipment they first attempted a dive at Ffynnon Ddu Resurgence (SN 848 153) during 1946. Unfortunately, they were not successful. However, of greater significance was the fact that during this meeting the Cave Diving Group was founded. This organization served to channel the activity of the individual cave divers.

Until 1960 no discoveries of 'dry' caves beyond sumps were made in South Wales, but the knowledge gained through these early years was to prove invaluable. Buxton and Wells, for example, commenced diving at the upstream end of Ogof Ffynnon Ddu I in 1958. An advance of about 30 m was made at this time in diving conditions among the finest in Britain. The potential for major discoveries was obvious in the Swansea Valley and numerous dives were also made in Dan yr Ogof (SN 838 160). Other caves examined with closed-circuit equipment included Llygad Llwchwr, White Lady Cave and Tunnel (now Cathedral) Cave (SN 839 161).

By 1961 locally based cave divers Charles George and Brian de Graff had acquired sufficient experience to extend explorations a considerable distance from their diving base. White Lady Cave, for example, was connected to Cwm Pwll y Rhyd

in June 1960, after a dive of 40 m. The same pair also discovered an air chamber (Number Five) in Llygad Llwchwr (SN 669 178) after a dive of 60 m and subsequently made a total penetration of 97 m. But by far the most significant exploration at this time was that at Ogof Ffynnon Ddu. By 1961 a distance of 146 m had been traversed in the Terminal Sump of Ogof Ffynnon Ddu I. The explorers were away from base for hours at a time and their explorations laid the basis for the later discovery of Ogof Ffynnon Ddu II (Fig. 9.1).

The advent of open-circuit, compressed air techniques heralded a golden era for cave divers. Early in 1966 Mike Wooding and John Sinclair passed the three short upstream sumps in Ogof Agen Allwedd (SO 188 158). Less than 45 m of diving gave access to 760 m of dry passages, terminating at Sump Five. But the greatest discovery of all was unquestionably that at Ogof Ffynnon Ddu in July 1966. Here, after a dive of 60 m, Charles George and John Osborne entered Dip Sump Series and opened the door to one of the longest and finest caves in Great Britain. In April 1967 the extensions were connected to the neighbouring cave, Cwm Dŵr (SN 857 156), allowing non-divers access to the vast series of passages associated with the major part of Ogof Ffynnon Ddu II.

Early in 1967 Oliver Lloyd and colleagues from Bristol took a keen interest in the Neath Valley, in particular the obvious potential at the head of the Nedd Fechan. Chris Gilmore quickly passed the downstream sump in Bridge Cave (SN 912 140) after an easy dive of 18 m. In February non-divers were allowed access to the new extension, Little Neath River Cave, by the opening of the notorious 'River Entrance' (SN 912 141) (see Chapter 14). In all >1220 m of fine stream passage was discovered and surveyed, terminating at Sump Two. Lying less than half-way to the resurgence the potential here was obvious. In March the assault was renewed. By July an additional 1525 m of passage had been found taking the divers to the New World Series. To reach this new extension divers now had to traverse 230 m of submerged passage. Non-divers have yet to gain access to this part of the cave. Despite some difficult terrain between and beyond the sumps exploration has now proceeded to within several hundred metres of the boulder-choked rising, Pwll Ddu. Sump Nine marks the

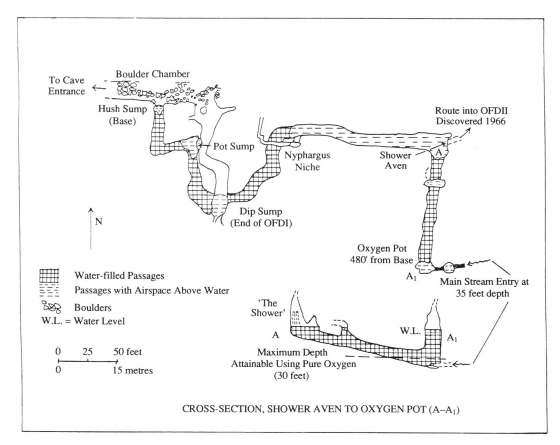

To Cave Entrance

Boulder Chamber

Hush Sump (Base)

Pot Sump

Nyphargus Niche

Shower Aven

Route into OFDII Discovered 1966

A

N

Dip Sump (End of OFDI)

Oxygen Pot 480' from Base

A₁

Main Stream Entry at 35 feet depth

Water-filled Passages
Passages with Airspace Above Water
Boulders
W.L. = Water Level

'The Shower'

A

W.L.

A₁

0   25   50 feet
0   15 metres

Maximum Depth Attainable Using Pure Oxygen (30 feet)

CROSS-SECTION, SHOWER AVEN TO OXYGEN POT (A–A₁)

Fig. 9.1. Sketch plan of the Ogof Ffynnon Ddu sumps.

current termination. This sump has been extended to considerable length and depth, but there appears little hope of establishing a physical connection with the rising in the foreseeable future.

Since 1966 virtually every known sump in South Wales has been examined. Contributing greatly to the explorations have been people such as John Parker, Roger Solari and Martyn Farr (1980). The major sites are described below.

Due to its ease of access Porth yr Ogof (SN 927 124) has received considerable attention, with John Parker taking much of the credit. This cave proved to be one of the first in the country to witness a cave-diving through-trip; the traverse between the Tradesman's Entrance and the upper, Hereford Entrance involves 210 m of continuous diving, broken by several small air bells *en route*. The upstream sumps here comprise a maze of underwater passages, a significant factor in the tragic loss of Paul Esser in February 1971. Most of this cave has now been surveyed showing that underwater surveying techniques can be just as accurate as those above (see Chapter 13).

Dan yr Ogof was likewise systematically explored in the late 1960s and early 1970s (see Chapter 16). Moon, Fairbairn and Arculus made the first long penetrations into the phreatic network of Mazeways in 1967 and 1968. Following long and awkward porterage a point over 150 m from diving base was eventually achieved. A protracted assault was mounted by Farr in 1971 and 1972. Dye-tests by Coase and others had by this stage established the existence of several interesting

hydrological connections, in particular that of Sink y Giedd, to the sumps of the lower series in Dan yr Ogof II. Efforts were rewarded by a relatively easy 118-m dive in the summer of 1972. This fell to Solari and Farr and yielded over 1600 m of inter-connected high-level passages (Fig. 9.2). The extension took the cave further southwest and to the south of the synclinal axis outlined by Coase & Judson (1977). It also proved to be the furthest upstream that it was possible to progress towards Sink y Giedd! From this point, unfortunately, virtually all routes, both above and below water, were blocked by boulder chokes, presumably due to the intensity of faulting in this area having fractured the rock more. A height of over 36 m above the water table was attained taking the cave into the upper limestones, through the easily recognizable Honeycombed Sandstone bed. A lengthy above-water digging operation yielded a further 430 m in 1978 (Mazeways Three) and in view of the draught circulating through this sector, future prospects must be reasonable – not only for a further advance to the west and northwest, but also for a bypass to the sumps.

Another site offering great potential to divers was Agen Allwedd. In 1971 Parker and J. Phillips pushed on through the terminal Sump Five in the Turkey Series and discovered over 460 m of new passages. In January 1973 Solari and Farr continued the explorations here when they passed a claustrophobic canal at the furthest extremity. Another 450 m of passage was found, giving a total of nearly 1 km, an extension which has been named the Remembrance Series. The

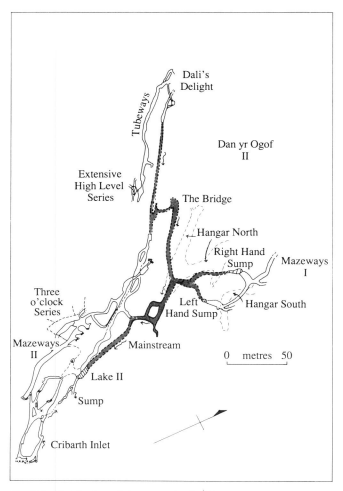

Fig. 9.2. Sketch plan of the sumps in the Mazeways area of Dan yr Ogof.

cave terminates at a massive fault-guided chamber >25 m high. A possible high-level continuation here has yet to be climbed.

Having exhausted the major potential at the upper end of the system searches were made at the Terminal Downstream Sump of Agen Allwedd. Survey and levelling techniques indicated that a height differential of >40 m existed between the Terminal Downstream Sump One and the Pwll y Cwm Rising in the Clydach Gorge. The horizontal separation was well over a kilometre. Parker made a tremendous advance here but it was left to Farr to make the breakthrough. After a dive of over 335 m he gained a major continuation in May 1974. After 600 m of streamway the passage once more ended at Sump Four. In June of the same year an ill-fated dive led to the exploration of the terminal sump for 180 m, reaching a depth of 22 m. Solari failed to return from this dive and a dynamic partnership was ended. The continuing sump was explored to 215 m in 1981, while more recently, in July 1987, Ian Rolland reached an air surface and short length of passage after a total dive of 285 m. Sump Five continues, and with a vertical interval of over 34 m to the level of the rising, further open streamway is almost certain to be reached in the very near future.

Diving at the major resurgence for the Agen Allwedd System was to prove, initially, very disappointing. The Pwll y

Cwm Rising (SO 215 125) in the bed of the River Clydach was found to be completely boulder-choked at a depth of 8 m. The neighbouring cave, Elm Hole, just a few metres away, proved more fruitful. Following a constricted, intimidating dive in 1974 the main subterranean flow from the caves of Llangattwg was encountered at a depth of 15 m. A second dive, the same year, achieved a depth of 22 m but the nature of the approach passage in Elm Hole was such that a continued assault was deemed to be too dangerous given the equipment then available.

The discovery of the major extensions in Ogof Daren Cilau (SO 206 153), in 1984 and 1985, presented cavers with the first real opportunity of a through-trip beneath Mynydd Llangattwg. In March 1985 the Terminal Downstream Sump was reached, over 4 km from the entrance and situated a mere 500 m from the Rising. Diving commenced less than two months later. Following three dives downstream in Daren and six upstream from Elm Hole the writer achieved a connection in July 1986. When the 625 m dive was undertaken from Daren Cilau to Elm Hole the following month a new record was established, for both the longest and the deepest caving through-trip in Great Britain.

Dive explorations are by no means complete in this area and a very interesting project has recently commenced. Divers of the Welsh Section of the Cave Diving Group are currently undertaking an excavation of the Pwll y Cwm Rising. At the present time a vertical blockage of about 5 m remains to be cleared. When access is gained to the large tunnel which runs beneath, a thorough exploration of the sump will be possible. Using large cylinders it will then be feasible to continue up the main underwater tunnel towards Agen Allwedd. A diving connection looks certain to be achieved within the next few years.

Another area of great potential lies in the Hepste Valley, northwest of Penderyn (see Chapter 13). The surface river here normally sinks into impenetrable fissures about 800 m upstream from the short cave, Tuck's Rift (SN 952 118). This point lies over 3 km from the rising (SN 936 098). Tuck's Rift drops directly to the water table and is associated with the main valley flow. Sumps here have not been conclusively explored and can be followed both up- and downstream to localized blockages of driftwood and cobbles.

The Hepste Resurgence System is a complex network, where numerous sumps have been passed, extending in a variety of different directions. In Ogof Afon Hepste, the largest of these caves, >900 m of passage is presently known, but comparatively little progress has been made with regard to a connection with Tuck's Rift. Quite apart from the main flow which is derived from the north, the resurgence is the focus of an extensive catchment area which extends for a considerable distance both east and west. The potential is obvious, but in view of the exceptional danger of flooding in this system drought conditions are essential for any exploration.

Another tantalizing hydrological connection is along the western flank of the Nedd Fechan, between Pant Mawr (SN 891 162) in the north and the risings 5 km away (see Chapter 14). These are situated a few hundred metres up the valley from Pwll Ddu. Associated with the risings is Ogof Cas (SN

912 124), which appears to be a flood overflow for the missing stream. A constricted boulder blockage must be negotiated about 80 m into the sump, aligned upon a major north–south fault. The continuation is constricted and to date exploration has only proceeded 30 m beyond the choke.

One of the long-standing mysteries in South Wales is that of Llygad Llwchwr (SN 669 178), near the village of Trapp (see Chapter 17). A major stream resurges here. Since the early 1960s a constricted 'slot', 110 m from diving base, has terminated all progress. This has been the scene of several underwater excavations. Progressive techniques could lead to further exploration of this site in the near future.

The southern edge of the limestone outcrop cannot be omitted from the overall picture. The Gower Peninsula, for example, presents considerable exploratory potential. Two adjoining systems, Llethryd Swallet (SS 531 912) and Tooth Cave (SS 532 911), drain an area of about 10 km² to Wellhead Resurgence (SS 539 897) at Parkmill (see Chapter 18). Both caves terminate in sumps. The resurgence has been penetrated for 8 m at a depth of 4 m. Llethryd Swallet terminates at an undived Sump Three, at a point >1600 m from the resurgence. The potential at Tooth Cave is >800 m in a straight line, but the terminal sumps here do not appear to lie on the main flow.

Another intriguing feature on the southern outcrop is the large Schwllt Rising (SS 884 768) near Ogmore. Water resurges within the intertidal limits and six million gallons per day are pumped off to supply parts of South Glamorgan. Diving has yet to be undertaken here. Again, the potential is immense.

Since the 1960s South Wales has witnessed tremendous discoveries in the diving sphere. Systems in the Swansea Valley, the Neath Valley and at Llangattwg testify clearly to the importance of diving in cave exploration. Considerable work remains throughout the area and with the inevitable advance in both equipment and techniques a series of major discoveries can be expected in the future.

References

Coase, A. C. & Judson, D. M. (1977). Dan yr Ogof and associated caves. *Transactions of the British Cave Research Association*, **4(1–2)**, 245–344.

Farr, M. (1980). *The Darkness Beckons*. London: Diadem Books.
Welsh Sump Index: Cave Diving Group 1986.

10

Limestones and caves of the Forest of Dean

DAVID J. LOWE

Lying between the limestone areas of South Wales and the Mendip Hills, between the valleys of the rivers Severn and Wye, is the deeply dissected plateau of the Forest of Dean. Within this region of mixed forest and farmland are large areas of outcropping limestone, although only rarely does true karst scenery make an appearance and there is little exposure away from quarries and river valleys. Caves are relatively common, ranging through large isolated 'fossil' passages to active swallets and resurgences in various stages of maturity.

The earliest serious examination of the caves, other than by miners in search of iron-ore, involved archaeological excavation of the rock shelters and caves south of Symonds Yat in the nineteenth century (e.g., Symonds, 1872) and later by the University of Bristol Speleological Society in the 1920s. More recently the obvious open holes have been explored (e.g., Cullingford, 1951; Standing, 1967), and the publications of the Hereford Caving Club (1956–60), Gloucester Speleological Society (1960–6) and Royal Forest of Dean Caving Club (1964–) provide detailed information.

Geology

The area (Fig. 10.1) is an undulating plateau between the incised valley of the River Wye and the River Severn. Rocks of Carboniferous age occur as outliers occupying a synclinal basin within the locally dominant Devonian rocks, the basin being the result of folding during the Hercynian orogenic episode.

Stratigraphy

General points regarding the various aspects of stratigraphy and terms in common usage are covered in Chapter 2. A similar chronostratigraphical framework is applicable in the Forest of Dean where, as on the Eastern Crop of South Wales, dolomitization of the lower part of the Dinantian sequence makes correlation of the various rock units highly conjectural and the placing of rock units into either biozones or regional stages equally uncertain. Table 10.1 shows a possible correlation of the major lithostratigraphical units to the regional stages and possible relationships of this modern

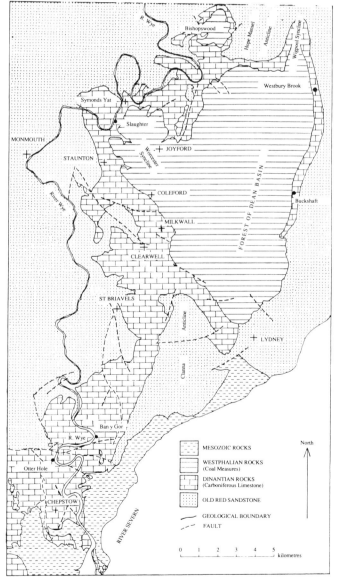

Fig. 10.1. Sketch map of the geology of the Forest of Dean.

Table 10.1 *Relations of new and old geological stratigraphical terminology*

Regional stages	Major cycle	Lithostratigraphy	Zones and subzones –	Series	Series
Holkerian	4	Upper Drybrook Sst Drybrook Limestone Lower Drybrook Sst	S_2	Upper Avonian	Viséan
Arundian	3	Whitehead Limestone	C_2S_1	,,	,,
Chadian	2	Crease Limestone	C_2	Lower Avonian	,,
Courceyan	1	Lower Dolomite Lower Limestone Shales Old Red Sandstone	C_1 Z K	,, ,,	Tournaisian

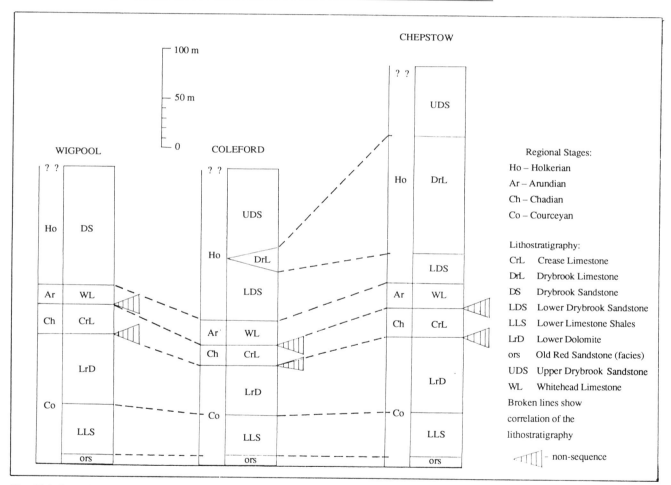

Fig. 10.2. Lateral variation of the strata of the Forest of Dean.

nomenclature to biostratigraphical terminology and other divisions which may be encountered in older literature or on geological maps.

As in South Wales rocks of Dinantian (Carboniferous Limestone Series) age follow conformably on Devonian rocks and the earliest Dinantian strata are of Upper Old Red Sandstone facies (George *et al.*, 1976). These essentially continental deposits pass upwards through continental–marine oscillations and eventually into an essentially marine sequence. Cyclic marine deposition, with non-sequences at certain levels (Table 10.1; Fig. 10.2) continued until at least the end of the Holkerian Stage (Ramsbottom, 1973). No younger Dinantian rocks are preserved in the area and the evidence of the change of depositional environment presented by the Drybrook Sandstone facies in the Holkerian would perhaps indicate that the area became positive in post-Holkerian times and remained so during the later part of the Dinantian. Likewise, no major spreads of Namurian age rocks (Millstone Grit) are preserved, although sandy deposits infilling impressive palaeokarstic features in the southern part of the area may be of Namurian age (Dixon, 1921) and indicate that more widespread deposition might have occurred, with the rocks subsequently being removed except in the palaeokarstic environment. Another possibility is that these sandy deposits belong to the early Westphalian (Coal Measures), there being some evidence to suggest that the upper part of the Drybrook Sandstone is of Westphalian age although mapped in continuity with Drybrook Sandstone of Dinantian age. Whatever the case, there is certainly a significant unconformity above the Dinantian part of the Drybrook Sandstone and, with the exception of the sandy karst-filling deposits and the thin early Westphalian beds mentioned, the next beds above the Holkerian rocks are of Westphalian C age (Upper Coal Measures).

The lithostratigraphic units comprising the Dinantian sequence and their predominant lithologies are as follows:

> Upper Drybrook Sandstone (part of); fine-grained quartz sandstone, red–off-white.
>
> Drybrook Limestone; bioclastic limestone, oolite and calcite mudstone, pale-coloured, thickening from <3 m in the Wigpool area to >100 m near Chepstow.
>
> Lower Drybrook Sandstone; medium- to coarse-grained quartz sandstone, locally pebbly, yellow–reddish-brown.
>
> Whitehead Limestone; mainly off-white calcite mudstone and dolomite mudstone with minor bands of multi-coloured shale and marl; massive 3 m bed termed 'Lidstone' at base.
>
> Crease Limestone; white oolitic limestone overlying

massive grey crinoidal limestone; dolomitized and hematitized locally.

> Lower Dolomite; generally uniform massive grey dolomitic limestone; relict part-dolomitized oolite locally.
>
> Lower Limestone Shales; variable sequence of argillaceous sediments with massive bioclastic and oolitic limestones.

Local variation of both the thickness and lithology of these deposits is common (Fig. 10.2) and local details are given by Welch & Trotter (1961).

Structure

The Carboniferous rocks form a synclinal basin, the centre filled by Coal Measures strata and rocks of Dinantian age around the margins, except where they are overstepped by younger beds (Fig. 10.1). Two main tectonic episodes are recognized: (a) an intra-Carboniferous episode during which Dinantian and older rocks were folded, faulted and possibly eroded prior to deposition of the Westphalian rocks; and (b) a pre-Triassic episode during which the Coal Measures were folded and the underlying strata refolded along slightly different axes to those of the first episode.

Within the broad synclinal structure numerous component folds are recognized and are described by Welch & Trotter (1961). Major faults affecting the Dinantian rocks are rare and, with the exception of the Lydbrook Fault (Fig. 10.3) they have had little effect on cave formation. Few minor faults have been recognized on the surface, but where visible underground they have exerted a strong influence on passage formation.

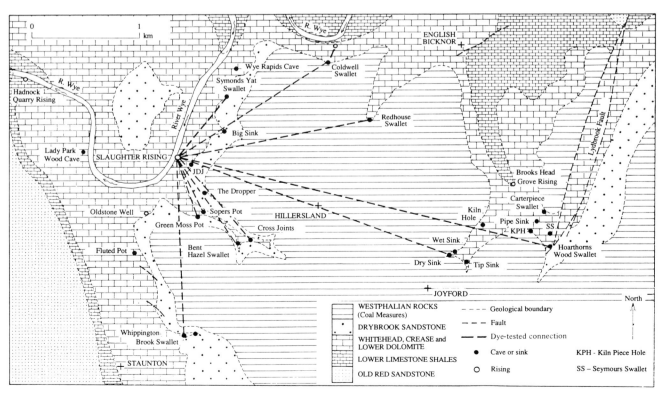

Fig. 10.3. Sketch map of the geology and drainage of the Slaughter area.

Joint-direction and density have not been studied in detail except in the more recently explored caves. Passage direction is locally influenced by joints, but more detailed examination is required before the interrelation of caves, joints and major structures can be fully understood.

Structure and stratigraphy combine with the present topography to provide an ideal setting for cave formation. Precipitation gathering on the higher ground of impermeable Coal Measures or Drybrook Sandstone strata channels onto the underlying limestones, usually to sink almost at the contact. Resurgence levels are controlled both by lithology, as is the case at junctions with the underlying Old Red Sandstone or within the Lower Limestone Shales, or by local saturation level, as along the banks of the Wye. A theoretical maximum thickness of limestone in excess of 250 m might be expected, but due to the effects of structure and topography, the maximum possible vertical separation of sink and rising is about 150 m. The maximum straight-line distance from sink to rising so far substantiated by dye-test is 3.6 km.

The caves

The outcrop of Dinantian limestones in the Forest of Dean is extensive (Fig. 10.1) and only a relatively small proportion has been subjected to thorough, systematic investigation. These better known areas are considered in detail below, followed by brief notes on the remaining areas.

The Slaughter area

This takes its name from a section of the east bank of the Wye (Figs. 10.1, 10.3) which has also given its name to the Slaughter Rising (SO 5554 1453), the largest resurgence in the northern part of the Forest. The Rising, which is fed by numerous sinks spread across a wide area (Richardson, 1924), comprises several discrete flows emerging from mudbanks and small, blocked culverts below a Forestry track and steep, wooded scree slope. The resurgent waters unite and flow through a larger culvert into the Wye. Richardson (1930) mentioned the resurgence and his description suggests that in high flood there is a connection directly into the river. This has not been proven but inputs to and output from the system suggest that there could be a normally undiscernible link to the riverbed. The precise stratigraphical level of the Rising has not been determined since all rocks in the vicinity are dolomitized, but it probably lies within the Lower Dolomite, fairly close to the top of the formation.

The caves draining to the Slaughter Rising are fed by streams sinking along the Drybrook Sandstone–Whitehead Limestone boundary above the Wye and by more remote sinks to the east where overstepping Coal Measures strata meet the Lower Dolomite. In this area the former boundary roughly follows the line of the cliff-top on the eastern side of the Wye at an average elevation of about 75 m above the Slaughter. It runs roughly north–south and comes within 300 m of the Rising at J.D.J. Hole (SO 5576 1453) (Fig. 10.3). Numerous small and several large streams flow onto the Whitehead Limestone and generally sink immediately. The area is pitted by innumerable 'shakeholes', probably subsidence dolines formed by collapse of the Whitehead Limestone beds into solutional chambers in

the underlying Crease Limestone. Collapse is not limited to the Whitehead Limestone, since many of the holes expose Drybrook Sandstone. Most impressive of these is the Dropper (SO 5516 1427) where a small stream sinks in a deep, quarry-like depression with walls of pebbly sandstone.

Coldwell Swallet (SO 5696 1554) (Fig. 10.3) is the furthest sink known to drain to the Slaughter from this boundary, locally modified by a tongue of Coal Measures strata overlapping the Drybrook Sandstone and Whitehead Limestone so that the entrance lies in the Crease Limestone. A steeply descending vadose streamway is solidly choked after 15 m, but water penetrating the choke has been traced to Coldwell Rising (see below) almost directly below. The original water route is along a small phreatic passage at higher level which leads to a joint-guided section trending down-dip towards the Slaughter Rising. This water route, which passes from the Crease Limestone into the Lower Dolomite, has been successfully traced to the Slaughter Rising. A final 25 m pitch leads to an impenetrable sump, which might be at local saturation level.

Further south along the Drybrook Sandstone–Whitehead Limestone boundary (Fig. 10.3) is Symonds Yat Swallet (SO 5602 1522) (Solari, 1974), where a small stream sinks in a large depression. An excavated route from an adjacent depression leads to a series of joint-guided rifts and chambers which penetrate the basal bed (Lidstone) of the Whitehead Limestone and give access to the Slaughter House. This large chamber in the Crease Limestone is of solutional origin, providing an interesting parallel with chambers in the local iron-ore mines. Several small streams sink through the chamber floor and can be followed down through boulders for about 8 m to a complete choke. Shower Passage, which leads from the chamber is also solidly choked but was probably the original main streamway of the system, now long abandoned.

Symonds Yat Swallet has many joint-guided passages, most noticeably within the Whitehead Limestone. Despite the presence of several marl bands in the upper levels, there is no evidence of bedding control. Slaughter House is totally out of context with the restricted passages which lead into it and only Shower Passage reaches a size corresponding to that of the chamber, suggesting the existence of a large, old system that has been invaded by the small streams of the upper, immature passages. Shower Passage is now occupied by an underfit stream and no continuation of the old route has been found. Slaughter House has been much modified by recent roof collapse, and it is possible that collapse of similar chambers back to the surface has formed the dolines previously mentioned.

Still further south (Fig. 10.3) Whippington Brook Swallet (SO 5563 1290) is one of the larger sinks in the area. A 16-m deep excavated shaft penetrates the oolitic facies of the Crease Limestone to enter a complex of small passages in the basal, crinoidal facies of the Crease. There has been much alteration and iron-staining of the rock and the area has an intricate geological history. The passages are typically phreatic, with roof-pendants, fretted rock surfaces and small solutional tubes. Larger, boulder-filled passages cut across the present open routes, but they are inactive and solidly choked. The cave ends in two sumps which are hydrologically linked. Why these sumps should occur 75 m above resurgence level is not certain. One possibility is the presence of a rock barrier associated with the lowest beds of the Crease Limestone, which at this locality are extremely dense and homogeneous. Another alternative is that the water follows a short phreatic riser on a minor fault to regain the Crease Limestone oolitic facies.

Intriguingly the sink in the valley bottom is just upstream of the outlet of a long drainage adit that was driven southeastwards to de-water Highmeadow Mine beneath Staunton. The scheme was a failure since the adit and mine workings did not connect. Even so the adit must collect substantial amounts of water from the ground it crosses, and this does not flow out of the adit mouth, which is blocked by collapse. The short section of explorable adit has intersected several small fragments of natural passage running along the Whitehead Limestone–Crease Limestone boundary, and more substantial natural passages are suspected beyond the current choke. Even more interesting is the old miners' motivation for commencing their adit in the vicinity of the Whippington Brook Sinks. Down-valley from Whippington Brook Swallet the diverted stream flows into a small culvert beneath the miners' spoil, and only a fraction of the flow reappears on the downstream side of the spoil. Possibly the intention was that the mine drainage would be into a pre-existing major sink, since obscured, the current Swallet being a more recent upstream sink, and one of several that exist in this stretch of the valley.

Between Symonds Yat Swallet and Whippington Brook lies the Mailscott Valley. Today the Mailscott Stream generally keeps to the surface across the limestone outcrop, to join the Wye about 300 m downstream of the Slaughter Rising. At least three sinks exist in the Mailscott Valley floor, each capable of transmitting the entire flow of the stream in normal conditions. Cross Joints Swallet (SO 5622 1385) (Fig. 10.4) (Lowe, 1981) is the only sink giving access to significant open cave passage and it provides a cave strikingly unlike the others currently known in the area. The entrance lies at the junction of two joints in the Whitehead Limestone, and drops into a low crawl a few metres long. A tiny vadose trench continues, dropping rapidly and passing through several small chambers before meeting a slightly larger passage carrying a stream from sinks up the valley. The combined streamway, still low, with pools, squeezes and numerous small formations, continues for more than 100 m before a way over a boulder-choke leads into a short rift passage. A few metres onwards the passage ends at the brink of a 10 m pitch (The Jump) which drops into Hyperspace, a chamber even larger than Slaughter House in Symonds Yat Swallet. From the foot of the pitch an excavated route leads down through boulders to a final blockage 35 m below the entrance.

The passages between the sink and Hyperspace are mainly vadose and immature, with fragments of small phreatic tubes preserved at roof level. Strong joint-guidance is apparent, but lithological or bedding control dominates as the cave trends essentially down-dip. Few nick points are present in the streamway and the passage is close to attaining a steady gradient on top of a relatively resistant bed, probably the Lidstone. A number of potholes or swirlpools have been cut

Fig. 10.4. Cross Joints Sink, Forest of Dean; a typical active swallet into the upper beds of the Whitehead Limestone (photo by D. J. Lowe).

into this bed, but generally the stream runs along its top. Hyperspace originated as a solutional chamber in the Crease Limestone, much as Slaughter House did, and likewise has been modified by blockfall of the more thinly bedded Whitehead Limestone above. The immature and recent streamway seems to have encountered the ancient system at the boulder choke above Hyperspace, but the origin of the boulders, probably an old abandoned passage, is not open, nor even apparent.

Up-valley (60 m) from Cross Joints Swallet and close to the top of the Whitehead Limestone is Mailscott Sink. The water sinking here is encountered in Cross Joints. A diversion ditch and dams have been constructed to send the stream past both Mailscott and Cross Joints Sinks, since if water flowed down either in flood conditions there would be no sanctuary for explorers in the streamway.

Downstream from Cross Joints a tributary valley joins the Mailscott valley from the south. A number of small sinks have been examined here, the main one being Bent Hazel Swallet (SO 5611 1384) (Fig. 10.3) comprising 6 m of tight vadose streamway to an impassable bend. Water from these sinks is assumed to reappear as an inlet at Lidstone level, high in the southwest wall of Hyperspace in Cross Joints Swallet, about 70 m away. Another inlet to Hyperspace, at the same level,

Fig. 10.5. Sopers Pot, Forest of Dean, looking up the 10 m shaft; the lower part is formed in the Crease Limestone and the upper in the basal Whitehead Limestone (photo by D. J. Lowe).

probably derives from leakage in the surface streambed almost directly above.

Beyond the valley junction the Mailscott stream, when flowing, covers about 400 m before reaching the third major sink, Sopers Pot (SO 5580 1408) (Fig. 10.5). Here an impressive fluted shaft penetrates the lowest beds of the Whitehead Limestone and the upper part of the Crease Limestone oolite. The shaft can transmit the entire stream, even in flood, but an apparently total excavation failed to locate a way on. Water sinking here, at Bent Hazel Swallet and at Cross Joints Swallet (and hence Mailscott Sink) has been dye-tested to the Slaughter Rising, all with rapid flow times.

Innumerable smaller sinks, named and unnamed, exist between the Drybrook Sandstone base and the Slaughter Rising. Green Moss Pot (SO 5585 1405 (Fig. 10.3) which swallows a stream on the hillside southwest of Sopers Pot, has been tested to the Slaughter, but the shaft is blocked at a depth of 8 m. Other sinks on the spur between the Whippington and Mailscott valleys have not been tested, nor is it known where Fluted Pot (SO 5518 1370) (Fig. 10.3), a short fragment of an abandoned stream cave in the Whippington Valley, originally resurged.

While discussing the sinks of this boundary it is convenient to consider two small associated resurgences. As previously mentioned, part of the water sinking at Coldwell Swallet

resurges at Coldwell Rising (SO 5701 1570) (Fig. 10.3). At the foot of the steep slope below Coldwell Rocks, possibly close to the Lower Dolomite–Lower Limestone Shales boundary, a strong stream flows from a short, choked cave. The flow is usually in excess of the stream sinking at Coldwell Swallet and since no other sinks are known to drain in this direction it is probable that much of the resurgent water derives from percolation.

Oldstone Well, or Whippington Brook Rising (SO 5532 1408) (Fig. 10.3), was once assumed to be the resurgence for Whippington Brook Swallet. Comparison of the meagre flow at the Rising with the large stream at the sink showed that the connection was unlikely and dye-testing proved that water from the Swallet resurged at the Slaughter. On the hill above Oldstone Well several small streams sink into the Whitehead Limestone or through collapses in the Drybrook Sandstone. These have not been dye-tested, but may supply part of the water to Oldstone Well, while an additional component is probably percolation water channelled by relatively imper- meable beds within the Whitehead Limestone.

To the east of this area the Drybrook Sandstone is uncon- formably overlain by late Westphalian Coal Measures. Run- off from these beds, in addition to augmenting the flow to those sinks already described, feeds several large sinks in the extreme north and east of the Slaughter Catchment, where the Coal Measures overstep onto the Lower Dolomite (Fig. 10.3).

Close to Hoarthorns Farm, Edge End, are two important sites; the active sink of Hoarthorns Wood Swallet (SO 5905 1376); and, further down the same valley, the fossil sink, Seymour's Swallet (SO 5904 1387), which is active only in very high flood. Hoarthorns Wood Swallet has a history of being excavated and reblocked by flood debris, but the original sink entrance is now bypassed by a short shaft which leads into the top of a steeply-hading rift 25 m deep. It was originally believed to lie on the Lydbrook Fault (Fig. 10.3) but sub- sequent survey has shown it to be perpendicular to the major structure and to have a small throw in its own right. There is no open way on at the foot of the rift; a steeply descending fissure joins at right angles and is probably a dry inlet.

A collapse area by the sink was excavated and gave access to a system of interconnected rift passages, the major one per- pendicular to the Lydbrook Fault. The system reaches a depth of >30 m before being blocked by a muddy pool. Soon after excavation the shaft became unstable and was backfilled, although access to the lower rifts is still possible by way of a blasted route from the original shaft.

Dye-tests have shown that water sinking at Hoarthorns Wood Swallet rises at the Slaughter in less than three days, and this has finally disproved the fallacy (e.g., Richardson, 1930) that the water emerged in the workings of Cannop Colliery. The remains of pipelines built and maintained at great expense to divert streams away from the sink and beyond Seymour's Swallet bear witness to the strength of earlier beliefs.

Seymour's Swallet is a complex system of rift passages guided by joints and a small fault parallel to the Lydbrook Fault. Generally it is inactive but for small trickles of percola- tion water. One of its most striking features is a 2 m diameter circular pot which is choked by silt. The main part of the cave lies across the top of the pot, but the hydrology of the system is obscure and it is not certain which is the main passage of the cave. There are several potential outlets, one of which is very close to the Lydbrook Fault and below a sloping aven. Much of the system is phreatic and careful study of wall-markings might provide the key to the problem. The lower parts of the cave are close to the base of the Lower Dolomite and therefore at about the same horizon or geologically lower than the Slaughter Rising, whilst topographically 100 m higher.

Water diverted past these sinks through the pipelines emerges in the dry valley slightly further north and sinks almost at once into Pipe Sink (SO 5890 1401) (Fig. 10.3). Its ultimate destination is unknown, but it is possible that together with water sinking at Carterpiece Swallet (SO 5895 1410) it resurges at Brooks Head Grove Rising (SO 5867 1436) on the junction of the Lower Dolomite and Lower Limestone Shales. Both sinks are untested and at present the origin of the water rising at Brooks Head Grove is unknown.

In the meadow south of Pipe Sink, at a higher level on the spur between the Hoarthorns and Joyford dry valleys, is Kiln Piece Hole (SO 5883 1393) (Fig. 10.3). A shaft of about 4 m, lined with oil drums, sidesteps into the top of a moderately large rift, probably formed on a minor fault, running north–south. At its northern end the rift is wide but choked, while to the south it is too narrow to penetrate, although seeming to

Fig. 10.6. Dry Sink, Joyford, in low flow conditions; an impressive rift swallet into the Lower Dolomite (photo by D. J. Lowe).

widen below. It is unlikely that a hole in this position can have swallowed a stream in recent times and the nature of the cave, if any, beyond its current limits, is problematical.

Further west, close to Joyford, a major stream flows north-wards to disappear in an impressive rift known illogically as Dry Sink (SO 5812 1373 (Fig. 10.3), a 10 m long and 7 m deep streamway with no open way on (Fig. 10.6). Another stream sinks close by in a steep-sided hole known as Wet Sink (SO 5820 1375), having flowed along the Joyford dry valley for the last part of its surface route. It is likely that the Wet Sink Stream was originally swallowed by Tip Sink (SO 5826 1368), an unimpressive hole nearby which has also been tested to the Slaughter Rising. In flood the stream might have gone on down the normally dry valley, to be eventually captured by the enlarging doline forming around Wet Sink. Beyond Wet and Tip Sinks a broad valley leads north to Brooks Head Grove, passing the old dig of Kiln Hole (SO 5840 1396). This hole swallows water in very high flood conditions and was excavated to a depth of about 25 m, where bedding development, possibly at the base of the Lower Dolomite was encountered. A number of depressions, probably of similar age to Kiln Piece Hole are located on the spur to the east of the Joyford Valley.

Another major sink exists to the northwest at Redhouse Swallet (SO 5737 1499). Here there is no open cave but the sink is in a classic blind-valley situation and has been successfully dye-tested to the Slaughter Rising. Other small sinks exist between Redhouse and the village of English Bicknor to the north (Fig. 10.3). No detailed examination has been carried out of the relationship of the sinks to each other and to local risings. Recently access has been regained to the Joyford Sinks and substantial discoveries should be expected.

The Bishopswood area

This area of Lower Limestone Shales lies in the northern part of the Forest of Dean Dinantian outcrop (Figs. 10.1, 10.7). A complex hydrology is evident despite the probable limiting of the caves to one band of limestone only a few metres in thickness (Solari, 1971, 1974). Numerous sinks and risings have been located, but little cave passage has been explored.

The major system is that which resurges at Lake Rising (SO 6036 1836) (Fig. 10.7). Several sinks are known or presumed to resurge here, the most notable being Dunderhole Farm Swallet (SO 6006 1943) which is a crawl 100 m long following the dip of the limestone. The potential length of this cave is in excess of 1 km and the depth is 60 m, but further progress will require massive removal of bedrock. Sixteen-acre Wood Sink (SO 5993 1877) has also been positively tested to Lake Rising, but no open passage has been explored. Both tests were positive in <27 h.

Further north Task Park Sink (SO 6098 1877) was positively traced to S.B. Rising (SO 6073 1938), with a fast throughput time. The drainage of this system is discrete from that of Lake Rising due to the presence of a belt of sharp small-scale folding with its axis passing between the two. A cross-section of this folding is well exposed in a track nearby (SO 6058 1942).

Many other small sinks and risings have been described in the area (Solari, 1971, 1974) but no further dye-tests have been

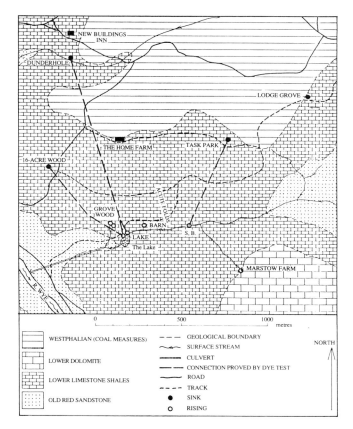

WESTPHALIAN (COAL MEASURES)

LOWER DOLOMITE

LOWER LIMESTONE SHALES

OLD RED SANDSTONE

– – – GEOLOGICAL BOUNDARY

⌒⌒ SURFACE STREAM

═══ CULVERT

—— CONNECTION PROVED BY DYE TEST

—— ROAD

– – – TRACK

● SINK

○ RISING

NORTH

Fig. 10.7. Sketch map of the geology and drainage of the Bishopswood area.

carried out. The process of opening and exploring the associated caves might present problems since most sites lie in open parkland, but the area of Lake Rising is well screened and digging here might open an extensive dendritic cave system.

The St Briavels area

Southwards from the Slaughter area the outcrop of the Lower Limestone Shales broadens around St Briavels (Fig. 10.1). Here again are numerous sinks and risings, the associated cave systems confined to single beds of limestone and most of the potential entrances being on shale–limestone boundaries.

Entry to St Ann's Well Cave (Clanna Resurgence) (SO 5798 0277) is from the rising. A low canal leads to a choke, whence an upward climb and a series of dry crawls and joint-guided rifts eventually lead to a large chamber with numerous fallen blocks. Here the stream is audible, but unreachable. From the chamber a low crawl leads over boulders which almost block a large passage, with swiftly flowing water audible at several points. Just before a boulder choke is reached the unseen stream crosses from left to right beneath the passage, and the remaining passage beyond the choke shows no sign of regaining the main water, although minor flows are visible. The survey shows the cave to follow a single bed of limestone up-dip from the rising and strong joint-guidance is evident throughout. Cullingford (1951) believed that the stream at St Ann's Well originated at sinks in the Aylemore Brook about 1 km to the west. The connection has not been confirmed and

would require the cave to make a 90° deviation from its present up-dip route. This could explain the loss of the main stream close to the end of the cave, the minor trickles beyond the choke being derived from percolation and accounting for the larger flow at the rising when compared to the sinking Aylemore Brook. Although only a short system, St Ann's Well Cave is important in that it demonstrates that passages of significant size can form within the Lower Limestone Shales.

In the area centred on Bearse Farm (SO 572 052) there are many depressions and deep, fluted shafts, often well hidden in the woods. Many have been lost to years of tipping or have been filled in for reasons of safety. Those remaining often reach 10 m in depth and some act as swallets for small streams, but little horizontal development has been located. The presence of an extensive underground drainage system can, however, be inferred from the presence of several sizeable resurgences.

At Slade Bottom, close to Bearse Farm, a large impenetrable resurgence – Slade Wood Rising (SO 567 055) – discharges from the valley side. Higher up the bank a narrow rift emits an impressive roar of water and strong draught. Close by is Dark Hill Cave (SO 568 054) which can be followed for about 20 m to a boulder blockage. To the south another resurgence cave, Slade Brook Rising (SO 568 053), ends after 6 m when water meets the roof.

Perhaps the most intriguing site in the St Briavels area is that noted by Cullingford (1951) at Highgrove Farm (SO 572 034). A shaft led into a chamber reported to be 30 m in diameter and 6 m high. The landing was on a huge debris cone of farm refuse and only a rapid exploration took place, revealing a 6 m long choked fissure. More recently access permission has been refused. Other shafts in the area were explored at about the same time (Cullingford, 1951) and some have been excavated since.

Considering the large number of shafts, sinks and risings, the area seems to be one of great promise. Significant caves ought to be present within the thicker limestone beds, forming relatively short independent drainage systems controlled by geological structure, lithology and the interaction of surface topography, but this remains to be confirmed by water-tracing and direct exploration.

Other areas

To the west of the Slaughter area at Hadnock Quarry a sizeable stream emerges from a choked hole in the bank of the Wye – Hadnock Quarry Rising (SO 5415 1540) (Fig. 10.3). The resurgence lies in the basal part of the Lower Limestone Shales, the sequence dipping eastwards at about 30° towards the axis of the Worcester Syncline (Fig. 10.1). No sinks have been positively tested to the Rising, but the drainage is assumed to derive from extensive limestone outcrops between the Wye and Staunton. Several streams run onto the limestone from the underlying sandstones and sink at various points according to flow-stage. Many depressions are also present in the area, but dense afforestation has thwarted any detailed systematic exploration.

West of Coleford (Fig. 10.1) is a wide outcrop of Lower Dolomite. Little speleological work has been carried out here since there are few signs of karst development at the surface. The presence of caves, possibly of great age, in the area has been shown by the location of truncated fragments such as Whitecliff Hole (SO 575 105), a small abandoned vadose streamway exhibiting a degree of joint-guidance. It seems fairly certain that in the Coleford area, and possibly in other parts of the Forest, important speleological sites have been lost to the march of civilization. A major stream flowing along the Whitecliff Valley, which once probably sank on meeting the limestone outcrop, is now culverted and the area covered by more than 5 m of fill.

The Drybrook Limestone is isolated from the main limestone sequence by the intervening Lower Drybrook Sandstone. In the north it is thin or absent (Fig. 10.2) and the first notable cave to be found is Cursits Cave (SO 653 111) near Ruspidge, 45 m long, 18 m deep and notorious for the extreme awkwardness of its entrance rift. Close by is Fester Dig (SO 652 111) which was abandoned when a connection with Cursits seemed inevitable. Milkwall Quarry Cave (SO 584 088) is a fragment of an ancient phreatic system with much infill of clay and calcite.

Southwards the Drybrook Limestone thickens appreciably at the expense of associated sandstones. Little work has yet been carried out along its outcrop, although small caves have been reported in cliffs above the Wye at Wintour's Leap (ST 542 964) and Ban y Gor Rocks (ST 545 971). Below Ban y Gor Rocks is a big resurgence known locally as the Great Well (ST 547 974) and potential for a large active system behind the rising is high, although the dip is away from the river and the active system might be flooded. Digging at a higher flood rising has revealed more than 400 m of tight passage above the active route, ending in a sump.

'Fossil' caves

Many abandoned caves unrelated to the present topography and active systems are found in the cliffs flanking the Wye from the Coldwell–Symonds Yat area as far south as Chepstow. Some were archaeologically excavated early this century, most of the effort being on the west bank of the river. The caves are of natural origin, although some have been modified by miners in search of iron-ore. Though all are blocked by clastic fill or ancient flowstone, draughts suggest that open passages exist beyond.

Lady Park Wood Cave (SO 5468 1465) (Fig. 10.3), southwest of Biblins, is over 300 m long, partly natural and partly mined, and was operated as a show cave from 1880 to 1910. The natural sections are essentially phreatic and complex, with passages blocked by mud, silt or mine waste. Scattered deposits of reniform and stalactitic hematite occur in both the mined and natural sections. The surface of the area above is pock-marked by depressions, some of which are obviously man-made.

Further north, in the cliffs below Symonds Yat, are many large cave entrances. Wye Rapids Cave, formerly designated C3 (SO 5611 1552) (Fig. 10.3), can be followed through several excavated sections for at least 200 m, with a strong draught throughout. At its furthest point the cave is trending towards the area of Symonds Yat Swallet. Other caves along the cliff,

designated C1 to C12, are of similar type, and all show signs of mining activity.

The absolute age of these fossil caves is unknown. Their occurrence, truncated and hanging above the Wye, suggests that they pre-date the downcutting of the river following post-Triassic, possibly mid-Tertiary, uplift. An even greater age for their inception can be postulated, however, on consideration that some of the passages might have been filled or partly filled by iron-ore. If this ore was contemporaneous in origin with the main Forest of Dean ore deposits, emplacement during the Triassic is a possibility. Further south rocks of Namurian (Millstone Grit) age, as well as Triassic rocks have been recognized, filling karst cavities in the Carboniferous Limestones, adding weight to the idea of a period of pre-Triassic or early Triassic cave formation in the Forest of Dean.

Those fossil caves are located at or close to the base of the Crease Limestone. Another series of abandoned caves is present at a higher level, around Coldwell Rocks near Symonds Yat in the north and at Wintour's Leap in the south. They probably lie at the top of the Crease Limestone, or even higher. The relationship of these passage fragments to those at lower levels, if indeed they are related, is obscure, and it seems that new long-ranging absolute dating methods must be developed before the early history of cave formation in the Forest of Dean can be elucidated.

The iron-ore mines

Reference has already been made to the iron-ore mines of the Forest. These are well known to cavers and industrial archaeologists alike. Iron-ore mining was certainly in progress during the Roman occupation and was carried on in much the same way, by excavation from outcrop, until the nineteenth century, when deep mining via shafts and levels commenced. This continued until about 1930 and there was a small revival during World War II. No ore is raised commercially today, but small quantities are mined at Clearwell for educational and experimental purposes. Some details of the history are documented by Hart (1953, 1971).

The ore is a mixture of hydrated ferric oxides (goethite and lepidocrocite) with anhydrous ferric oxide (hematite). Its origin and mode of emplacement are controversial, but in the light of modern research here (Lowe, 1974; Solari, 1974) and at Llanharry in South Wales (Gayer & Criddle, 1970), Triassic emplacement seems most likely. Parts of the mines have a natural appearance and this simple observation is crucial to the overall concept of ore genesis.

Conclusions reached by Gayer & Criddle (1970) can largely be applied in the Forest, while the observations made locally by Lowe and Solari remain valid. Gayer and Criddle suggested an elegant theory whereby acidic iron-bearing solutions corroded cavities in the limestone and in so doing were neutralized, causing precipitation of iron compounds which filled the newly formed voids. Peripheral to this cavity filling was a process of alteration or hematitization of the wall rocks. In the Forest, in those places where ore is still visible, there is generally a marked boundary between granular ore, which presumably came out of solution into a cavity, and hematitized bedrock, while the zone of ferrification is almost always thin

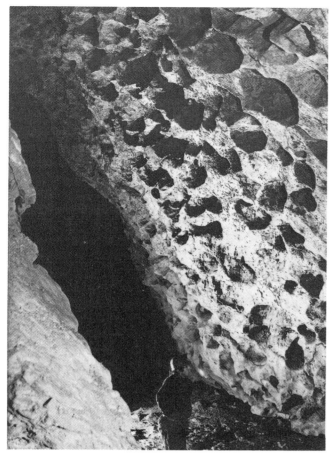

Fig. 10.8. Honeycomb Rift in Westbury Brook Iron Mine; a steeply dipping natural rift along bedding exhibiting intense solution features in the base of the Whitehead Limestone (photo by D. M. Judson).

Fig. 10.9. The floor of Honeycomb Rift, Westbury Brook Iron Mine, showing desiccated iron-rich mud, thought to mark the limit of inundation by the iron-rich solutions responsible for ore deposition (photo by D. M. Judson).

and transitional into a dolomitized zone away from the cavity. This observation implies that the iron-rich solutions were transmitted through the limestone along pre-existing conduits and did not penetrate the entire mass in any 'saturation' or 'water table' sense. Indeed, the ore is known to fail at depth and it seems possible that this failure is somehow associated with a different chemical environment in a zone of complete saturation. The presence of ancient conduits is difficult to prove, but there are several features which could suggest pre-hematitization cavities.

Some of the passages encountered in the mines were void of ore and showed no sign of iron-staining. If the Gayer and Criddle mechanism is correct, this should not be the case unless the cavities pre-dated iron solution inundation. The top part of Honeycomb Rift in Westbury Brook Iron Mine (SO 6626 1648) (Fig. 10.8) is a fine example with a 'tide mark' of iron-rich mud deposits marking the upper limit of inundation (Fig. 10.9). Less spectacular examples occur in Wigpool Iron Mine (SO 65 19) and Buckshaft Scowles (SO 65 11). One passage in the Columbus Pit area of Wigpool Iron Mine possesses an active stream and there is no sign of ore having been present or removed in the area. In Buckshaft Scowles there are many solutional rifts with smoothly polished phreatic surfaces showing no sign of iron-staining. Locally these rifts

are endowed with a heavy and apparently very old layer of flowstone, which might pre- or post-date the ore deposits. Either possibility implies a cavity that was devoid of ore and hence not formed at the time of hematitization.

Significantly the more intricate and natural-appearing mines lie close to the base of the Crease Limestone or at the Whitehead Limestone–Crease Limestone boundary, the preferred geological level for the fossil caves previously described. To the caver the question of whether the mine passages were present before ore deposition is of academic interest. What is important is that a vast network of essentially natural cavities remains. If it were possible to connect the various workings, many of which intersected during the mining period, very long systems indeed would result. Two areas in particular offer link-ups of this type, near Clearwell (Fig. 10.1) on the Western Outcrop, where several mines are already connected, and a long stretch of the Eastern Limestone Outcrop southwards from Wigpool.

Whereas the old miners concentrated on emptying the cavities by following the natural line of the infilled cavities, the miners of the nineteenth century, with power tools and explosives, were able to sink deep shafts and levels for haulage and drainage. Generally these man-made passages are blocked, filled-in or below water level today, but those which remain open allow access to fragments of natural cave which would otherwise be unknown, including some parts of the Buckshaft and Westbury Brook systems. It is, however, the shallower workings, usually entered from crop, which present the most fascinating insight into the ancient systems of the Forest, being characterized by three-dimensional phreatic networks of rifts and tubes connecting large solutional chambers, locally termed churns. The discovery of similar chambers in active systems, Slaughter House in Symonds Yat Swallet and Hyperspace in Cross Joints Swallet, was one of the linking factors in constructing the theory of ore deposition outlined above. The obvious corollary is that active systems similar to the passages preserved in the mines, but free of iron-ore infill, might be present in the Crease Limestone of the Forest of Dean.

Conclusion

Most of the active caves described are relatively small and immature, except where major solutional chambers in the Crease Limestone are intersected and in the Lower Dolomite on the edge of the Slaughter drainage area. The major passages in Otter Hole (Chapter 11) on the Welsh side of the Wye are also formed in the Lower Dolomite, illustrating that the formation is potentially cavernous. In the Forest of Dean similar conditions of geology and hydrology suggest that comparable development might be present behind the Slaughter Rising.

Acknowledgements

Many cavers have contributed to the exploration of the area as a whole and to the individual caves and drainage systems described. The publications of the local caving clubs have been an invaluable source of information. The unpublished results of work by the Cave Projects Group are also incorporated. In particular the wide-ranging work of the late Roger Solari has greatly influenced the direction of speleological research in the Forest. Thanks to Ian Standing of Coleford for suggesting additions to update this chapter.

References

Cullingford, C. H. D. (1951). The limestone area of the Forest of Dean. *British Caver*, **22**, 48–55.

Dixon, E. E. L. (1921). The unconformity between the Millstone Grit and the Carboniferous Limestone at Ifton, Mon. *Geological Magazine*, **58**, 157–64.

Gayer, R. A. & Criddle, A. J. (1970). Mineralogy and genesis of the Llanharry iron ore deposits, Glamorgan. *Ninth Commonwealth Mining and Metallurgical Congress 1969, Mining and Petroleum Geology Section*, **Paper 15**, 605–26.

George, T. N., Johnson, G. A. L., Mitchell, M., Prentice, J. E., Ramsbottom, W. H. C., Sevastopulo, G. D. & Wilson, R. B. (1976). A correlation of Dinantian rocks in the British Isles. Geological Society of London, Special Report No. 7, 87 pp.

Gloucester Speleological Society Journals, 1960–6.

Hart, C. E. (1953). *The Free Miners*. Gloucester: British Publishing Co.

Hart, C. E. (1971). *The Industrial History of Dean. With an Introduction to its Industrial Archaeology*. Newton Abbot: David & Charles.

Hereford Caving Club Newsletters, 1956–60.

Lowe, D. J. (1974). The geology of the Forest of Dean limestones. *Cave Projects Group Newsletter*, **5**, 73–9.

Lowe, D. J. (1981). Cross Joints Swallet. *Caves and Caving*, **14**, 4–5.

Ramsbottom, W. H. C. (1973). Transgressions and regressions in the Dinantian: a new synthesis of Dinantian stratigraphy. *Proceedings of the Yorkshire Geological Society*, **39**, 567–607.

Richardson, L. (1924). Swallow holes in the Forest of Dean, Gloucestershire. *Proceedings of the Cotteswold Naturalists Field Club*, **22(1)**, 29–31.

Richardson, L. (1930). *Wells and Springs of Gloucestershire*. Memoirs of the Geological Survey of Great Britain. HMSO.

Royal Forest of Dean Caving Club Newsletters, 1964–.

Solari, R. A. (1971). Hydrology of Bicknor Parish/Grove Wood Rising. *Cave Projects Group Newsletter*, **4**, 15–22.

Solari, R. A. (1974). Hydrology of the Slaughter Rising. *Cave Projects Group Newsletter*, **5**, 54–65.

Standing, I. J. (1967). Speleology in Gloucestershire. *Cave Research Group Newsletter*, **108**, 4–6.

Symonds, W. S. (1872). *Records of the Rocks*. London.

Welch, F. B. A. & Trotter, F. M. (1961). *Geology of the Country around Monmouth and Chepstow*. Memoirs of the Geological Survey of Great Britain. HMSO.

11

Otter Hole, Chepstow

CLIVE D. WESTLAKE, JOHN V. ELLIOTT and MARK E. TRINGHAM

The discovery of Otter Hole between 1973 and 1976 and the investigations that have followed, prove the existence of a previously unsuspected extensive and well-developed karst region between the Forest of Dean and the main limestone outcrops rimming the Coalfield. The cave potential here is by no means fully realized. Apart from being a sizeable cave many kilometres from any comparable system, Otter Hole is particularly notable for its tidal resurgence and its magnificent stalagmite formations (Elliott, Westlake & Tringham, 1979).

The northern limit of the Otter Hole catchment is clearly defined, with the Carboniferous Limestone faulted down against Old Red Sandstone for several kilometres west of the River Wye. Streams flowing southwards off the Old Red Sandstone from Gaer Hill, Chepstow Park Wood and Itton are engulfed by swallets along this boundary at elevations between 110 m and 130 m O.D. (Fig. 11.1). Although the limestone extends far to the southwest towards Llanfair Discoed and Penhow, the rocks there generally dip southwards carrying the sub-surface drainage away from Otter Hole. Likewise the southern limits of the area are not clearly defined but are probably to the north of Mounton Brook. The deep gorge of the Wye marks the eastern edge of the catchment.

There are many notable landforms, karstic and otherwise. The Wye Gorge is in places more than 100 m deep with impressive limestone scars such as the Wynd Cliff of Lower Dolomite (Courceyan) and the Piercefield Cliffs of Crease (Chadian) and Whitehead (Arundian) Limestones. The Lancaut and Piercefield meanders are deeply incised while the tidal reaches of the Wye exhibit extensive muddy alluvium at low water. The land surface above Otter Hole and its assumed continuation westwards differs from most other Welsh karst in its wooded pastoral aspect, although there are dry valleys, such as those running south from Rogerstone Grange, Howick and Llanquilan (Fig. 11.1). There are some shallow dolines and most of the swallets are situated in pronounced depressions. The Mounton Brook is notable for disappearing into its bed at various points dependent upon flow.

The main exploration of Otter Hole began in 1974 and by September of that year 270 m of passage had been found, leading to a lake which was below the level of the Wye and was only passable for seven hours out of every twelve. In bad weather the sump did not open at all because of flooding from within the cave. A streamway was explored beyond for nearly 400 m and divers passed three short but constricted sumps to explore 120 m of large stream passage before another sump. A high-level choke yielded to two months of digging in December 1975 and in the next few weeks 2000 m of the most spectacular passages were explored. Calcite formations were so profuse that they seriously hampered progress.

Otter Hole has provided the key to understanding the sub-surface drainage pattern north of Chepstow, although there remain many years' work to solve the details, with the likelihood of a lot more cave to be found.

Description of the cave

From its entrance upstream of the resurgence a few metres above the high water mark Otter Hole begins with constricted muddy bedding-plane passages (Fig. 11.2). The first 70 m of cave descend in a southerly direction to a region which is at the level of the highest tides and may become impassable, especially during the winter. From this point the trend is generally westwards in somewhat roomier passages, which are initially joint- and later fault-controlled. The mud becomes increasingly pervasive until the spacious chamber housing the Tidal Sump One is reached some 300 m from the entrance (Fig. 11.3). Beyond the sump a fixed ladder is ascended, the First Boulder Choke negotiated and the streamway reached. This is pursued in a northwesterly direction in a fault-determined passage up to 2 m wide and 15 m high, punctuated by various boulder obstacles and larger rift chambers. Upstream of the First Boulder Choke (300 m) is the Second Choke, the far side of which is descended by a 7 m pitch. A westward-trending passage of 100 m leads to Sump Two. Although this is <1 m long it is soon followed by further sumps of 3 and 14 m. After 160 m of large streamway the 30 m Sump Five is passed, only to be succeeded by Sump Six which is 40 m long, followed by 55 m of canal ending in Sump Seven, which has been penetrated for 30 m.

The route to the upper passages of the Extension starts 50 m before Sump Two. A confusing series of crawls, climbs and

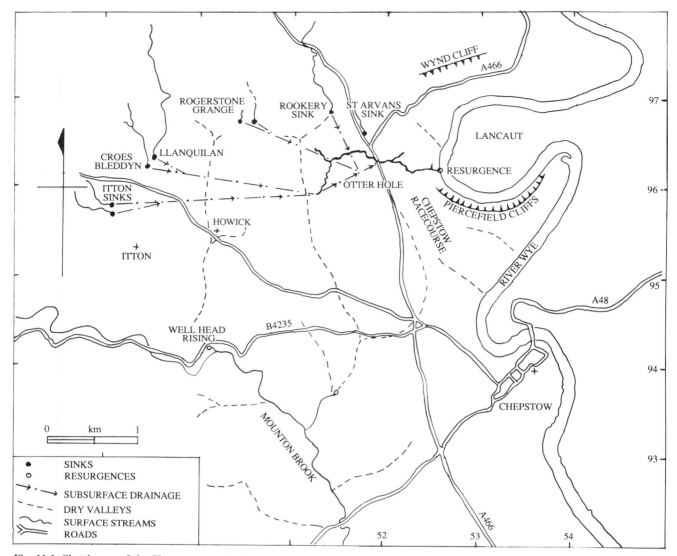

Fig. 11.1. Sketch map of the Chepstow area showing the topographic and hydrological relationships of Otter Hole.

squeezes lead up to more comfortable passages approximately 4 m wide by 2 m high at a level some 20 m above the stream. The west-trending passage becomes larger and begins to be filled with speleothems of a profusion and magnificence previously unimagined in the British Isles. A further 250 m into the Extension, the Hall of the Thirty is encountered (Fig. 11.4). The sloping floor of this chamber is covered with massive stalagmites up to 4 m high and the flat roof is resplendent with huge stalactites.

The next 100 m are notable for the variety and multicoloured quality of the speleothems, and lead to Camp One where a high aven provides one of the few sources of water in the dry passages of the Extension. Shortly beyond Camp One the fault-determined passage turns northwest and few speleothems are encountered for the next 200 m. When the passage turns WSW, again on well-defined faults, it is up to 20 m high and contains spectacular stalactites, notably in the breath-taking Long Straw Chamber (Fig. 11.5). Bedding control predominates in the next section where there are superb stalagmites. At Tunnels Junction the main route is Tunnels

Left, a southwest-trending bedding passage some 500 m long where phreatic features are encountered (Fig. 11.6). Towards the end sand-floored crawls linking high cross rifts descend to the Far Streamway which sumps downstream. To the south is a boulder choke with a sizeable rift chamber above, whilst the source of the stream is a flooded bedding plane not far upstream. A rift passage can be followed westwards to a sump which discharges only a trickle where divers have entered 100 m of large underwater passage. The boulder choke has been recently passed to 90 m of restricted and flood-prone passage.

The main side passages are Crystal Ball Passage and Tunnels Right. The former is reached through a boulder choke at the beginning of the Extension and, although the first part of this northeast-trending, fault-controlled passage is unusual in being undecorated, the far end is resplendent with curtains, flowstone and crystal formations. Tunnels Right is 150 m long, trending northwest again with magnificent stalactites. The present length of Otter Hole approaches 3500 m.

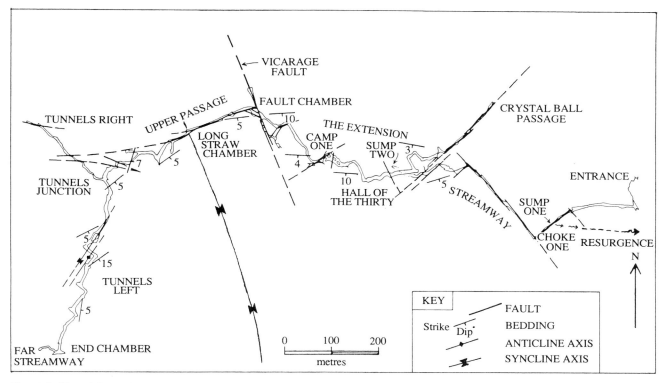

Fig. 11.2. Plan of Otter Hole showing principal geological features
in the cave.

Fig. 11.3. The Tidal Sump in Otter Hole at low water: at high tide
both the arch behind the caver and the eye-hole above it are
submerged.

Fig. 11.4. Hall of the Thirty, Otter Hole.

Fig. 11.5. Long Straw Chamber, Otter Hole.

Fig. 11.6. The phreatic tube of Tunnels Left, Otter Hole.

Regional hydrology

The hydrology of Otter Hole is best considered in relation to the drainage of the whole Carboniferous Limestone outcrop (Fig. 11.7). Along the 8 km junction between the underlying Devonian Old Red Sandstone and the limestones,

a number of sinks are encountered. From east to west the more important are St Arvans Sink (The Rookery), then the two Rogerstone Grange Sinks, Llanquilan Quarry Sink, Croes Bleddyn Sink, North and South Itton Sinks, Cas Troggy, Wentwood Reservoir Sink and Whitebrook Sink.

Before the discovery of Otter Hole, Cas Troggy, which lies about 6 km WSW of the cave, had been proved to flow to the Severn Tunnel Great Spring (Drew, Newson & Smith, 1970). On this evidence and that of discharge measurements it was postulated that the Great Spring drained the 20 km^2 west of the Mounton Brook, a hypothesis supported by recent investigation. That work coincided with conditions of low stage perhaps resulting in an underestimate of the significance of some of the swallets. In particular the suggestion that Mounton Brook Rising has its origins in sinks further upstream is somewhat misleading when the sites are examined in wetter conditions. Although it is possible that some of the large volume of water

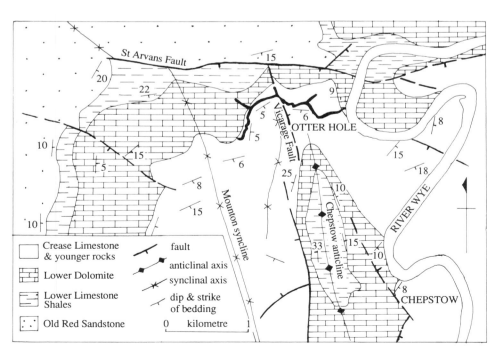

Fig. 11.7. Sketch map of the geological relationship of Otter Hole.

sinking in the bed of Mounton Brook may resurge at the comparatively small rising, the majority must flow elsewhere, probably to the Severn Tunnel Great Spring.

Within Otter Hole there are two sizeable streams. The Main Stream flows from Sump Seven to the Tidal Sump One and on to the resurgence. When the cave is accessible the estimated discharge is in the order of 100–200 l s^{-1}. The Far Streamway at the end of Tunnels Left flows at about half this volume. It is clear that winter discharges are significantly higher. Dye-testing has established that the Far Streamway joins the Main Stream at some point upstream of Sump Seven. The remaining half of the Main Stream is largely derived from St Arvans and Rogerstone Grange Sinks, tests proving positive at Sump Two but negative in the Far Streamway. The origins of the latter are further to the west, tests from Croes Bleddyn and Itton North both proving positive. The catchment areas of these, and other immediately adjacent untested sinks, such as Llanquilan Quarry Sink, total about 8 km^2 which is appropriate for a resurgence the size of Otter Hole's. Thus it is unlikely that the other sinks further west drain in this direction, but are more likely to join Cas Troggy in resurging at the Severn Tunnel Great Spring.

Otter Hole has already gained notoriety for its unique tidal sump – Sump One. In the region of Aust and Chepstow the River Severn has a tidal range of about 14 m at spring tides and about 7 m at neap tides. Although Otter Hole is some 8 km up the Wye from its confluence with the Severn, there is still a considerable tidal range. While the first 200 m of the cave are above the level of all but the very highest tides, Sump One is situated between the low and high tide marks. The hydrology is also complicated by the flows of the Main Stream and the River Wye which are largely, but not entirely, seasonal. During the summer months the sump is usually passable for three hours before and three hours after low water, when the Wye is below the level of the floor. The Main Stream sinks here to reappear at the resurgence on the river bank between high and low water marks. At mid-flood tide the sump can be observed filling up, and the through-route becomes impassable in about 30 minutes. Conversely, at mid-ebb tide the sump drains out. For about three hours before and three hours after high water the 10 m long sump remains submerged and impassable. During the winter months these fluctuations continue but the sump is not likely to open. Whether this is a result of the higher stage of the Main Stream or of the River Wye or both is unknown. There is doubtless a density differential between the fresh-water Main Stream and the brackish estuarine waters of the Wye which further affects the behaviour of the sump.

Geology

Around Chepstow the Carboniferous Limestone is represented by a number of limestone and sandstone formations, which can be correlated to those of the Forest of Dean with only minor thickness changes (Fig. 11.8). However, in contrast to the Forest of Dean, of the five limestone formations present, only one, the Lower Dolomite, is known to contain a significant cave.

The Lower Dolomite comprises a 110 m thick monotonous series of well-bedded grey, sparry dolomites and dolomitic

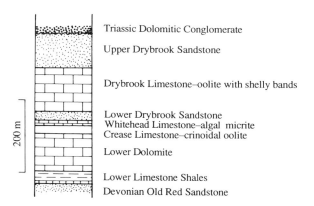

Fig. 11.8. Stratigraphic sequence in the Otter Hole area.

limestones, in which the original sedimentary features such as ooliths and fossils have mostly been obliterated by the dolomitization. Bedding surfaces are commonly undulating or stylolitic, with thin partings of red or green clay. Around the edge of the Carboniferous outcrop, surface streams flow off the surrounding Old Red Sandstone and sink into a number of swallow holes in the basal limestone of the Lower Limestone Shales, here separated from the Lower Dolomite by 30 m of shale. The outcrop pattern of the Carboniferous Limestone is affected by folds and faults of Hercynian age, with the main fold axes NNW–SSE, perpendicular to the regional trend. The folds generally plunge south and are asymmetric with steeper westerly dips (Fig. 11.7). The major faults include one set of east–west trend, such as the St Arvans Fault which bounds the Carboniferous outcrop north of Chepstow; another set trends NNW–SSE, parallel to the folds.

The present landscape and drainage has evolved by progressive river downcutting during falls in sea level up to the end of the Devensian. On the Wye, pauses in the river downcutting are represented by four terrace levels between heights of 3 m and 60 m O.D., of which only the second terrace is recorded in the Chepstow area, at 25–30 m O.D. A river channel buried beneath estuarine alluvium at Chepstow records a subsequent rise in sea level of at least 15 m (Welch & Trotter, 1961).

Otter Hole lies within the Lower Dolomite mostly beneath a capping of Crease Limestone at a depth of about 80 m beneath the surface. The cave crosses a branch of the gentle Mounton Syncline and skirts the northern end of the laterally impersistent Chepstow Anticline. The Vicarage Fault lies close to the Chepstow Anticline axis, and is a wrench with an apparent small downthrow to the east; it is seen in the cave at Fault Chamber (Fig. 11.2). The cave lies approximately beneath the outcrop of the boundary between the Crease Limestone and the Lower Dolomite, the main high-level passage following a single stratigraphic horizon within the Lower Dolomite. In many parts of the cave there is a close relation between the location and type of cave passage and particular geological structures (Fig. 11.9). Overall, the nearly horizontal high-level cave follows the curving strike of the Dolomite, but Tunnels Left is located along a zone of jointing on two minor fold axes (Elliott et al., 1979).

A number of passages follow faults, in many cases for distances exceeding 200 m. The faults are all either vertical or

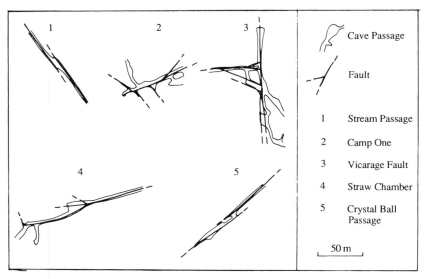

	Cave Passage
	Fault
1	Stream Passage
2	Camp One
3	Vicarage Fault
4	Straw Chamber
5	Crystal Ball Passage
	50 m

Fig. 11.9. Detailed relationship of some Otter Hole passages to faults.

steeply inclined, and in some cases zones of fractures occur rather than single breaks; the Vicarage Fault is one such zone on which a complex of rift passages has formed. The faults contain spectacular breccias up to 5 m or more in thickness, where crushed fragments of the Lower Dolomite are cemented in calcite, dolomite and clay. The adjacent rock is in places strongly veined by calcite, eroded into a distinctive boxwork surface. Some fault surfaces bear horizontal slickensides, indicating wrench-type displacements.

The inherent mechanical weakness of the breccias has in places caused roof collapse with large boulder piles. Where two or more faults meet this collapse is particularly evident and near Camp One an aven which marks the highest point in the cave has formed at such a position. Where faults are traced along their length they are frequently seen to branch, or to form an en echelon arrangement: in many cases this has yielded slightly offset cave passages which are joined by chambers. Water percolating down through the fault zones has in many places produced spectacular red and black formations of iron-rich calcite.

Speleomorphology

Otter Hole falls into two main parts; a Lower active streamway, and an Upper Passage which is mostly dry. The Upper Passage forms the larger part of the cave and consists of one main route, and one tributary, Tunnels Right. It is nearly horizontal, and most of the passages have been extensively modified by collapse of walls and roof and by speleothem infilling, so that the original passage morphology is infrequently seen. Passage cross-sections, where preserved without breakdown, as at Tunnels Junction, are either circular or elliptical and in places large roof domes and rifts occur on faults and joints. Where present, scallops all indicate an original water flow direction from the end of the cave towards Crystal Ball Passage. This phreatic cave, with its contributory Tunnels Right, acted as a route for water draining towards the Wye, when the water table was well above its present level. In

Tunnels Left flood markings of red clay coat the walls and formations up to a height of about 33 m O.D., indicating a relatively recent flooding which would have drained downhill towards the end of the cave and then along the present stream route.

The Lower active stream passage is presently partly phreatic and partly vadose and presents the unique feature of also being partly intertidal. The stream flows from a height of 21 m O.D. at Sump Three down to the intertidal parts of the cave at Sump One. The stream passage, for the most part, consists of a straight rift which follows a major fault. Enlargement up joints and small faults indicates that part of the passage development could have been phreatic. Upstream from the rift passage is a short section of anastomosing bedding cave. A similar section of bedding passages has developed at the entrance of the cave, and both these bedding cave-networks are largely of phreatic origin.

Geomorphological evolution

Otter Hole records phases in its development which correlate with the evolution of the River Wye. The first phase is that of a phreatic drainage route whose height correlates with the second terrace of the Wye at 25–30 m O.D. The Upper Passage probably formed just below a water table which sloped down towards the Wye. Enhanced solution at this level formed roof domes and rifts by water mixing. Doubts exist concerning the age of the Upper Passage, because the timing of river downcutting and terrace formation is itself in doubt. Devensian, Ipswichian or earlier ages are possible, and, as pointed out by Smart & Statham (1984), speleothem isotope dating would be of particular use here.

The second phase in the cave development was initiated by capture of the stream flow to its present lower route. The capture may have first occurred near Crystal Ball Passage, but now occurs further upstream. The streamway has developed under both vadose and phreatic conditions, during progressive downcutting of the Wye to below its present level. A third phase in the cave development is represented by the inundation of the lowest parts by tidal water and alluvial mud. This has resulted from the approximately 20 m rise of the Wye which accompanied the widespread rise in sea level during the Flandrian.

A number of geological controls have been important in locating the cave. Firstly, the cave as a whole has formed at a particular stratigraphic position in the folded Lower Dolomite, and secondly, individual passages have been located along structural features such as faults and joints. Faults are of the greatest importance, forming lines of high permeability and mechanical weakness, while joints related to minor folding are of lesser importance.

References

Drew, D. P., Newson, M. D. & Smith, D. I. (1970). Water-tracing of the Severn Tunnel Great Spring. *Proceedings of the University of Bristol Spelaeological Society*, **12(2)**, 203–12.

Elliott, J. V., Westlake, C. D. & Tringham, M. E. (1979). Otter Hole, near Chepstow, Wales. *Transactions of the British Cave Research Association*, **6(4)**, 143–58.

Smart, P. L. & Statham, I. (1984). A note upon a fossil cave feature in Chepstow. *Proceedings of the University of Bristol Spelaeological Society*, **17**, 71–80.

Welch, F. B. & Trotter, F. M. (1961). *Geology of the Country around Monmouth and Chepstow*. Memoir of the Geological Survey of Great Britain. HMSO.

12

The Mynydd Llangattwg Cave Systems

PETER L. SMART and CLIVE G. GARDENER

The Llangattwg area, with over 62 km of known cave (Fig. 12.1), rates as one of Britain's major caving areas. Furthermore, with the recent spectacular extensions to Daren Cilau and water-tracing evidence for substantial as yet unknown systems in the Dowlais Limestone, the prospects for further major finds are excellent. Indeed, this text may well prove incomplete on publication. In writing this chapter we hope to stimulate both this exploration and further scientific study.

The Mynydd Llangattwg Cave System currently comprises three major caves: (Ogof) Agen Allwedd, Ogof Daren Cilau and Ogof Craig a Ffynnon, although many smaller caves such as Eglwys Faen and Ogof Pen Eryr are also known (Table 12.1). These caves are developed under Mynydd Llangattwg, a gently sloping peat-covered plateau, which falls from an elevation >520 m O.D. in the north to <400 m O.D. some 3.5 km to the south. To the north and east the plateau is bounded by steep screes and extensively quarried crags developed in the limestone (Fig. 12.2), which give way below to more gently pastured slopes on the Old Red Sandstone of the Usk Valley. The Usk, a major river with an upstream catchment >500 km², meanders gently, falling from 100 m O.D. in the northwest at Bwlch to <60 m O.D. in the south. At Gilwern, the River Clydach enters from the west through a steep rocky gorge incised into the plateau margin and leading the Heads of the Valleys road up to Brynmawr. In the floor of this gorge the three major resurgences of the Llangattwg Cave System are found: Ffynnon Gisfaen (277 m O.D.), rising at the top of the limestone; Pwll y Cwm (207 m O.D.), the resurgence for Agen Allwedd and Daren Cilau, rising in a deep pool near the base of the limestone; and Ogof Capel (220 m O.D.) issuing from bedding planes in the walls of the gorge downstream from Pwll y Cwm. A further large resurgence is known about 11 km to the southeast at Pontnewynydd, where the limestone is exposed in the valley floor, but this has not been traced and its relation to the Llangattwg caves remains unknown. However, a borehole drilled at Pen-ffordd-goch (SO 2553 1058), near Keeper's Pond and not far from the summit of the Blorenge, intersected a 4 m void with 2 m of sand on the floor at a depth of 120 m.

Only one major stream sink, Llangattwg Swallet (Fig. 12.3), is known, although Crochan Sion Hopkin, Pwll Coedog and the Cascade Stream Sink all take a considerable amount of water in wet conditions. The remaining water from the plateau either drains into numerous small sinkholes, cascades down the crags to the north into the Cwm Onneu-fâch or flows southwards across the Coal Measures to feed the reservoirs above Brynmawr. Water-tracing by Gascoine and others (see Chapter 4) has demonstrated that the underground catchment extends under Mynydd Llangynidr, the westward continuation of Mynydd Llangattwg, to Ogof Cynnes 7.25 km northwest of the resurgence at Ffynnon Gisfaen. It is clear from the tracing that the feeders of Ffynnon Gisfaen form a separate system to the known caves. A large phreatic conduit intersected in the Carno Adit, a tunnel excavated through the limestone of Llangynidr for water supply, may eventually provide access to this system.

In October 1959, the Craig y Cilau National Nature Reserve was established to protect the alpine flora of the limestone crags. Over 250 species are known, including the rare least whitebeam (*Sorbus minima*) (first identified by Reverend Augustine Ley in 1895), several lesser known hawkweeds, alpine enchanter's nightshade (*Circae alpina*) and angular Solomon's seal (*Polygonatum odoratum*). In 1986 the major caves of Llangattwg were incorporated into a much larger Site of Special Scientific Interest. Agen Allwedd is now probably the largest hibernation site of the lesser horseshoe bat (*Rhinolophus hipposideros*) in Britain and these are being studied by Sean Heaver of Chelsea Speleological Society (Heaver, 1985). Small numbers of Daubenton's bat (*Myotis daubentoni*), long-eared bat (*Plecotus auritus*) and Natterer's bat (*Myotis nattereri*) are also found and the colony appears to be thriving. Access and conservation are currently controlled by the Mynydd Llangattwg Cave Management Committee and similar protection is planned to extend to the other Llangattwg caves.

The Llangattwg caves

Agen Allwedd

Agen Allwedd was, for a considerable period, all that was known of the Llangattwg System. Entered from a small excavated entrance at the base of the crag, the restricted entrance series fed by several percolation streams leads to a boulder-choke connection with a larger dry fossil passage – the 1.2 km long Main Passage (Fig. 12.4). This is a portion of a much larger complex of high-level passages now extensively modified by breakdown and choked by sandy sediments. A short way along Main Passage a substantial streamway (Main Stream Passage) (Fig. 12.5) can be followed for 1.0 km via a second boulder choke to the Northwest Junction where water from the entrance series, Meander Passage, Flood Passage, Keyhole Inlet and other small percolation inlets meets a second major stream passage, Turkey Streamway. Upstream this leads to the high-level passages of Summertime, where several circular tours are possible. Turkey Streamway itself leads to inlets from the Remembrance Series, accessible only to divers. Downstream, Main Stream Passage can be followed through three of the five major boulder chokes in the cave (Fig. 12.6), the last being bypassed by a high-level route through Biza Passage to Lower Main Stream Passage (Fig. 12.7), the Terminal Sump and Maytime Series. An important but irritatingly small tributary stream (Southern Stream Passage) leads back to the central section of the Main Passage, close to the Main Passage termination near the Cliffs of Dover (Fig. 12.8). Trident Passage offers the strongest possibility for making the elusive link east to Ogof Daren Cilau. Recent discoveries leading to the continuation of High Traverse Passage via the southern end of Gothic Passage may well provide a dry bypass to the downstream sumps and similar good potential for an eventual connection with Ogof Daren Cilau. In total the Turkey and Main Streamways are over 6 km long.

Eglwys Faen

About 0.5 km east of the Agen Allwedd entrance lies Eglwys Faen, an enigmatic system with an initial passage of considerable size but little length despite the determined attention of several generations of cave diggers. A line of boulder chokes met at the end of the Inner Chamber, beyond the Seven Tunnel in the St Patrick's Series and at the present end of the Hereford Dig in the high-level Warren Series, almost certainly follows a fault. This has yet to be breached to realize the substantial potential for large well-developed cave passage beyond. The boulder choke on the far side of Glump Sump in the Western Series may well play an important part in achieving this highly sought after breakthrough.

Ogof Craig a Ffynnon

More success was attained at Pope's Hole, a small draughting entrance above the Rock and Fountain public house, which eventually gave its name to Ogof Craig a Ffynnon. This interesting and, in places, beautifully decorated system (Figs. 12.9, 12.10) also features five major boulder chokes which presented significant barriers to exploration. Exceptionally linear passages of moderate dimensions interconnect two substantial but unfortunately isolated segments of major trunk passage. An understorey of active streamways is also present (Fig. 12.11). After an initial period of rapid exploration further finds in this system have proved elusive, despite several years of persistent effort.

Ogof Daren Cilau

Commonly known to cavers simply as Daren Cilau, this cave system was for long the *bête noire* of Llangattwg, infrequently visited owing to its tight and awkward entrance series. However, tantalizing possibilities were suggested by the size of the Old Main Chamber, and the breakthrough was made through a maze of boulders to Jigsaw Passage. This leads into over 1 km of substantial fossil trunk passage, Epocalypse Way, whose grandeur is well matched by the delicacy of the aragonite formations. A major breakdown passage entering a short distance prior to the final chokes, Antler Passage, commences with antler-like helictites (Fig. 12.32) and terminates at a boulder choke, which, when passed, may well provide an alternative entrance in the Daren Cilau Quarry or a possible connection with Ogof Pen Eryr. Epocalypse Way terminates in three major chokes, which will require considerable ingenuity to pass, yet probably offer most hope of an extension towards Ogof Craig a Ffynnon.

At the upstream end of Epocalypse Way the small, high-level Eglwys Passage leads via an area of extensive collapse, in St Valentine's Chamber, to the continuation of the Jigsaw Passage stream in Preliminary Passage. Skilful climbing was necessary to enter the high-level routes leading to the downstream continuation (Fig. 12.12), White Passage, and the Time Machine, the largest section of cave passage yet encountered in Britain (Fig. 12.13). Downstream the high-level routes separate from the active streamway and there is considerable potential for substantial extensions towards the Clydach Gorge. The Bonsai Streamway can be followed to the Terminal Sump and St David's Streamway. The Terminal Sump is 625 m from the Pwll y Cwm resurgence (Fig. 12.14) and has been linked by diving to the constricted overflow sump at Elm Hole. Upstream, open passage is again encountered beyond St David's Sump, leading towards the Maytime Series of sumps in Agen Allwedd. Borrowed Boots Streamway is a significant tributary from the north in this area, possibly bringing water from Llangattwg Swallet into the system. Digging and diving exploration are currently proceeding at a rapid pace, the most recent extensions leading north from the lower Bonsai Streamway–King's Road junction to some 1.2 km of cave, terminating in a large passage (12 O'clock High) leading westward towards Agen Allwedd, here only 500 m distant. Further substantial additions to the system can be expected. The extensions to Ogof Daren Cilau are without doubt the most significant find in British caving in the last 10 years.

Further information on the caves can be obtained from the following: Leitch (1960) and Jenkins (1963) for Agen Allwedd, Parker (1978) and Gascoine (1979) for Ogof Craig a Ffynnon, and Gardener (1984, 1985, 1986) and Farr (1985, 1986) for Ogof Daren Cilau.

History of exploration

The quarries

Access to the Llangattwg caves from Brynmawr makes use of the course of the old tramway built by the Bailey brothers from 1828 to 1831 (Rattenbury, 1980) (Fig. 12.15). While the majority of the quarrying took place in the early

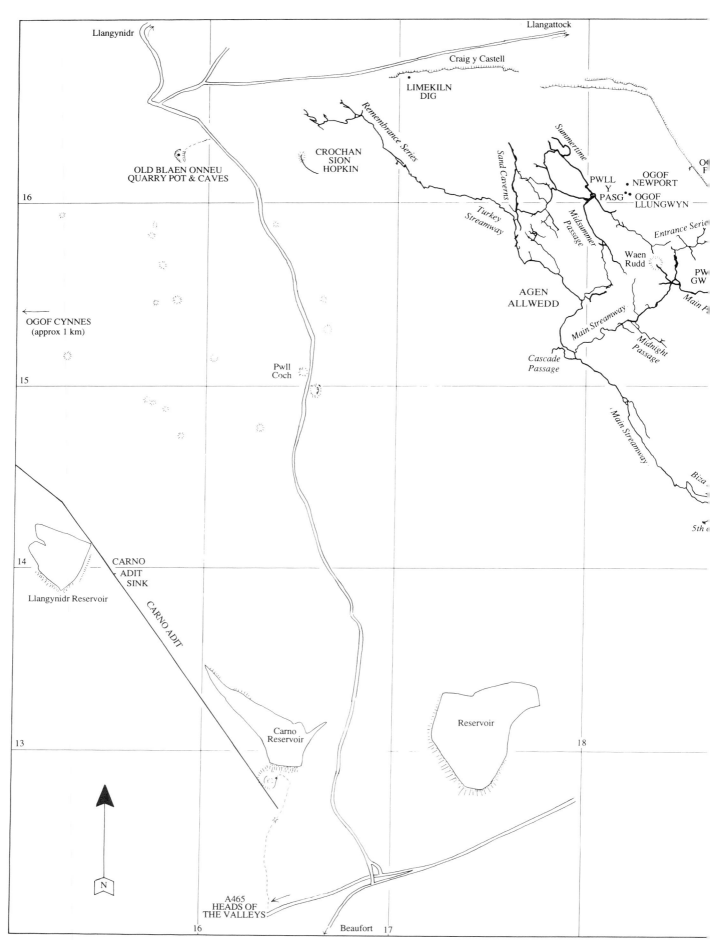

Fig. 12.1. Compilation plan of the caves of Mynydd Llangattwg.

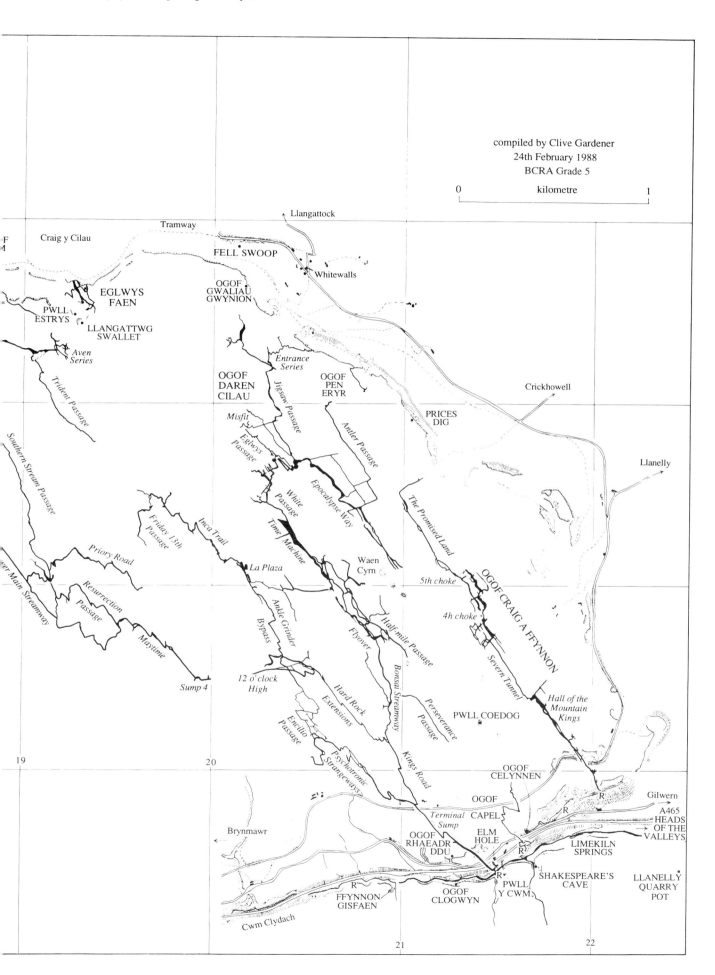

compiled by Clive Gardener
24th February 1988
BCRA Grade 5

0   kilometre   1

Llangattock

Craig y Cilau   Tramway

FELL SWOOP

Whitewalls

OGOF
GWALIAU
GWYNION

EGLWYS
FAEN

PWLL
ESTRYS

LLANGATTWG
SWALLET

Aven
Series

Entrance
Series

Crickhowell

OGOF
DAREN
CILAU

OGOF
PEN
ERYR

Jigsaw Passage

Misfit

PRICES
DIG

Eglwys Passage

Antler Passage

Trident Passage

Llanelly

White Passage

Epocalypse Way

Time Machine

Southern Stream Passage

The Promised Land

Friday 13th Passage

Inca Trail

La Plaza

Waen
Cym

5th choke

OGOF CRAIG A FFYNNON

Priory Road

4h choke

Resurrection Passage

Ankle Grinder Bypass

Flyover

Half-mile Passage

Severn Tunnel

Maytime

Bonsai Streamway

Hall of the
Mountain
Kings

Main Streamway

12 o'clock
High

Hard Rock
Extensions

Perseverance Passage

PWLL COEDOG

Sump 4

Encilio Passage

Kings Road

Psychotronic
Strangeways

OGOF
CELYNNEN

Gilwern

19

20

OGOF
CAPEL

R

A465
HEADS
OF THE
VALLEYS

Brynmawr

Terminal
Sump

ELM
HOLE

R

LIMEKILN
SPRINGS

OGOF
RHAEADR
DDU

R

SHAKESPEARE'S
CAVE

LLANELLY
QUARRY
POT

FFYNNON
GISFAEN

OGOF
CLOGWYN

PWLL
Y CWM

Cwm Clydach

21

22

Table 12.1. *Length and depth of significant caves of the Llangattwg area, and the elevation, location and geological horizon of their entrances*

Cave (English version)	Length (m)	Depth (m)	Entrance NGR	Altitude (m)	Geological horizon at entrance
Llangattwg Swallet or A.1. Pot	—	—	SO 19310 15169	490	Millstone Grit
Pwll Coedog (sink with a rowan tree)	15	15	SO 2141 1327	414	
Crochan Sion Hopkin (John Hopkin's Pot)	6	4	SO 16508 16282	463	
Ogof Cynnes (warm cave)	914	c. 25	SO 1408 1540	524	
Blaen Onneu Quarry Pot (Ashtrees at the head of the valley)	76	17	SO 15840 16260	498	
Pwll y Pasg (Eastertime pot)	299	34	SO 18199 16054	478	Dowlais Limestone
Ogof Newport (Newport cave)	36	12	SO 18204 16100	479	
Ogof Llungwyn (Whitmonday cave)	56	16	SO 18214 16043	479	
Pwll Gwynt (Windy Pot)	269	28	SO 18770 15665	443	
Pwll Estrys (Ostrich Pot)	c. 10	c. 10	SO 1926 1548	400	
Ogof Rhaeadr Ddu (Black waterfall cave)	230	40+	SO 2127 1253	252	
Waterfall Cave	105	9	SO 2127 1250	248	
Ogof Clogwyn (Rock face cave)	165	6	SO 2129 1238	235	
Llanelly Quarry Pot (After Llanelly village)	91	36	SO 2246 1245	265	Gilwern Oolite
Price's Dig (After Brian D. Price)	76	5	SO 21059 14919	412	
Limekiln Dig or Ogof Caci (muddy cave)	6	2	SO 17050 16680	405	
Ogof Pen Eryr (Eagle head cave)	427	8	SO 20741 15201	409	
Ogof ar Olygfa Braf (The cave with a fine view)	30	—	SO 19904 15807	384	
Ogof Daren Cilau (Rock of refuge)	21 000	190	SO 20512 15295	401	Blaen Onneu Oolite
Channer's Dig or Ogof Gwaliau Gwynion (Whitewalls cave)	34	2	SO 20165 15645	395	
Ogof Ffw (Badger sett)	24	2	SO 18518 16175	389	
Eglwys Faen (Stone church)	1 240	19	SO 19233 15654	366	
Agen Allwedd (Keyhole cleft)	31 000	160	SO 18730 15870	367	
Ogof Gam (Twisting cave)			SO 18738 15842	364	

Table 12.1. *Continued*

Cave (English version)	Length (m)	Depth (m)	Entrance NGR	Altitude (m)	Geological horizon at entrance
Ogof Craig a Ffynnon (Rock & fountain cave)	8 000	150	SO 22011 12873	259	
Pwll y Cwm (pool in the valley)	—	9	SO 21487 12438	207	
Elm Hole (Wych elm)	70+	25	SO 21501 12457	209	
Devil's Bridge Cave	18	c. 2	SO 2150 1246	210	Pwll y Cwm
Ogof Capel (Chapel Cave)	c. 650	?	SO 21656 12593	220	Oolite
Ogof Celynnen (Holly Cave)	c. 20	c.2	SO 21679 12612	221	
Shakespeare's Cave (Midsummer Night's Dream inspiration?)	366	—	SO 21710 12483	207	
Fell Swoop or Mud Mine (After Jack Fell)	91	3	SO 2013 1586	344	Lower Limestone Shales
					Old Red Sandstone

Fig. 12.2. View westwards along the Llangattwg escarpment; the ledge below the lower cliff is the Tramroad with the entrance to Agen Allwedd near its end in the upper centre of the view (photo by C. Howes).

nineteenth century and horse-drawn trams were initially used to transport broken limestone rubble from the Pant y Rhiw Quarry to the iron and steel works at Nantyglo and Beaufort, records exist from about 1840 of the locomotive haulage of coal from mines in Monmouthshire to the Llangattwg wharf.

The initial construction on the Llangattwg side (1814–15) included a tramroad and double incline to remove limestone for burning from the Pant y Rhiw Quarry. The lime was distributed to local farms using the Brecon and Abergavenny Canal which had opened in 1797. In 1827 permission was given to extend the tramroad west to further quarries situated in an area that is now within the Craig y Cilau nature reserve. This

was shortly followed by the completion of the Nantyglo–Llangattock incline tramway in mid-1831.

By the 1840s a network of tramways had been established, eventually linking the Daren Cilau Quarry to the Nantyglo tramway. The oldest tramway (constructed before 1818) ran from the Nantyglo works to the Daren Disgwylfa Rocks and was, in land leases of 1870, referred to as an iron-ore tramway. Although this is shown as active tramroad on the 1890 1st edition of the 25 inches:1 mile Ordnance Survey map (Sheet XLI.II), it had been dismantled along with the other tramroads by 1900. The last fires burnt at the Llangattwg Wharf kilns in 1920 but quarrying continued at Blackrock until 1970 and only recently ceased at Blaen Onneu New Quarry. A massive rock-fall during a severe storm in 1948 substantially altered a section of the face of the far western Pant y Rhiw Quarry. As late as the early 1960s an attempt was made to reopen the Daren Cilau Quarry, but this idea was abandoned. It is perhaps ironic that the practice of quarrying which frequently destroys caves has been the most important factor in providing access at Llangattwg.

The caves

Eglwys Faen is the oldest cave recorded in historical texts of the Llangattwg area and is known to have been used as a place of worship and refuge in Cromwellian times (Jones, 1911). Present-day exploration started in 1946 when Brian Price of the South Wales Caving Club (SWCC) excavated the first duck in Ogof Gam and the entrance to Agen Allwedd. Three years later, on 24 December 1949, he was able to shift the remaining boulder from Agen Allwedd, penetrating as far as Sally's Alley (from the nickname of his fellow-digger, David

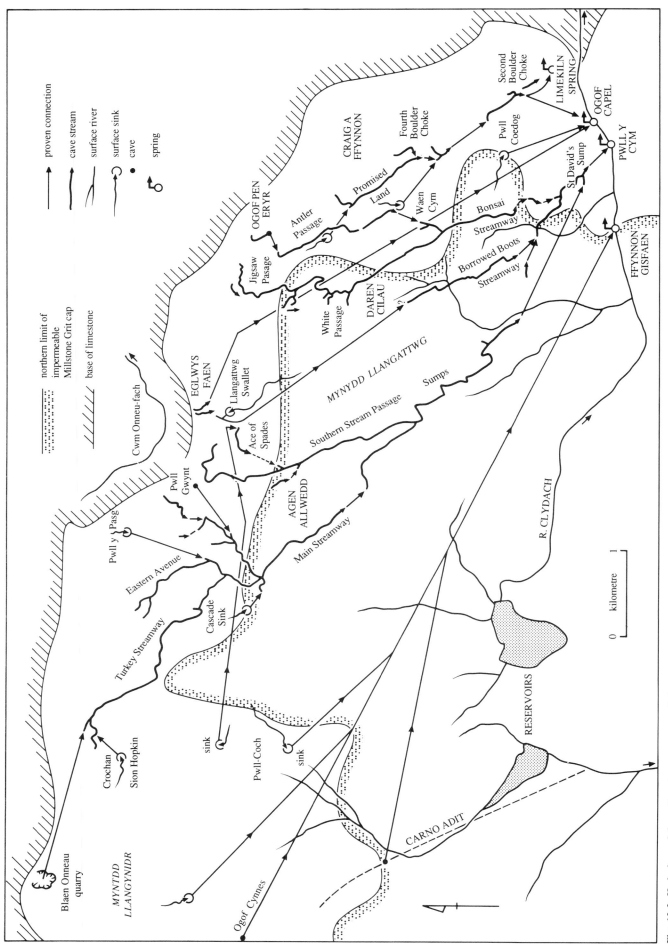

Fig. 12.3. Hydrology of the active stream passages of the Llangattwg System based on Gascoine (Chapter 4) and other dye-tracing tests (based on Welsh Water Authority information 1980).

Fig. 12.4. The Main Passage of Agen Allwedd (photo by
P. R. Deakin).

Fig. 12.5. The Main Stream in Agen Allwedd (photo by
A. C. Waltham).

Fig. 12.6. The First Boulder Choke in Agen Allwedd (photo by
A. C. Waltham).

Fig. 12.7. Approaching Sump One, lower main stream passage,
Agen Allwedd (photo by C. D. Westlake).

Seagrave). Six days later, progress was made as far as the First
Boulder Choke.

Attention then shifted to Eglwys Faen with Mel Davies and
other members of the British Nylon Spinners Speleology
Section (BNSSS) digging through the crawl at the end of Main
Chamber to discover both New Cavern (The Inner Chamber)
on 24 February 1956 and, after the subsequent climbing of a
9 m aven, the High Level Series. Ogof Daren Cilau was first
noticed on 6 July 1957 by Vic Howells and Mel Davies as a low
silted arch with a powerful draught. This was entered on 2
November 1958 when two scouts working with Brian Price

succeeded in passing the muddy entrance pool. In the same
quarry BNSSS commenced exploration of Ogof Pen Eryr in
1957, making limited progress with the assistance of
explosives.

The biggest discovery of the era was made in Agen Allwedd
on 13 October 1957 by members of Hereford Caving Club
(HCC) who poked upwards through the boulders into Main
Passage. An initial trip was made to the Cliffs of Dover and
North Wing, and although exploration rapidly followed to the
Second Boulder Choke in Main Stream Passage, this obstacle
was not passed until 2 February 1958 by an HCC party led by

Fig. 12.8. Thick-bedded sediments at the cliffs of Dover, Agen Allwedd (photo by T. D. Ford).

Fig. 12.10. Helictites in Ogof Craig a Ffynnon (photo by C. D. Westlake).

Fig. 12.9. Hall of the Mountain Kings, Ogof Craig a Ffynnon (photo by C. D. Westlake).

Fig. 12.11. Gasoline Alley, Ogof Craig a Ffynnon (photo by I. Davinson).

Fig. 12.12. The large vadose canyon passage just before the Time Machine (photo by C. D. Westlake).

Fig. 12.13. The Time Machine – the largest cave passage in Britain (photo by C. D. Westlake).

Fig. 12.14. The Pwll y Cwm resurgence in the bed of the Clydach River; the submerged pothole is in the pool in front of the figures.

inflatable dinghy. Many hundreds of man-hours went into the passing of the Third and Fourth Boulder Chokes, the latter being reached on 23 June 1960. However, the passing of this second major barrier would have to wait a further 12 years. Upstream exploration of Turkey Series by Kingsley Hawkins and members of HCC led via a collapsing choke (Hawkins' Horror) into the Summertime Series on 19 April 1959, the first day of British Summer Time. By June, HCC had completed the Inner Circle route. Exploration and work on the survey continued with minor passage discoveries in Summertime, culminating in the opening of the 2.5 km Outer Circle by Harold Lord and Bob Toogood of the Technical Speleology Unit (TSU) in April 1963.

In 1962 Ken Pearce and a British Speleological Association (BSA) party dug through the blockage in Southern Stream Passage to enter the impressive Lower Main Stream Passage and its associated high levels. The first breakthrough trip was halted by the Lower Main Streamway in full spate; however, at Whitsuntide, exploration to Terminal Sump and the Fifth Choke was complete – the subsequent ascent into Biza Passage being achieved by Bob Toogood in May.

Excavations at Price's Folly in 1961 employed the first British cave monorail to remove the total mud fill. More success was achieved by Mike Boon and Fred Davies on 12 April 1963, when the two blasted the terminal choke in Daren Cilau and discovered over 500 m of passage – ending in a chamber whose size gave an indication of the further potential. A long working trip by Tich Morris and Boon saw the entire cave surveyed in 17 h. From 1963 to mid-1965 blasting work by David Dadley and members of the Llangattwg Cave Group gave a 120 m extension to Ogof Pen Eryr in the same quarry.

The Agen Allwedd resurgence was confirmed as Pwll y Cwm by BNSSS members using dye and absorbent detectors made of nylon yarn in possible resurgences. Elm Hole was explored following an HCC dig through the roots of a wych elm tree on 22 July 1961. Diving also began in Ogof Capel and the Pwll y Cwm pothole in the River Clydach (see Chapter 9).

In March 1966 M. J. Wooding and J. F. Sinclair dived the upstream Turkey Sump and discovered over 700 m of new passage beyond Sump Three. Encouraged by the subsequent diving efforts of Martyn Farr and Mike Ware of SWCC in July

Paul Hartwright. On the latter occasion the far end of the Keyhole Passage traverses was reached, enabling a subsequent descent to the streamway and exploration to Deep Water the following month. Also in March 1958 BNSSS were invited to take part in the exploration and made the first visits to the nearer reaches of Southern Stream Passage, Cascade Passage and Turkey Streamway as far as the junction with Coal Cellar Passage. A month later, BNSSS explored to the 'Coal 'ole' and followed Turkey Streamway until they were stopped by Turkey Pool. This was later negotiated, using ex-RAF lifebelts, by a combined group of HCC and BNSSS cavers reaching Terminal Chamber at the top end of Turkey Stream-way later the same year.

Downstream, Deep Water was crossed in November 1958 when members of Chelsea Speleological Society (CSS) used an

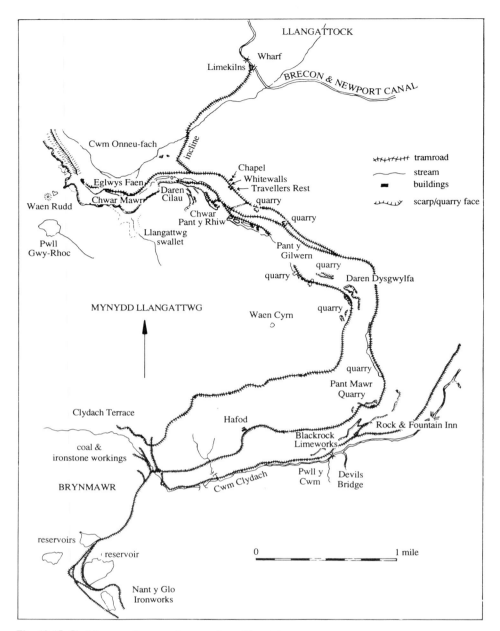

Fig. 12.15. Sketch map of quarry tramroads on Mynydd Llangattwg (modified after Rattenbury, 1980).

1971, John Parker took over and on his third dive in Sump Five he and Jeff Phillips entered the Remembrance Series on 18 December.

An important step forward was made when Clive Westlake and Paul Deakin of Eldon Pothole Club visited Biza Passage on 13 April 1971 and obtained a radio contact with the party who had entered via the Third Boulder Choke. Five trips and 435 man-hours later, Dave Gill joined with these two in removing the final boulder, allowing the first 5 km traverse of the Grand Circle on 20 May 1972. This was soon followed by the first Outer Circle Figure of Eight trip by Cambridge University Cavers, who completed their tour in 8.25 h.

On 2 April 1972, and with portering assistance from the Eldon diggers, John Parker dived the Agen Allwedd main downstream sump. By 22 July he had pushed over 220 m into

Sump Three where, unknowingly, the line was left just 30 m short of the end of the sump. This was eventually passed by Martyn Farr on 4 May 1974 when over 450 m of passage was explored. A month later an attempt on Sump Four ended in tragedy when diver Roger Solari failed to return. Despite more than 2760 man-hours spent in mounting a rescue he was not found.

At the Pwll y Cwm resurgence, a large underwater tunnel entered via Elm Hole was partially explored by Farr in April 1974. The main difficulty was the constricted nature of the initial part of the sump which limited the size of the diver's air cylinders.

Also in the Clydach Gorge, early digging work in the quarry above the Rock and Fountain public house had been started in 1954 by Phillip Jones, Mel Davies and members of BNSSS.

Further efforts, commencing on 7 August 1966, led to the discovery of 10 m of passage in Pope's Hole – initiated by Russell Pope. However, most of this early digging concentrated on a lower hole where a collapsed passage with stalagmited boulders was entered and led to a small chamber.

Members of Cwmbran Caving Club (CCC) took over the site in 1973 but it was not until the summer drought of 1976 that an old mud plug cracked and a draught rekindled interest in Pope's Hole. John Parker, Jeff Hill and Ann Franklin successfully dug through the mud to discover the initial section of Ogof Craig a Ffynnon, to the First Boulder Choke, on 11 August. On 26 September a heavy flood from the entrance completely blocked the Heads of the Valleys road. Gasoline Alley was discovered on 19 November and the Second Boulder Choke passed on 10 July 1977 to enter the Hall of the Mountain Kings. Progress through the Third and Fourth Boulder Chokes followed and the lower series was also explored.

Following the discovery of Pwll y Pasg and Ogof Newport early in 1962, high on the mountain scarp, Pwll Gwynt was found in March 1981 by Trevor Knief of CSS. This cave is unusual for Llangattwg comprising six high avens, some of which are roofed by the Millstone Grit. It has been found to be hydrologically linked to an inlet below Keyhole Chamber in Agen Allwedd. On 30 August 1982, Clive Gardener and Dave Ramsay of CSS made the breakthrough which led to a major extension of Trident Passage in Agen Allwedd. This 600 m crawl enters a blank area on the survey and takes almost 2 h to traverse. The CSS discovery of Northern Stream Passage followed in October 1983 and after being trapped by a flood in Turkey Chamber on New Year's Day, Martyn Farr discovered the 1984 Series off Shattered Passage.

There can be no doubt that the most celebrated discovery in British caving during the last decade has been the major extensions in Ogof Daren Cilau which began in the summer of 1984. On the August Bank Holiday, three cavers from Chelsea Speleological Society (Clive Gardener, Jock Williams and Joe Goodall) started a search for a route through to the floor of the Old Main Rift, obscured by an unknown depth of boulder breakdown. This was found in the form of a surprisingly wide open crawl which for some reason had not been the subject of serious earlier exploration.

After digging through boulders, they met an impasse but this was overcome by following the direction of a strong draught blowing out from a bedding crawl. Excavations led via a tight 45° squeeze down to two levels and a crystal pool. With Martyn Farr and Paul Tarrant the breakthrough was on 8 September 1984 and they followed Jigsaw Passage to a short calcite constriction, later called 'The Wriggle'. The mud coating on boulders and floor resembled melted chocolate, totally unblemished by human or animal footprints. In places the deposition covered an even older surface of hollow calcite. Sections of this fragile, wafer-thin false floor crunched beneath cavers' boots.

Beyond 'The Wriggle' the roof aven to The Highway and the door-like entrance to Misfit Passage stood out as enticing leads for a return trip. The jumble of boulders at the start of the Big Chamber–Nowhere Near the Entrance gave images of an immense terminal boulder choke looming close ahead, yet upwards the chamber opened out and a view was gained of a long steeply sloping scree ramp. Shattered rocks underfoot slipped away with alarming ease. First footfalls were made into the Upper Loop Route where small mounds of decomposed calcite, resembling spent carbide, were seen sprinkled over boulders. An ominous choke waited to greet the party a short distance ahead. Miraculously, a small hole led through an already open ('Anti-Gravity') tunnel to a lower more impressive passage with enormous breakdown boulders standing like monoliths.

Eglwys Passage was entered next and led to a series of sandy crawls culminating at a terminal choke, conjectured to be close to the Eglwys Faen System. However, hardly any time could be allowed to lever the boulders ahead for, on the return, side leads to the left of the Big Chamber needed closer examination. Epocalypse Way unexpectedly turned out to be the biggest and most impressive trunk route of the new find and offered the finest collection of speleothems in the cave at 'The White Company'.

Further on, beautiful selenite crystals, up to 15 cm long, grew on the walls of an old dry streambed while the floor was littered with hundreds of crystal-coated mud balls. A pagoda of unique character appeared isolated on the mud floor, a perfect match for its Ogof Craig a Ffynnon counterpart. Around it lay odd broken fragments of thick calcite which appeared to have fallen from the roof. Beyond this a new stream trickled across the passage floor.

The anticipated connection down into Ogof Craig a Ffynnon was, disappointingly, not found and instead, at the end, chokes called a conclusion. The Lower Streamway was entered by a ladder on the subsequent trip, but also soon terminated in a large impassable boulder choke. The entrance rift to Antler Passage was glossed over on the first visit, yet John Cooper later pushed on beyond a squeeze and discovered the remarkable Antler formations and the passage continuing for 700 m.

Urchin Oxbow was not initially reported as having any special quality, yet photographers were soon returning with spectacular reports of aragonite formations, including a representation of the 'Silver Goddess' Rolls-Royce figurine. Man in the Roof and Aqueous Choke were rapidly entered. This latter major lead, taking more water than any other stream in the first breakthrough series, remains to be fully explored.

A relatively short excavation for the original diggers, Clive Gardener and Jock Williams, led in February 1985 up through a small choke in Eglwys Passage to the unexpected view of a large black void – St Valentine's Chamber. A slow methodical dig in boulders flooring this chamber enabled the two cavers to descend through large wedged blocks to the first sight of Preliminary Passage, which was then followed to a boulder choke.

On a subsequent visit, Steve Holmes and Tony White explored climbing leads at the end of this passage. Ascending and traversing to a dead-end alcove, Tony decided to complete a 20 m climb to a promising-looking roof aven. After exploring passages at the top level (Higher Things) he made a makeshift solo descent to the floor of White Passage (Fig. 12.16). Proceeding to the top of the boulder slope at the entrance to Red

Fig. 12.16. The 70-foot pitch in Ogof Daren Cilau (photo by C. D. Westlake).

was quite a surprise. The obvious conclusion, that a new water route had been located, was indeed correct, despite the fact that the Bonsai Streamway water had passed underneath The King's Road and was now also issuing from the right-hand wall upstream. Downstream, green water spouted from a fissure in the passage wall. This was the first underground sighting of water from the Llangattwg Swallet stream sink, which had been dyed the previous day.

In addition to the downstream sump, the upstream route, which had to lead to the Llangattwg Swallet Streamway, was guarded by first a lake and then a 40 m long sump. Although the lake was easily crossed by the wet-suited cavers on the first trip, this St David's Sump would later be passed by Martyn Farr on 27 April – leading to Psychotronic Strangeways, which ends at the dark and foreboding Gloom Room Sump. Ian Rolland and Rob Parker subsequently discovered a further 2 km of influent streamway (Borrowed Boots Streamway) terminating in the Seventh Hour Sump, on 23 November 1985. Work continues in the cave, with the UC4 group and Bristol Exploration Club active in the Hard Rock Extensions, the task now only being possible by using elaborate underground camps. Their organization and digging effort has presently yielded in excess of 1200 m of new passage. Members of Birmingham University Speleological Society, University of Leeds Speleological Association, Chelsea and Gloucester Speleological Societies are currently engaged in successful excavating in Gothic Passage in Agen Allwedd and have so far

River Passage, and with no sign of a further choke ahead, he decided to retrace his steps and carry out the full exploration one week later, in the company of his companions! Thus, on 2 March 1985, seven people made their way down the tall impressive canyon of White Passage until a boulder slope guided the climb to the start of the Time Machine. That first glimpse suggested an expansive chamber with shadowy voids off to distant walls. The bypass to the end choke of the Time Machine was entered and along the new smaller active stream passage the outstanding delicacy of intricate helictites was quickly noted. Not one but several 'Bonsai trees' adorned ledges aside the passage walls and long thin calcite straws hung as silent companions to the tumbling waters. A snake of cavers followed the stream's irreversible route through the heart of the mountain. A large inlet streamway entering from a wide passage on the left suggested that the water lost at the end of Epocalypse Way had once again been regained in the new passage. Later dye-tracing was to confirm this. A short distance ahead, a large fossil passage gaped onto the streamway in the form of a flyover at high level. Anticipation of a possible boulder choke round the next corner proved a lucky omen in prolonging the traverse of the passage. Then a short climb and The King's Road was entered for the first time. Here the blackness of the rock and stillness of the air came in stark contrast to the bubbling, lively Bonsai Streamway.

The water had been lost on the left-hand side of The King's Road and therefore, after the prolonged steep descent, to come across a T-junction with an inlet stream from the right

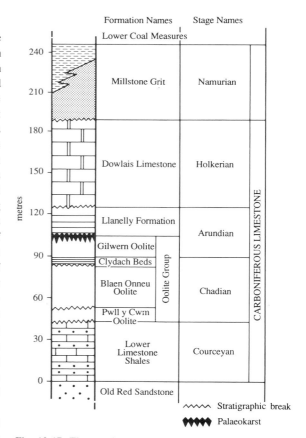

Fig. 12.17. The stratigraphic sequence of Carboniferous Limestone beds in the Mynydd Llangattwg area.

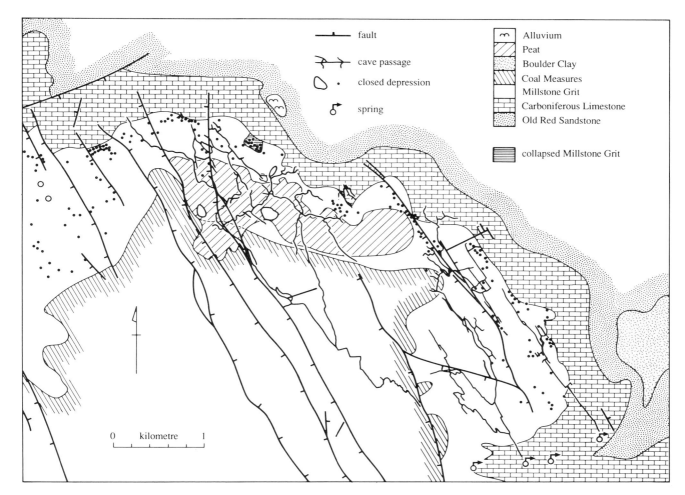

Fig. 12.18. Geological sketch map of Mynydd Llangattwg showing the relationship of caves to geology and the distribution of surface depressions (adapted from a Welsh Water Authority plan 1980).

entered 1600 m of new passage trending southeast towards Daren Cilau and Maytime (Agen Allwedd).

Divers have also begun to clear the Pwll y Cwm pothole of debris to provide a direct upstream access to St David's Streamway. From March 1986, Martyn Farr pushed both the Terminal downstream sump in Daren Cilau and Elm Hole resurgence sump to connect the diving lines on 3 July 1986. The first through trip from the Daren Cilau quarries to the Clydach Gorge was made on 11 August 1986. This three-mile underground route with a descent of 190 m has paved the way for a future link between Agen Allwedd and its Clydach resurgence, and the ultimate realization of the 'Mynydd Llangattwg Cave System'. Ogof Capel's sump and unstable boulder choke were passed in July 1987 to yield some 500 m of stream passage, but as yet no link with the other caves.

Geology

The Carboniferous Limestone of the Llangattwg area is <150 m thick and forms the tapering north edge of a thick wedge of carbonates developed between fluviatile Old Red Sandstone (Devonian) below and the deltaic Millstone Grit (Upper Carboniferous) above. Important discontinuities are present in the upper part of the limestones which generally thin

to the north and are more dolomitized to the south. The stratigraphic subdivisions are shown in Figure 12.17 and their distribution in relation to the caves in Figure 12.18. The limestone outcrop is relatively restricted, being confined to the impressive quarried scarp face of the Usk Valley, where exposures are abundant. The caves are therefore overlain by Millstone Grit to the north and east and in the case of Agen Allwedd also by the Lower Coal Measures. The regional geology was described by George (1970), while a detailed account of the Carboniferous Limestone of the area was presented in George (1954), and more recent work describing the depositional environments has been conducted by Wright (1981, 1986) and Wright, Raven & Burchette (1981).

The Lower Limestone Shales

The Lower Limestone Shales form a transitional unit between the fluviatile clastic deposits of the Old Red Sandstone and the shelf carbonates of the Lower Carboniferous. The base is diachronous, younging from south to north and in the Clydach Gorge a thin basal conglomerate is present. Three cycles of regression and transgression were recognized by Burchette (in Wright et al., 1981), but the lowest is absent at Llangattwg, where the formation has a maximum thickness of

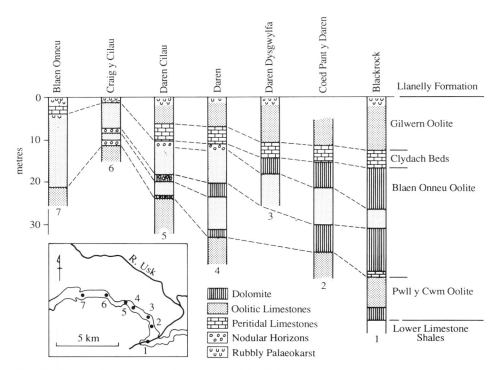

Fig. 12.19. Lateral variations in the members of the Oolite Group along a north–south transect (after George, 1954).

27 m. Coarse oolitic calcarenites deposited on shoal bars form the base of the sequence and are replaced upward by fine-grained lagoonal limestones with a restricted brachiopod fauna. Following the second transgression, blocky mudstones containing numerous thin bioclastic and silty limestones were deposited on the deeper shelf. This unit, which is 13 m thick in the Clydach Gorge, forms an important barrier to water movement in the area. There is a gradational transition to the overlying Oolite Group representing progradation of the shoreline towards the south.

The Oolite Group

The major known caverns of Llangattwg are all developed in limestones of the Oolite Group. In the Clydach Gorge, the Group is some 52 m thick and comprises five sub-members (Fig. 12.17), but to the north erosion prior to the deposition of the Llanelly Formation has caused overstep of the top two members. At Daren Cilau, the Gilwern Oolite is reduced to <2 m in thickness, while near Blaen Onneu the Llanelly Formation rests directly on the Blaen Onneu Oolite; the Clydach Beds having being completely removed (Fig. 12.19). George (1954) has ascribed this to local tectonic uplift along the Neath Disturbance which runs just to the north (Fig. 12.20). This uplift has also been responsible for periodic emergence during deposition, with soil formation and dolomitization and general reduction of the thickness of individual members to the north (Fig. 12.19).

The basal unit is the Pwll y Cwm Oolite, which is about 11 m thick in the Clydach Gorge, the lowest 3 m comprising a coarse ferroan dolomite. Above, coarse bioclastic limestones grade upwards into massive, black, well-sorted oosparites which weather to a light grey. At Daren Ddu, the Blaen Onneu

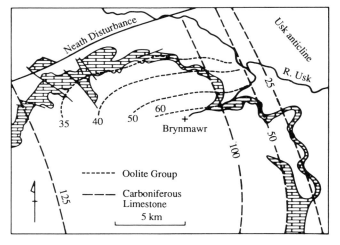

Fig. 12.20. Isopach map of the Carboniferous Limestone and of the Oolite Group in the Mynydd Llangattwg area (isopachs in meters).

Oolite, which is lithologically very similar to the Pwll y Cwm Oolite, is sandwiched between thinly bedded shales and dolomites. To the north and west the dolomites and shales thin and are replaced by prominent nodular horizons (Fig. 12.21). The 'nodules' are contained within grey, green and (rarely) red mottled clays, have sharply defined margins and may reach 40 cm in diameter. Some comprise fibrous calcite, others are micritic and may incorporate coarse spar-filled cracks. They are similar to the calcrete palaeosols described by Wright (1982b) from the Llanelly Formation and suggest that periods of emergence give rise to palaeosol development to the north, with dolomitization further to the south, perhaps in a peritidal environment. The periods of emergence alternate with trans-

Fig. 12.21. The Nodular Bed in the roof of Washout Chamber, Agen Allwedd.

gressions shown by developments of bioclastic shore-face or beach deposits, followed by the development of well-sorted oolite shoals (Wright, 1986).

The palaeosols at the top of the Blaen Onneu Oolite were curtailed by the onset of a transgression giving rise to a complex peritidal sequence of interbedded pelsparites, micrites and thin shales about 12 m thick, the Clydach Beds (formerly the Marker Beds of George, 1954). The highest member of the Formation is the Gilwern Oolite (maximum 12 m thick) which may be distinguished from the Lower Oolites by its light grey colour. It is an impure poorly sorted oosparite which contains abundant bioclastic debris, especially towards the base, and intra-clasts of oolite fragments up to 10 cm in diameter. Palaeokarst development has affected the top three members of the Formation progressively further north giving a distinctive rubbly appearance in outcrop.

Llanelly Formation

The Llanelly Formation (previously the calcite–mudstones group of George, 1954) has been the subject of detailed study by Wright (1981). The Formation reaches a maximum thickness of 20 m, thinning to the west as a result of overstep by the Dowlais Limestone. It comprises four members which show pronounced lateral variations in both thickness and lithology, although several persistent beds are readily recognizable (Wright et al., 1981). The lowest unit, the Clydach Halt Member, overlies a spectacular palaeokarst developed in the top of the underlying Oolite Group. This comprises a rubbly horizon up to 5 m thick with intense solutional piping and fissuring, filled with green and red mottled clay. It has been interpreted by Wright (1982a) as relict kavernossen karst. The Clydach Halt Member (0–6 m thick) includes siliceous conglomerates, laminated sandstones and thick green claystones deposited on a semi-arid floodplain. Limestone bands and calcite and dolomite nodules are interpreted as calcrete palaeosols and playa deposits (Wright, 1982b). The Cheltenham Limestone Member is 1.75–8 m thick and is predominantly composed of dark well-cemented peloidal and bioclastic limestones with beds up to 1 m thick. Thin green clays and laminated mudstones <0.5 m thick also occur and represent the final members of a regressive sequence

from open bays through lagoonal, intertidal and supratidal environments to terrestrial soils with calcretes. The overlying Penllwyn Oolite (3–5 m thick) shows a return to marine conditions with well-sorted cross-bedded oolitic limestones. The final member is >7 m thick and comprises mottled red–green clays with platy limestone calcretes overlying a palaeokarst surface deposited in a similar environment to the Clydach Halt Member.

These lithologies are replaced laterally by up to 10 m of mainly pale grey and buff fluvial orthoquartzites with some conglomerates and thin mudstones which are well exposed in Blaen Onneu Quarry. Towards the south, pale grey–green limestones also occur and the unit is only some 2 m thick at Daren Cilau. This unit has been mapped as a separate formation, the Carn Laws Sandstone, by Barclay & Jackson (1982) who considered it to be Holkerian in age. These siliciclastic deposits are the result of the southward progradation of a fluvio-deltaic complex onto the Dinantian carbonate shelf and to the west extensive erosion of the Llanelly Formation has occurred.

Dowlais Limestone

Much less geological information is available for the Dowlais Limestone Formation. In the Carno Adit it is seen to comprise 87 m of tabular, medium–thick beds of dark grey, fine- or medium-grained partly dolomitized bioclastic limestones with dark grey shale interbeds. Brachiopods, coral debris and gastropods are abundant and massive oolitic limestones are also found (Barclay & Jackson, 1982). The carbonates are predominantly peritidal, built up behind a major oolite shoal complex in the area of the present Vale of Glamorgan (Wright, pers. comm.). The Formation thins eastwards due to overstep by the Millstone Grit (Wright et al., 1981), being <55 m thick in the Clydach Gorge. The junction is unconformable and represents a major break, with development of an irregular erosional palaeokarst surface with pebble conglomerates in the Millstone Grit, locally including clasts of Dowlais Limestone (Owen & Jones, 1961; Barclay & Jackson, 1982).

Millstone Grit

The Millstone Grit was previously subdivided into the Basal Grit, Shale Group and Farewell Group units (George, 1970), but more recent mapping has demonstrated that such units cannot be recognized laterally due to abrupt facies variations, and they are also diachronous (Barclay, pers. comm.). Furthermore, on the basis of the fauna present in several marine bands the Farewell Group is now considered to be part of the Lower Coal Measures. In the Brynmawr area the Millstone Grit is 50–60 m thick, thinning markedly towards the Usk anticline from a thickness >2000 m near Swansea. The coarser rocks are wedge- and current-bedded, white–yellow well-sorted quartz conglomerates. These coarse sediments comprise the upper part of a poorly developed rhythmic sequence commencing with fine-grained grey–black highly fossiliferous marine shales (often with pyrites) through progressively coarser more terrestrial units. The sequence may be capped by a thin coal with a root bed. The deposits show a marked contrast to the underlying carbonate rocks and represent terrigenous sedimentation under estuarine conditions on the margin of a landmass exposed to the north (George, 1970). Progressively more terrestrial conditions developed into the Coal Measures above, which are not described in detail here although they overlie the limestones in the southwest of the area.

Structure

Detailed structural information is not available but the beds dip generally southwest at an angle of about 2–3°. Minor flexures are present with dips up to 15° and the bedding may locally depart from the general trend as can be observed in the roof of Antler Passage (Daren Cilau) where a small monoclinal fold is present. The major regional structure is the Neath Disturbance running some 5 km to the north, which has been active since Carboniferous times. It is associated with the Dinas sinistral tear fault with a trend from WSW to ENE. This trend is poorly developed in the faults mapped from the surface of Llangattwg, but observations underground indicate that several minor faults belonging to this orientation are important in guiding the cave passages. Both the Fifth Boulder Choke in Agen Allwedd and Man in the Roof Passage in Daren Cilau are examples which may even be developed on the same fault, in this case seen to have a throw of 4 m at Waterfall Chamber in Southern Stream Passage. The main surface fault set trends NNW–SSE, the downthrow being generally to the west. While this trend parallels several of the major cave passages, there is no conclusive proof that faults offer preferential routes for cave development compared to joints. The lower part of Jigsaw Passage and Lower Loop Route may well be controlled by faults mapped at the surface. Leitch (1973) has demonstrated that in Agen Allwedd both the NNW–SSE and NNE–SSW joint sets are both well developed but the former guides more passages as it is oriented down the preferred drainage direction to the resurgence. The statistical validity of this work does, however, require further consideration (see also Cousins, 1977).

Regional geomorphology

The traditionally accepted views on the origin and development of the relief of Wales are embodied in the classic work of Brown (1960). He identified a series of extensive planation surfaces developed during the episodic uplift of the Welsh Massif in late Tertiary times. Mynydd Llangattwg formed part of the High Plateau surface, while the Middle Peneplain was identified near Brynmawr at an elevation of 420–430 m O.D. The lowest surface, the Low Peneplain, occurred only in the major valleys and was well developed in the vicinity of Langorse Lake between the Wye and Usk valleys. The Great Escarpment running to the north of the Llangattwg area from the Black Mountain to the Brecon Beacons was considered both to separate the High Plateau from the Middle Peneplain of Mynydd Eppynt (Brown, 1960) and to result from differential weathering of the Upper Devonian sandstones and conglomerates and the Lower Devonian marls (George, 1970). More recently Battiau-Queney (1980, 1984) has suggested that the scarp is the result

of differential uplift controlled by ancient structures within the basement, a suggestion reflecting the earlier thinking of Jones (1951), who believed the Plateau surface fell from over 550 m O.D. in central Wales to 60 m O.D. at the South Wales coast. However, Battiau-Queney (1980) has suggested that the uplift involved both flexure and movement along pre-existing faults, the earlier surfaces being destroyed by this differential movement at the start of the Pleistocene.

Prior to this differential elevation, Battiau-Queney (1980) argued that much of upland Wales formed a low-relief tropical landscape with deep weathering and a balance between uplift and the rate of bedrock weathering controlled by regolith depth – a dynamic equilibrium landscape as described by Hack (1960). She cited evidence of relict weathering forms preserved under residual clay-rich regolith on Mynydd Llangynidr (Trefil Quarry) and only recently exposed, but there is little evidence of the true age of these minor landforms. More compelling is the extensive dissolution and rotting of the Millstone Grit, first reported by Thomas (1974) which gives rise to closed depressions and zones of foundered strata (Thomas, 1954, 1959). Battiau-Queney (1980) suggested that this resulted from extensive dissolution of silica under a deep regolith with accumulation of core-stones and leached silica grains in a residual ferralitic matrix. Dissolution of the limestone did not occur until uplift opened fissures and caused circulation through rotted zones. The latter is in contrast to the views of Thomas (1974) and Burke & Bird (1966), who considered that even minor leakage of aggressive water through fissures in the Millstone Grit causes limestone dissolution and cave development with eventual collapse of the overlying Millstone Grit. The evidence from caves such as Pwll y Pasg, Siambre Ddu and Ogof Newport supports the Burke–Bird model but there is no doubt that relict Tertiary weathering forms are also present.

Remarkably, these products of ancient weathering appear to have survived in the landscape despite the Pleistocene glaciations. A late corrie glaciation occurred on northeast-facing slopes (Walker, 1984), possibly including that below Craig y Cilau (Leitch, 1973). Mynydd Llangattwg was, however, completely covered by ice from the local Welsh ice-caps developed during the last glacial maximum (Newer Drift) and at least one other glaciation (Older Drift), probably to ice thicknesses well in excess of 100 m (George, 1970; Bowen, 1973; Boulton et al., 1977). This is supported by the presence of both glacial striae and till near the Carno reservoir. In addition glacial striae can clearly be seen on the Llangattwg Plateau above Ogof Craig a Ffynnon and Eglwys Faen. However, erosion appears to have been concentrated into the faster moving ice-streams occupying the pre-existing valleys such as that of the Wye. Indeed, Williams (1968) suggested that the southward flow of ice from central Wales may have overridden the crest of the major Brecon Beacons–Black Mountain escarpment causing its breaching near Bwlch and the capture of the Wye headwaters southwards into the Usk.

This view is at variance with the radial development of the regional drainage pattern favoured by Brown (1960) in which the Usk was considered to be the downstream part of a Wye–Usk system, the northern part of which was captured only

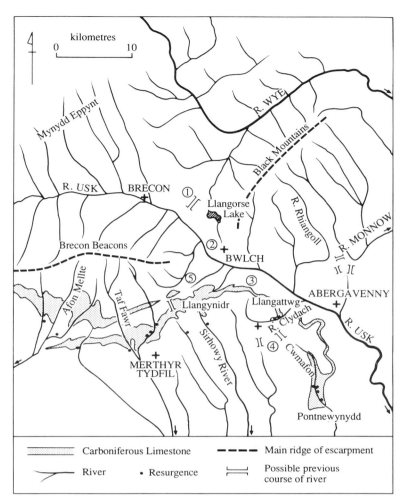

Fig. 12.22. Regional drainage of the Llangattwg area. (1) Previous course of the Upper Usk to Wye, or, conversely, Wye to Usk (Brown, 1960). (2) Breaching of the Great Escarpment dividing • rivers draining southeast from those draining northeast to the Wye, or, ancient path of the Usk (Brown, 1960). (3) Headwaters of the Usk prior to breaching of the Great Escarpment impinge on the Carboniferous Limestone. (4) Incision of the Clydach after capture of the Upper Usk at Bwlch caused beheading of southeast-draining rivers. (5) Breaching of the escarpment causes beheading of SE-draining rivers, diverting them away from the limestone.

recently, in geological time, into the northeast-trending lower Wye along the escarpment. The available evidence is limited and details of the early drainage pattern remain mere conjecture. However, there is some support for a substantial reorganization of the drainage pattern in the Llangattwg area (Fig. 12.22) with rivers such as the Clydach cutting headward in response to the incision of the Usk Valley, following capture of the Wye headwaters. Such a radial reorganization would cause substantial changes in the karst drainage system of the area, a possibility considered further below.

Cave development

Leitch (1960, 1973), Jenkins (1963) and Cousins (1973) have previously discussed the development of Agen Allwedd but nothing has been written on Craig a Ffynnon or Daren Cilau. This is also the case with the cave sediments which have been studied by Bull (1976a,b) and Noel, Retallick & Bull

(1981). Only a limited amount of additional work has been undertaken in preparing this chapter but a detailed study of the Llangattwg System is now long overdue, particularly in view of the recent major extensions which have substantially improved our knowledge of the System.

Present hydrology

The outcrop area of the Carboniferous Limestone in the Pwll y Cwm – Ogof Capel catchment is relatively small (2.3 km^2) and is concentrated around the margin of Mynydd Llangattwg. The recharge is thus via widely distributed percolation inlets and seeps which gather in relatively small branching passages such as those in the Entrance Series of Agen Allwedd and Ogof Daren Cilau and feed larger streamways deeper in the system. This small catchment area is augmented by leakage of water through the Millstone Grit which overlies much of the caves. Such recharge is often

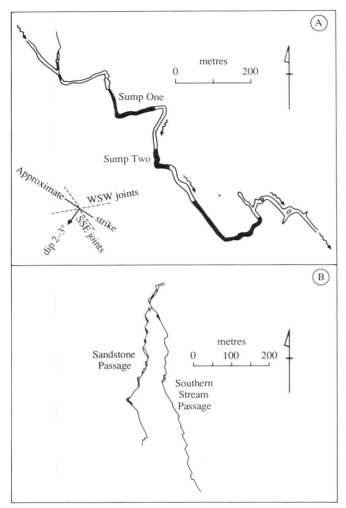

**Fig. 12.23. Control of cave passages by geological structure.
(a) Lower Main Stream Passage and Terminal Sumps, Agen
Allwedd, showing relationship of sumps to down-dip loops.
(b) Southern Stream Passage, Agen Allwedd, showing relationship
to NW–SE joints.**

limestone. In the case of a sink near Blaen Onneu, the water
reappears at Ace of Spades inlet, having crossed above Turkey
Passage, Main Stream Passage and Southern Stream Passage
before finally emerging in the cave. This suggests that lithological perching may well occur on the thick shales of the Clydach
Halt Member of the Llanelly Formation. Further support for
this suggestion is provided by the completely separate nature
of the Ffynnon Gisfaen and Pwll y Cwm catchments as defined
by tracing studies (Fig. 12.3). The former is confined to the
Dowlais Limestone and the latter to the Oolite Group which
are separated by the Llanelly Formation. Where exposed in
the caves, the distinctive thick green shales of the Clydach Halt
Member show no evidence of either active seepage or penetration by solutional cave development.

Recharge from both limestone and Millstone Grit is directed
into a series of sub-parallel vadose streamways within the
caves. These streamways have a pronounced S–SE orientation
aslant the dip in contrast to County Clare, Ireland, where
similar parallel vadose caves are strictly dip-oriented (Self,
1981). This difference is due to the non-orthogonal relation
between dip and joint directions in Llangattwg, the streamways being controlled by the well-developed joint set parallel
to the direction of dominant faulting in the Mynydd
Llangattwg area.

However, down-dip passage segments are also present.
Close inspection shows that in Southern Stream Passage, for
example, segments controlled by the main S–SE joint set are
interspersed with much shorter sections oriented more down-
dip, although still joint-guided (Fig. 12.23b). Furthermore,
both Main Stream Passage upstream of Northwest Junction in
Agen Allwedd and the adjacent streamway leading to the Coal
Cellar show strong down-dip orientation. The overall course of
Turkey Passage and Main Streamway is to the southeast. It
therefore has a greater strike component than the S–SE-
draining sub-parallel streamways which are tributary to it and
give a substantial increase in discharge downstream. An
understanding of the interactions between the structural controls on passage orientations and the overall geometry of the
cave is crucial in explaining the development of the system and
this is discussed further below.

Development of the early cave system

There is a clear disparity between the size of the active
stream passages in the Llangattwg System and the size of the
major fossil trunk passages such as Main Passage (Agen
Allwedd) and Epocalypse (Daren Cilau); only the downstream parts of Main Stream Passage approach their dimensions. In Agen Allwedd these major passages can be followed
upstream to within 200 m of the present northern outcrops of
the limestone where they all terminate in boulder chokes,
probably caused by the proximity to the scarp. The large
passage size combined with the localized input contrasts
markedly with the present active passages, and even though
the most upstream parts of the System have undoubtedly been
truncated by scarp retreat, it must be concluded that a concentrated input from a major stream sink was present (others
may also have been present, but are not so clearly indicated).
The current Pwll y Cwm – Ogof Capel catchment is about

concentrated at discrete points where small streams gathering
on the impermeable peaty uplands drain into closed depressions, formed where prominent joints allow water to penetrate
into the limestone. Rarely, open caves may be entered at these
sites (often after considerable excavation) and usually comprise unconformity bedding-caves below which fluted vadose
shaft complexes may penetrate 35 m into the Dowlais Limestone before becoming constricted or blocked by sediment and
boulders. Pwll y Pasg is one good example. The Millstone Grit
waters appear as discrete streams running in narrow vadose
canyons within the major caves such as the Ace of Spades inlets
in Agen Allwedd. In Ogof Craig a Ffynnon, water derived
from the Millstone Grit also enters through the massive
boulder chokes which are such a feature of the cave. These
contain blocks of foundered grit suggesting that the shaft
development mechanism envisaged by Burke & Bird (1966)
and Bull (1977) has reached an advanced stage in the east of
the area.

Tracing work (Fig. 12.3) has also established that not all flow
from the Millstone Grit is essentially vertical through the

10.0 km² of which over 3.5 km² is tributary to the Main Stream Passage. It is thus very unlikely that these early fossil passages were formed by the upper Usk whose catchment currently exceeds 550 km². This confirms the suggestion made above that the Usk only occupied its present course rather later in the evolution of the System following breaching of the Old Red Sandstone escarpment.

Figure 12.22 suggests two possibilities; first, that the Rhian-goll followed the regional pattern to the south and was engulfed where it intersected the limestones. Its present catchment area is some 49 km² and this may be rather large even for the biggest passages in Summertime. The second possibility is that this stream fed into the headwaters of the developing lower Usk and that a rather small stream drained from the escarpment in the area now occupied by the link with the upper Wye. The catchment area (>15 km²) seems a good fit but detailed measurements of the hydraulic geometry of the cave passages are needed to test these suggestions and examine the possibility that further inlets were present, for example, to feed Jigsaw Passage in Daren Cilau.

The course of the lower Usk probably traversed the limestone outcrop in the vicinity of Abergavenny much as today but was also fed by rather more water from the north, now captured by the River Monnow. Assuming that at 2–3° the dip of the limestone was greater than the gradient of the river (highly probable), the limestones would have been exposed in

the valley floor first in the north, with continued incision causing migration of the outcrops southwards with time (Fig. 12.24). Thus, while the point input to the limestones was at first fixed, the lowest outlet migrated down-dip and southwards in the Usk Valley through time. Hence the initial drainage had a relatively large strike component (>4 km), while the extent of down-dip flow was more limited, perhaps <1 km, due to the proximity of the Old Red Sandstone escarpment to the north.

The Cave System could therefore be modelled as an essentially vadose down-dip segment to the zone of saturation, followed by a major strike-oriented phreatic conduit which also gained stratigraphic elevation, the input being at the base and the outlet towards the top of the limestone (Fig. 12.24). The highest and therefore oldest strike passage in this sequence is the Old Main Chamber of Daren Cilau which discharged to the southeast through the fault-guided Old Main Rift Passage, the downstream continuation now being choked by a flowstone blockage. The major Inner Circle to Main Chamber route in Agen Allwedd is probably the best preserved representative of these early strike-oriented conduits but the upstream segments have been destroyed by erosion.

Within the Llangattwg System it is apparent that many passages have developed immediately below the nodular shaly bands towards the top of the Oolite Group. These are particu-

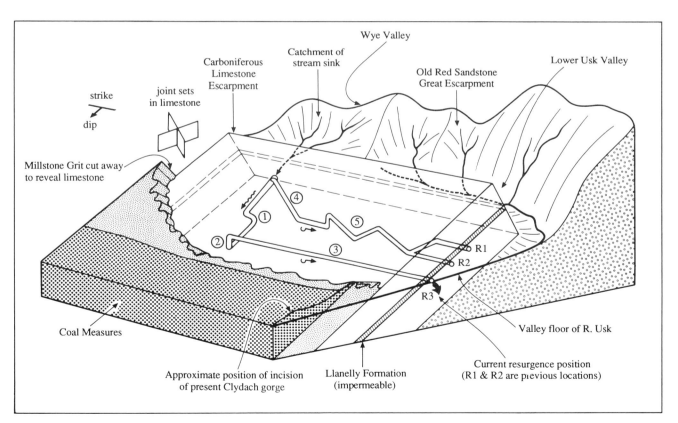

Fig. 12.24. Block diagram of the early development of the Llangattwg caves. R1, R2 and R3 represent three successive positions of the resurgence controlled by the incision of the River Usk. (1) Down-dip flow in the vadose zone. (2) Lifting segment taking water from the base of the limestone up to beneath the nodular band. **(3) Strike flow in the phreatic zone. (4) Strong SSE-joint control of vadose passage. (5) Phreatic strike conduit developed on WSW–SSE joint set giving looping form. Note: the R3 resurgence is idealized, whilst R1 and R2 are more realistic.**

Phreatic passage forms

Vadose passage forms

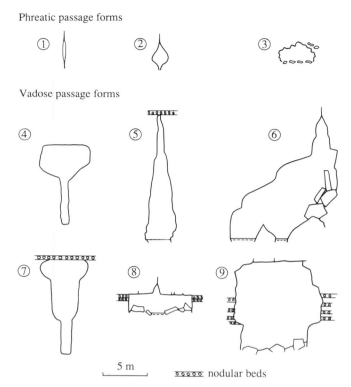

5 m

○○○○○ nodular beds

Fig. 12.25. Typical passage cross-sections in the Llangattwg caves. (1) Initial phreatic 'rift' on joint or fault (Man in Roof Passage, Ogof Daren Cilau). (2) Development of phreatic tube (Misfit Passage, Ogof Daren Cilau). (3) Phreatic tube and spongework developed along bedding plane (Upper Loop Route, Ogof Daren Cilau). (4) Single vadose trench below phreatic passage (Keyhole Passage, Agen Allwedd). (5) Widening vadose trench below phreatic passage (White Passage, Ogof Daren Cilau). (6) Single vadose trench below phreatic passage with meander undercuts (Epocalypse Way, Ogof Daren Cilau). (7) Multiple vadose trench below phreatic tube (No Name Canyon, Ogof Daren Cilau). (8) Collapse of roof controlled by nodular band in vadose passage (Jigsaw Passage, Ogof Daren Cilau). (9) Extensive collapse causing upward stoping of roof in vadose passage (Time Machine, Ogof Daren Cilau). Note: all the above types of passage may be subject to infilling to any depth by clastic sediments or, less commonly, by speleothems.

larly well developed to the north and acted as aquicludes, preventing water from attaining higher beds. In some cases entirely separate passages developed, isolated by the nodular bands, for example, Man in the Roof Passage above and Epocalypse Way below a prominent nodular band in Daren Cilau. Elsewhere, they were penetrated: thus Preliminary Passage in the same cave is roofed by one band through which a phreatic opening at the top of the 20 m pitch connects into Higher Things Passage, which is roofed by a second higher nodular shale band. The Llanelly Formation with its thick shales was, however, much more effective and no passages in the System are known to penetrate this aquiclude (Eglwys Passage and Old Main Rift Passage are both capped by this formation, while in Agen Allwedd it has been exposed in the roof by collapse as, for example, in the dome of St Pauls and in Erse Passage). While spongework and bedding-plane anastomoses are observed in these phreatic passages formed under the nodular horizons, the majority are guided by joints and the initial opening takes the form of a narrow phreatic rift

up to 20 cm wide and sometimes several metres in height (Fig. 12.25). Thus, close inspection of the strike phreatic passages shows that they are in fact often guided by a long segment formed on the S–SE joint set and another formed on the W–NW-oriented set. This is clearly seen in the lower Agen Allwedd Streamway where, because the former set leads the passage down-dip, the W–NW set lifts the water back up to the previous level, giving a classic looping phreatic form. In fact ponding in the bottom of these loops is currently responsible for Sumps One, Two and Three in Agen Allwedd (Fig. 12.23a). To the south and east the nodular shale horizons are replaced by dolomites whose role in controlling cavern genesis is less clearly understood; a good example is The King's Road, Daren Cilau.

An additional structural control of some significance is the effect of the faulting which is generally downthrown to the west on the dominant S–SE set. Thus, water confined beneath a particular nodular horizon and under hydrostatic pressure in the phreatic zone may lift across the fault plane before again being confined by the nodular shale horizon at a higher level. This may also permit a down-dip deflection of the passage. The faults are also significant in allowing water to penetrate between the various nodular horizons within the Oolite Group. The phreatic precursor to Jigsaw Passage, for example, lifts up a minor fault and discharges down the well-marked tube in Misfit Passage rather than following the lower part of Jigsaw Passage.

Evolution of the early Cave System
   Many of the major passages in the Llangattwg System are clearly two-phase, with an initial phreatic passage modified by a vadose trench of comparable dimensions. Main Passage in Agen Allwedd and Epocalypse Way in Daren Cilau are of this type. This suggests that as the base level controlled by the River Usk fell, passages which were initially in the shallow phreatic zone developed a free air surface and trenches were incised in their floors. Many of the larger vadose passages in the caves show either a marked widening towards the base or wide-sweeping undercuts of which good examples are seen in the first part of Epocalypse (Fig. 12.25). This morphology suggests that incision was not rapid and that sediment provided a protective cover on the floor causing at least comparable, if not greater, wall than floor erosion. Detailed observation may demonstrate that some of these vadose canyons are actually incised into the crest of major phreatic loops and are thus isolated by phreatic sections. This can be clearly observed in Main Stream Passage of Agen Allwedd where at Deep Water the phreatic passage is barely trenched but both upstream and downstream the height of the canyon passage increases to in excess of 15 m.

In addition to the slow vadose modification of existing passages, new phreatic passages continued to develop at lower levels and were eventually able to capture the major flow causing abandonment of the higher level passages. Again, careful examination of the evidence in Summertime may enable a detailed sequence to be reconstructed but accurate levelling of this part of the System will be needed. In Daren Cilau, the sequence is more clear (Fig. 12.26) although boulder

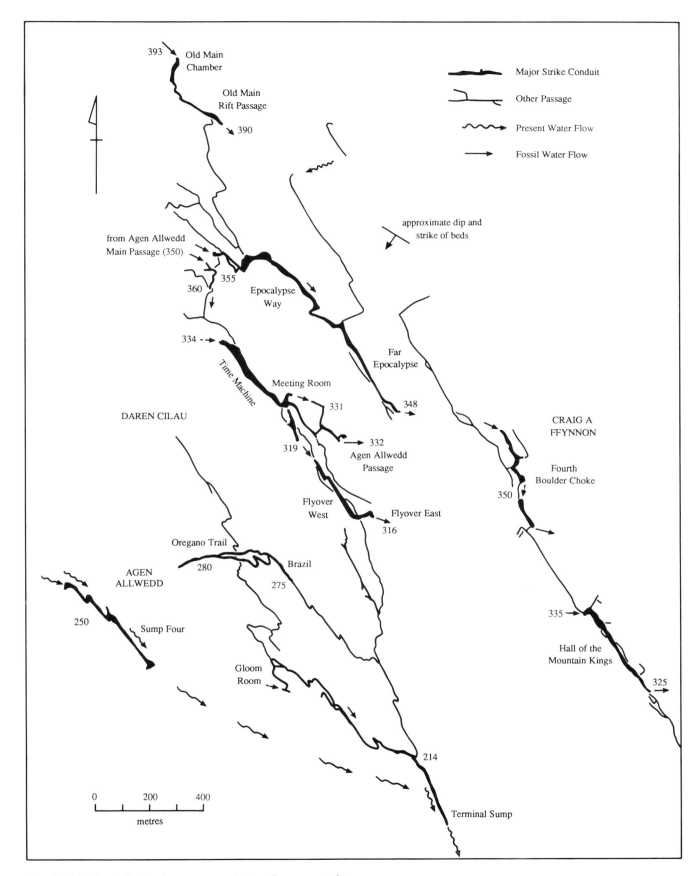

Fig. 12.26. Major strike trunk passages and later vadose passages in Ogof Daren Cilau, Ogof Craig a Ffynnon and the Terminal Sump of Agen Allwedd (numbers are spot heights in metres).

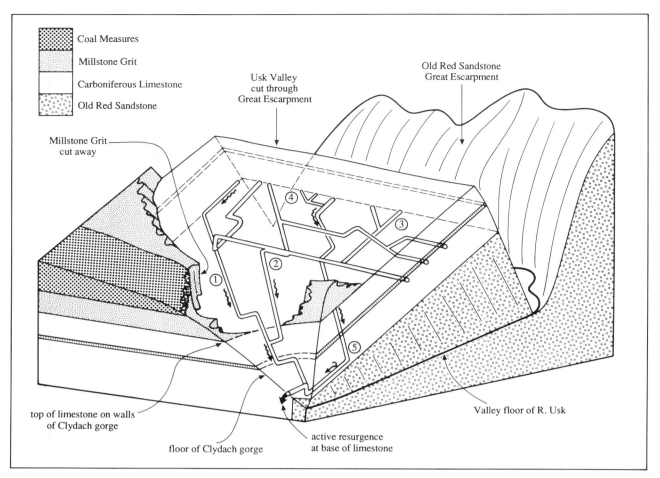

Fig. 12.27. Block diagram of evolution of the early cave system following the breaching of the Old Red Sandstone Great Escarpment by the River Usk. (1) Active drainage reoriented onto SSE joints to give larger down-dip component. (2) Previous passages utilized by vadose streams. (3) External invasion trench derived from concentrated recharge from Millstone Grit or glacial sources. (4) Limestone percolation water collects below outcrops in small vadose passages. (5) Limited phreatic development near resurgences.

chokes currently prevent mapping of the strike continuations; here the influence on passage orientation of the S–SE joints and faults is much greater than in Agen Allwedd. Further exploration will clarify this sequence which at present shows several considerable gaps, for example between the Flyover and Gloom Room levels, although the recently entered Oregano Trail leading to 12 O'clock High may be important here. These may simply result from the effects of essentially random geological controls on the elevation of new phreatic passages giving irregularity in the sequence of falling passage elevations, but another possible explanation is considered below.

Breaching of the Old Red Sandstone escarpment and the connection of the upper and lower Usk drainage routes have had a substantial effect, both in increasing the rate of incision of the lower Usk and in causing diversion of the major stream sinks into the surface drainage. Headward erosion of the steepened River Clydach caused capture of the headwaters of the southward-flowing surface streams near Brynmawr (Fig. 12.22) and the positions of the resurgences were rapidly localized within this valley near the base of the limestone (Fig. 12.27), perhaps explaining the Flyover–Gloom Room capture mentioned above. The down-dip component of the under-

ground drainage therefore increased and there was a significant reorganization of the drainage (Fig. 12.27). The effects are apparent when it is recognized that the area occupied by the caves is much wider to the east than the west and that the major ancient streamways (e.g., St Pauls–Main Passage–Epocalypse) have a more westerly orientation than the present Main Stream Passage drainage. The incision of the Usk, both before and after capture, caused the development of an extensive vadose zone within the limestones. Much of this was fed by percolation at the outcrop, as described for the present active drainage, giving numerous small vadose trenches. In some cases, these occupied down-gradient segments of pre-existing passages, sometimes reversing their phreatic flow direction and often being significantly smaller in size than the formative stream. This produced the distinctive keyhole-shaped cross-section with a narrow trench typified by Keyhole Passage in Agen Allwedd, where incision reached 10 m (Fig. 12.25) (Agen Allwedd is Welsh for keyhole cleft). Where the guiding phreatic passage turned along the strike or up-dip a new vadose passage developed following the limited phreatic openings along the major joint network, as in the case of Southern Stream Passage.

In some cases, the vadose passages attained a significant size, perhaps due to concentration of inflow on the Millstone Grit, as at the present Llangattwg Swallet. These developed major new routes towards the trunk streamway, cutting across the predominantly strike-oriented ancient passages. This is clearly illustrated by the Jigsaw Passage streamway, which has reversed the floor gradient of the Epocalypse Way trunk route in the St Valentine's Loop Route segment, before carving a deep exit trench into the lower part of the system (Time Machine) via Preliminary and White Passages. Because the majority of the phreatic openings guiding these passages were developed below the nodular horizons near the top of the Oolite Group, trench development was not limited by the dolomitization and impurity of the basal limestones. Thus, in Agen Allwedd, trenches >20 m deep have formed spanning a substantial thickness of the Oolite Group. In fact the present active stream downstream of Northwest Junction is now cutting into the Lower Limestone Shales at several points, an indication of its prolonged activity. Main Stream Passage upstream of Northwest Junction and Southern Stream Passage also cut deeply into the Lower Limestone Shales.

This continuity of vadose erosion and the effects of many generations of cave streams is also responsible for the substantial size of passages such as the Time Machine. Here, the original phreatic route has hosted all subsequent vadose activity, including that of the present Bonsai Stream. Undercutting of the walls and solutional removal of the resulting

collapse debris continues today and the effects are easily seen just below the entry to White Passage. Downstream of the Time Machine, the routes diverge, with many different passages occupied at various stages in the caves' evolution, none of which attain a comparable size (Fig. 12.28). It should be possible to define a sequence of previous water-table levels within the System by the levels to which the various trenches are graded, and at which they pass downwards into phreatic passages. At present, while good-quality survey information is becoming available, insufficient detailed fieldwork has been conducted to make this possible.

The nodular shaly bands are mechanically weak and frequent breakdown of the roof and walls of the passages has occurred at this level. This may give rise to the distinctive rubbly notches in passage walls (as in the Time Machine) but more usually produces an angular flat-roofed passage, as in the upper part of Jigsaw Passage, Daren Cilau (Fig. 12.25). Continued breakdown of the roof may also occur (e.g., Main Chamber of Agen Allwedd) and eventually the distinctive thick green mudstones of the Llanelly Formation may become exposed. These are very blocky and usually a more stable dome-shaped roof is formed (e.g., St Pauls, Agen Allwedd) following development of a downward bulge, as seen in the Big Chamber–Nowhere Near the Entrance (Daren Cilau), which may subsequently collapse. Some of the boulder chokes which are such a characteristic feature of the Llangattwg caves are thus derived wholly by roof collapse, particularly at the inter-

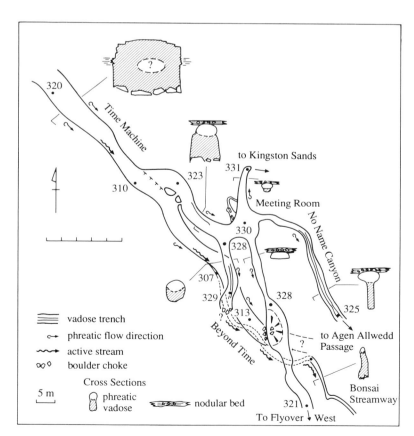

Fig. 12.28. Divergence of phreatic routes downstream of the Time Machine in Ogof Daren Cilau and comparison of the extent of vadose modification (numbers are spot heights in metres).

section of ancient high-level and more recent vadose passages. There is no reason in this case to expect any topographic expression on the surface (Bull, 1977). This type of choke contrasts markedly with that derived by development of a sub-Millstone Grit solution pipe along some major fracture line. In the former case penetration over the top of the collapse offers good potential, while in the latter better progress may be made by following the rock wall along the side of the choke (cf. Eglwys Passage Choke, Daren Cilau, and Third Boulder Choke, which is associated with a surface depression, Craig a Ffynnon). Boulder chokes may cause considerable ponding in the cave passages which may make interpretation of vadose trench sequences and phreatic water-table levels more difficult.

Sedimentary infill and the effects of glaciation

A particular feature of the caves of Llangattwg is the deposits of sand and silt which infill many of the passages. They are particularly abundant in the northern parts of Agen Allwedd giving rise to names such as Sand Caverns Passage. Sedimentary structures indicate flow was from the northwest and the coarse grain size suggests unimpeded flow throughout the upper System, although there is some evidence of ponding in the Cliffs of Dover area (Fig. 12.8). This may relate to multiple inputs as suggested by Bull (1976a) or be caused by a boulder choke beyond the present end of the cave. It is clear that many of the passages in Agen Allwedd have been completely filled to the roof; the declining size of Midsummer Passage to the southeast is almost certainly due to the increasing depth of fill in this direction. The sediments also overlie breakdown in passages which show a complete suite of developmental forms. There is thus little doubt that the fill is relatively recent and not derived from the caves' evolutionary phases.

In a detailed study, Bull (1976b) has demonstrated that quartz grains from the basal parts of the sediments show distinctive glacial surface morphology when inspected under the scanning electron microscope. Higher in the sequence these features are absent and the grains are remarkably fresh suggesting limited transport and derivation from the overlying Millstone Grit. Bull (1976b, 1980) suggested that they entered the caves under a periglacial regime but it is equally possible that they could be derived from stagnant ice or névé accumulations on the plateau where the high pressures needed to cause grain crushing would not occur. Given that the last glaciation persisted for a minimum of 10 000 y and that there is clear evidence for the existence of sediment-transporting streams entering the caves, some significant morphological modifications would be expected. One possibility is that the extensive fill allowed rejuvenation and trenching of high-level passages following blockage of the lower routes. Such an explanation could, for example, be applied to explain passages like Preliminary Passage, described previously. Many of the smaller passages, such as the entrance series in both Agen Allwedd and Daren Cilau, must certainly have been enlarged at this time by the much greater flows which were supplied from the margins of the Usk glacier. Perhaps it is a pity the ice did not remain longer! In the Daren Cilau entrance passages, at least

two separate phases of vadose excavation have occurred and there is a gravelly fill overlain by several tens of centimetres of flowstone.

In contrast to Agen Allwedd the other caves appear to have finer-grained sediment fills implying a proximal–distal relationship and a flow to the southeast, as under non-glacial conditions. Fine sands are known, however, as for example in Flyover West. In many places breakdown boulders are coated with mud and quite thick laminated sequences are also found, as near the Third Boulder Choke in Craig a Ffynnon. In Epocalypse the removal of such a fill has produced large numbers of mudballs but there is remarkably little earlier coarse sediment flooring the passage. In Agen Allwedd the thick sandy deposits are capped by a mud deposited in ponded conditions which endured sufficiently for consolidation to occur prior to retreat of the water and desiccation of the sediment. Bull (1980) has shown that much of this sediment entered the cave by a translatory flow mechanism via solutional openings from the surface, giving a truly remarkable regularity in the thickness of the alternating fine and coarse couplets which make up the laminated mud deposits along the whole length of Main Passage (Fig. 12.29). Subsequent

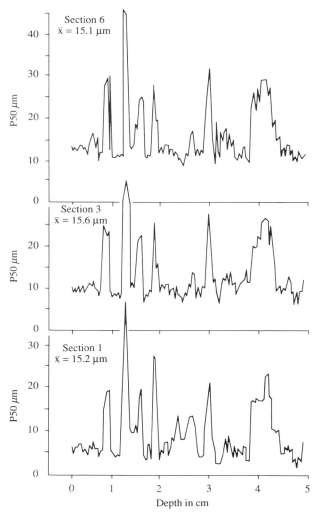

Fig. 12.29. Median grain size curves for three sections of laminated cap-mud in Main Passage, Agen Allwedd (after Bull, 1980).

palaeomagnetic studies have confirmed this interpretation (Noel *et al.*, 1981; Noel 1983, 1986). This ponding may well represent the glacial maximum when, as Ford (1979) envisaged, hydraulic heads in the cave were low because it was completely overriden by ice. This suggestion, whilst attractive, does not completely conform with modern studies of subglacial hydrology (Smart, 1986) and the ponding may be a local event within Main Passage (or the upper part of the System) due to some essentially random blockage caused by the prior influx of coarse sediment.

Substantial speleothem deposits in the Llangattwg caves are largely limited to the marginal zone where the limestone is exposed at the surface and soil bacterial processes are more active, generating elevated carbon dioxide partial pressures in the soil atmosphere. Here significant deposits may accumulate as at the stalagmite squeezes in Daren Cilau. Below the Millstone Grit, many waters remain aggressive where they enter the cave and carbonate speleothems are less common. This lack of secondary cementation by calcite may explain why preservation of ancient sediments is limited. Gascoine (1982) has observed that soft black secondary deposits are developing beneath the gritstone inlets and provide a useful guide to the water source. On analysis these were found to be iron and manganese hydroxy-oxides, precipitated because of the change from partially reducing conditions in the shales and peats of the Millstone Grit capping to the oxidizing conditions of the fissures. There is little organic matter present in these deposits.

The shales of the Millstone Grit are pyritiferous and this may well provide the source of sulphate for the development of the gypsum deposits in the caves. These are found growing from the surface of desiccating mud banks, from shales, from stalactites (Figs. 12.30, 12.31) and much more rarely, direct from the limestone bedrock. Typically the habit is bladed or acicular, often with twinning but pure white fibrous masses are also found. Very beautiful forms of aragonite (Kendall, 1988), are present in Urchin Oxbow and there are also aragonite overgrowths on ancient and massive fractured calcite formations in Epocalypse. The delicacy of many of these forms is mirrored by the calcite helictites (Fig. 12.32), widely found in the System and best known as the Bonsai Tree in the stream-

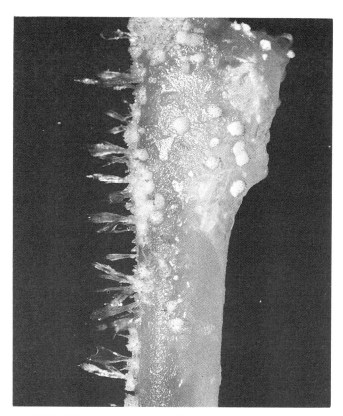

Fig. 12.31. Selenite (gypsum) crystals <1 mm long growing on the side of a straw stalactite, Ogof Craig a Ffynnon (photo by C. Howes).

Fig. 12.32. 'The Antlers' helictites in Ogof Daren Cilau (photo by J. J. Rowland).

Fig. 12.30. Gypsum crystals on mud block, Main Passage, Agen Allwedd (photo by T. D. Ford).

way of the same name. These and the 'popcorn' and crystal wall coatings suggest that significant intergranular or microfissure seepage must occur but these depositional processes are poorly understood and await detailed study.

Conclusions

Cave exploration on Llangattwg has currently received a major impetus with the discoveries of Daren Cilau and the water-tracing work of Gascoine. This and the geomorphological analysis presented above clearly indicates that connections between the three major Systems under Llangattwg are probable and merely require determined exploration. It is also apparent that while the extensive sediment fills have proved a great hindrance, the cross-cutting of the ancient strike-oriented sink fed systems with the later, vadose, SSE–oriented, joint-controlled system feeding the Clydach provide opportunities for linking genetically separate segments of the System. This leaves an enigma as to where the strike-oriented Systems had their resurgences, as the presently active resurgences in the gorge are there only because of the recent incision of the gorge. Nevertheless, the key to continued exploration under Llangattwg remains the negotiation of the frequent boulder chokes, both internally and externally derived, which bring exploration of many of the main passages to a premature halt. Perhaps in the near future such exploration, and the survey and mapping to a high standard which now routinely follow, will enable a more complete and detailed account to be made of the development and evolution of the caves in this classic area.

Acknowledgements

We would like to thank the many people who have contributed both directly and indirectly to the final production of this chapter. Peter Bull and John Parker deserve special mention for their initial work, while Mark Noel was significant in getting the thing moving. Paul Wright proved patient and helpful in sorting out the geology, and greatly improved the standard of this section. Bill Gascoine readily provided details of his tracing work in the area and also kindly read through the draft text. Many cavers, including John Hunt, John Cooper, and the Cambrian Cave Registry, have contributed to the cave surveys and details presented here, but the work of Harold Lord, Pete Cousins and Ian Penney (Agen Allwedd), John Parker (Craig a Ffynnon) and especially Dave Ramsay (Daren Cilau) are greatly appreciated. References for the area history were helpfully provided by Doug Muir at Crickhowell Public Library. Simon Godden translated the draft figures into neat diagrams.

References

Barclay, W. J. & Jackson, D. I. (1982). The geology of Carno adit, Ebbw Vale, South Wales. *Report of the Institute of Geological Sciences*, **82(1)**, 41–4.

Battiau-Queney, Y. (1980). *Contribution à l'Etude Géomorphologique du Massif Gallois*. Unpublished Ph.D. thesis. L'Université de Bretagne Occidentale, France.

Battiau-Queney, Y. (1984). The pre-glacial evolution of Wales. *Earth Surface Processes and Landforms*, **9**, 229–52.

Boulton, G. S., Jones, A. S., Clayton, K. M. & Kenning, M. J. (1977). A British ice-sheet model and patterns of glacial erosion and deposition in Britain. In *British Quaternary Studies*, ed. F. W. Shotton, pp. 232–46. Oxford: Clarendon Press.

Bowen, D. Q. (1973). The Pleistocene history of Wales and the borderland. *Geological Journal*, **8**, 207–24.

Brown, E. H. (1960). *The Relief and Drainage of Wales*. Cardiff: University of Wales Press.

Bull, P. A. (1976a). *Cave Sediment Studies in Agen Allwedd*. Unpublished Ph.D. thesis. University College of Wales, Swansea.

Bull, P. A. (1976b). An electron microscope study of cave sediments from Agen Allwedd, Powys. *Transactions of the British Cave Research Association*, **3**, 7–14.

Bull, P. A. (1977). Cave boulder chokes and doline relationships. *Proceedings of the 7th International Congress of Speleology, Sheffield*, pp. 93–6.

Bull, P. A. (1980). Towards a reconstruction of timescales and palaeoenvironments from cave sediment studies. In *Timescales in Geomorphology*, ed. R. A. Cullingford, D. A. Davidson & J. Lewin, pp. 177–87. London: Wiley & Sons.

Burke, A. R. & Bird, P. F. (1966). A new mechanism for the formation of vertical shafts in Carboniferous Limestone. *Nature, London*, **210**, 831–2.

Cousins, P. R. (1973). The geomorphology of Agen Allwedd. *Journal of the Wessex Cave Club*, **12**, 282–8.

Cousins, P. R. (1977). Survey work in Agen Allwedd. *Bulletin of the British Cave Research Association*, **18**, 9–11.

Farr, M. J. (1985). Daren Cilau – how Britain's most spectacular passage was discovered. *Descent*, **64**, 12–15.

Farr, M. J. (1986). Through Llangattwg Mountain. *Descent*, **72**, 19–26.

Ford, D. C. (1979). A review of alpine karst in the Southern Rocky Mountains of Canada. *Bulletin of the National Speleological Society*, **41**, 53–65.

Gardener, C. G. (1984). Daren Cilau – the route to the heart of the Llangattwg Mountain. *Descent*, **61**, 17–24.

Gardener, C. G. (1985). Daren Cilau – a new 5 km extension and the largest underground passage in Britain. *Caves and Caving*, **28**, 6–11.

Gardener, C. G. (1986). Cascading water and echoes of David Bowie from the depths of Daren Cilau. *Caves and Caving*, **33**, 20–6.

Gascoine, W. (1979). Ogof Craig a Ffynnon. *Journal of the Cambrian Caving Council*, **5**, 66–9.

Gascoine, W. (1982). The formation of black deposits in some caves of southeast Wales. *Transactions of the British Cave Research Association*, **9**, 165–75.

George, T. N. (1954). Pre-Seminulan Main Limestone of the Avonian Series in Breconshire. *Quarterly Journal of the Geological Society, London*, **110**, 283–322.

George, T. N. (1970). *British Regional Geology: South Wales*, 3rd edn. London: HMSO.

Hack, J. T. (1960). Interpretation of erosional topography in humid temperate regions. *American Journal of Science*, **258A**, 80–97.

Heaver, S. (1983, 1984, 1985, 1986). *Bat Investigations on Mynydd Llangattwg*. Private publications from S. Heaver, Chelsea Speleological Society.

Jenkins, D. A. (1963). Notes on the geology of Agen Allwedd and Mynydd Llangattwg. *Proceedings of the British Speleological Association Annual Conference*, **1**, 49–65.

Jones, O. T. (1951). The drainage systems of Wales and the adjacent regions. *Quarterly Journal of the Geological Society*, **107**, 201–25.

Jones, T. (1911). *A History of the County of Brecknock, III*. Brecon: Blissett Davis & Co.

Kendall, A. (1988). Aragonite in Ogof Daren Cilau. *Cave Science*, **15**, 83–4.

Leitch, D. E. (1960). Ogof Agen Allwedd in relation to the Mynydd Llangattwg. *Cave Research Group Publication*, **10**.

Leitch D. E. (1973). A review of the speleogenesis of Agen Allwedd. *Hereford Caving Club 21st Anniversary Publication*, 7–43.

Noel, M. (1983). The magnetic remanence and anisotropy of

susceptibility of cave sediments from Agen Allwedd, South Wales. *Geophysical Journal of the Royal Astronomical Society*, **72**, 557–70.

Noel, M. (1986). The palaeomagnetism and magnetic fabric of cave sediments from Pwll y Gwynt, South Wales. *Physics of the Earth and Planetary Interiors*, **44**, 62–71.

Noel, M., Retallick, W. G. & Bull, P. A. (1981). Further palaeomagnetic studies of sediments from Agen Allwedd. *Transactions of the British Cave Research Association*, **8**, 178–87.

Owen, T. R. & Jones, D. G. (1961). The nature of the Millstone Grit–Carboniferous Limestone junction of a part of the north crop of the South Wales coalfield. *Proceedings of the Geological Association*, **77**, 187–98.

Parker, J. (1978). Ogof Craig a Ffynnon – 4 miles of new cave found in South Wales. *Descent*, **38**, 16–26.

Rattenbury, G. (1980). *Tramroads of the Brecknock and Abergavenny Canal*. Hartland: Railway and Canal Historical Society.

Self, C. A. (1981). *Caves of County Clare*. Bristol: University of Bristol Speleological Society.

Smart, P. L. (1986). Origin and development of glaciokarst closed depressions in the Picos de Europa, Spain. *Zeitschrift für geomorphologie NF*, **30(4)**, 423–43.

Thomas, T. M. (1954). Solution subsidence outliers of Millstone Grit on the Carboniferous Limestone outcrop of the North Crop of the South Wales coalfield. *Geological Magazine*, **91**, 220–6.

Thomas, T. M. (1959). The geomorphology of Brecknock. *Brycheiniog*, **5**, 129–36.

Thomas, T. M. (1974). The South Wales inter-stratal karst. *Transactions of the British Cave Research Association*, **1**, 131–52.

Walker, M. J. C. (1984). Craig-y-Fro. In *Field Guide to Wales: Gower, Preseli, Fforest Fawr*, ed. D. Q. Bowen & A. Henry, pp. 91–5. Quaternary Research Association.

Welsh Water Authority (1980). *Southeast Wales Groundwater Study*. Unpublished Report by Howard Humphreys.

Williams, G. J. (1968). *Contributions to the Pleistocene Geomorphology of the Middle and Lower Usk*. Unpublished Ph.D. thesis. University of Wales, Cardiff.

Wright, V. P. (1981). *The Stratigraphy and Sedimentology of the Llanelly Formation between Blorenge and Penderyn, South Wales*. Unpublished Ph.D. thesis. University of Wales, Cardiff.

Wright, V. P. (1982a). The recognition and interpretation of palaeokarsts: two examples from the Lower Carboniferous of South Wales. *Journal of Sedimentary Petrology*, **52**, 83–94.

Wright, V. P. (1982b). Calcrete palaeosols from the Lower Carboniferous Llanelly Formation, South Wales. *Sedimentary Geology*, **33**, 1–33.

Wright, V. P. (1986). Facies sequences on a carbonate ramp: the Carboniferous Limestone of South Wales. *Sedimentology*, **33**, 221–41.

Wright, V. P., Raven, M. & Burchette, T. P. (1981). *A Field Guide to the Lower Carboniferous Rocks Near Abergavenny*. Cardiff: Geology Department, University College Wales.

13

The caves of Nant y Glais, Vaynor

TREVOR D. FORD

A small group of caves lies within the flanks of the Nant y Glais Valley, west of Vaynor village. The Nant y Glais is a tributary of the Taff Fechan with headwaters on the southern flanks of the Vaynor Moors; largely drift-covered these are composed of Upper Old Red Sandstone and the Lower Limestone Shales. The middle reaches of this river cross the Carboniferous Limestone along the line of a major NNW–SSE fault downthrowing eastwards. Subsidiary faults have given rise to numerous fractures parallel to the river and it is in these shattered limestones that the caves are developed.

The Nant y Glais River either disappears underground or runs on the surface through a small gorge where it crosses the limestone, according to the stage of flow. Some of the sinks are into tight joint passages and others are poorly defined soakaways in the riverbed. Resurgences are similar. No contribution has been proved to come from sinks in the adjacent moorlands (see Chapter 4), although it is likely that some percolation water reaches both main caves, possibly from doline-type sinks. The stream in Ogof y Ci is fed from sinks in the riverbed upstream from Blaen y Glais Farm (Fig. 13.1).

The gorge and caves are developed in the lower part of the Dinantian sequence, in the Abercriban Oolite where it rests on dolomitized limestones. Owing to faulting the walls of the gorge tend to be in different beds, those of the east bank being somewhat higher than the west. The caves, however, have utilized the same horizon so that the passages of Ogof Rhyd Sych lie somewhat lower than the riverbed.

The caves
Traversing downstream the caves are as follows:

Ogof Robin Goch (SO 0392 1076). Barely 100 m in length, a small hole in the east bank takes flood water into a bedding tube leading to a moderate-sized chamber with a boulder choke almost beneath a doline in the field above. The choke is apparently the upstream side of the terminal choke in Ogof Rhyd Sych.

Ogof y Ci (SO 0403 1051). This cave lies closely parallel to the gorge beneath its west bank, and has three entrances, the upper two being dry and the third wet. Totalling some 1600 m in length, Ogof y Ci is mostly a single active stream passage of

varying height with several tubes developed on cross-joints all becoming too tight to negotiate after a few metres. The upstream end of the cave splits into two closely parallel joint-controlled passages, with the easterly of these close to the river and sinks therein. The downstream end of Ogof y Ci is close to river-bank resurgences, where the water sinks again almost immediately and reappears in Ogof Rhyd Sych.

Ogof Dŵr Dwfn (SO 0415 1022). This is a minor resurgence cave in the west bank about 150 m downstream from the Ogof y Ci risings, and a few metres upstream from Ogof Rhyd Sych.

Ogof Rhyd Sych (SO 0416 1021). This is the second main cave of the Nant y Glais area. A stream issues from the small entrance in the east bank which leads into a substantial stream passage (Fig. 13.2). This soon closes down and most of the cave is a series of tight rifts in joints parallel to the gorge and linking crawls. Totalling about 2000 m in length, the cave trends generally northwards and ends in a boulder choke close to Ogof Robin Goch. The stream from Ogof y Ci reappears from tight joints half-way through the cave. The system is susceptible to flooding as it carries most of the Nant y Glais flow, even in normal conditions.

Ogof Pysgodyn Gwyn (SO 0416 1016). This is a resurgence cave in the west bank of the Nant y Glais some 25 m downstream from Ogof Rhyd Sych. It is a phreatic tube subject to repeated silting-up which leads through a series of sumps to a short canal, some 30 m long. It is fed by sinks in the stream course approximately 50 m upstream.

Ogof Jonny Bach (SO 042 100). This is a bedding-plane cave in the west bank which takes water, only to resurge about 20 m downstream.

Geomorphology
Lying close to the walls of the Nant y Glais Gorge the caves are in an interesting position which can be interpreted as potentially leading to collapse and hence widening of the gorge. While both Ogof y Ci and Ogof Rhyd Sych are largely vadose streamways today, they have been relatively little modified from the network of phreatically enlarged joints resulting from faulting. As both caves have their floors close to the relatively insoluble dolomite, their passages tend to have

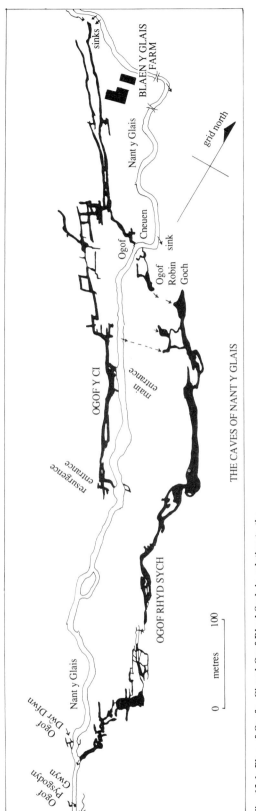

THE CAVES OF NANT Y GLAIS

Fig. 13.1. Plan of Ogof y Ci and Ogof Rhyd Sych in relation to the Nant y Glais River course.

Fig. 13.2. The Main Streamway of Ogof Rhyd Sych; note the large flowstone banks in the left centre (photo by C. D. Westlake).

triangular cross-sections with flat floors. The profile of Ogof y Ci is close to that of the riverbed in the gorge while the main passage of Ogof Rhyd Sych is 5–10 m lower. Speleothems are largely restricted to the middle sections of both caves where vadose features are best developed. The close relationship to the valley floor suggests that the caves are mainly of post-glacial origin in this tributary valley sheltered from active glacial scour by the Vaynor Moors. However, there is ample scope for further study.

14

The caves of the Mellte and Hepste Valleys area

TONY WALTHAM and D. G. EVERETT
(largely based on material by the late Oliver C. Lloyd)

The limestone outcrops at the head of the Vale of Neath are noted for their structural complexity and they contain some extensive caves. The largest cave systems are on the Afon Nedd Fechan – the Little Neath River. These were described in Chapter 13. The two major tributary rivers, the Mellte and the Hepste, also have important cave systems, and there are various other caves and karst features in the areas around Ystradfellte and Penderyn.

Geology

As is clear from Figure 14.1, the overall geology of the area is dominated by the gentle southerly dip of the limestone and the major offsets by faulting associated with the Neath Disturbance. The southerly dip of the limestone is generally

around 5°, and the overlying conglomerates, grits and quartzites of the Millstone Grit Series form upland immediately south of the limestone outcrops. This simple structure is broken by numerous small faults orientated NNW–SSE, but the Neath Disturbance is a far larger structural break aligned ENE–WSW. It is a major fault zone containing zones of tight complex folding of the Carboniferous rocks, and accounts for the repeat of the limestone outcrops, firstly through the Hepste and Mellte basins, and then again through the Penderyn area.

The primary drainage of the area is just east of south, from the high slopes of the Brecon Beacons into the heart of the South Wales Coalfield. The surface river courses progressively climb the stratigraphic succession, as dip is steeper than regional slope. Drainage has, however, been complicated by a

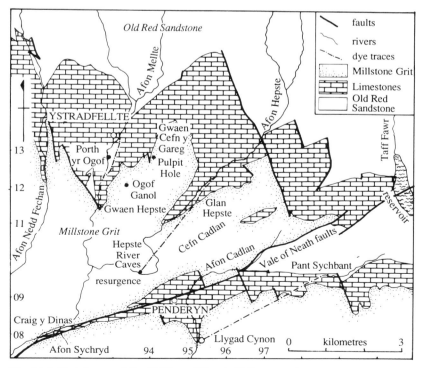

Fig. 14.1. Sketch map of the Mellte and Hepste Valleys area to show the relationship of rivers and caves to the geology.

succession of river captures, the most important being that by the River Neath, working headwards along the weakness of the Neath Disturbance. This youthful capture means that sections of many of the rivers are deeply entrenched – notably the Mellte, Hepste and Sychryd. The stages of capture were probably accompanied by glacial diversions, and the southwards flow of the main ice sheets was itself complicated by diversions around the grit escarpment and into the Vale of Neath.

Porth yr Ogof

The flow of the River Mellte is underground for much of its course across the limestone outcrop, and part of its route is through the well-known cave of Porth yr Ogof (SN 928 124). Located a kilometre south of the village of Ystradfellte, the cave has the largest entrance in Wales where it swallows the flood course of the river (Fig. 14.2). The underground course is open to the resurgence, but the cold water and deep lakes, which sump in wet weather, make the through-trip hazardous, and ill-equipped cavers have drowned in the past. Porth yr Ogof has been known for centuries, but the first systematic exploration was in 1936. Since then the cave has been further explored by divers, and has also been resurveyed (Standing & Lloyd, 1970; Lloyd, 1980).

Fig. 14.2. The river entrance of Porth yr Ogof (photo by C. Howes).

Cave morphology

The cave has 15 entrances (Fig. 14.3) of which the largest is the Main Entrance, where the wet-weather flow of the River Mellte goes underground. This lies at the downstream end of a short, steep-sided, rocky gorge, at the foot of a 15 m high cliff. The boulder-strewn riverbed sweeps into the Entrance Chamber, over 15 m wide and 5 m high. Just at the limit of daylight the full width of the passage is occupied by the White Horse Pool; Porth yr Ogof is sometimes known as the White Horse Cave from the fancied resemblance of a calcite vein in the wall over the pool.

The permanent flow of the River Mellte also enters this pool, from the Upper Stream Passage, and then flows on into the Great Bedding Cave with the impressively wide span of sloping bedding roof (Fig. 14.4). The main streamway continues in fine style, with the wide tunnel floored by either deep

pools or major shingle banks. Down a few rapids, the final pool is deep and long, with a low airspace through the narrow canal to the joint-aligned resurgence. Oxbows lie on both sides of the main streamway, and on the western side these expand into the complexities of the Right Hand Series and the three phreatic levels of the Maze. From the southeast, the Cwm Porth Inlet Passage is a mostly flooded inlet draining from sinks on the adjacent valley flank.

The upstream parts of Porth yr Ogof carry the baseflow of the Mellte, but have been explored for less than one-third of the way to their source at Church Sink. Most of the main stream route is flooded, until it feeds into the Upper Stream Passage; this fine square canyon carries the water east of the Entrance Chamber and is broken by several short sumps before draining into the White Horse Pool. The upstream cave has various phreatic loops in its long sumps, and divers have also explored a complex of high-level fossil phreatic passages.

Hydrology

The regime of the Mellte River is very flashy, with short periods of rapid run-off interspersed with long periods of quiet conditions. In summer, the bed of the river is usually dry between Church Sink at Ystradfellte and Porth yr Ogof. In wet weather, however, the sink soon fills up and the river flows on along the surface to the Main Entrance of the cave. Mellte means lightning and is an appropriate name, for after heavy rain the river can rise with alarming speed. This fact is not at all surprising, for the catchment area covers over 35 km² of the central Fforest Fawr Mountains, which reach a maximum altitude of 730 m.

The Main Entrance of Porth yr Ogof makes an impressive spectacle when the river is in spate. In high flood the approach ledges to the entrance become submerged and at the resurgence water gushes up under a considerable hydrostatic head. After exceptional floods, water may even reach the roof of the Main Entrance, and much of the cave becomes submerged. Under such conditions, a large amount of debris is washed in and large tree trunks are often found in the cave. Also, when there is snow on the southern slopes of Fforest Mountains and when the Mellte is frozen, meltwater can create a mild flood in the cave by rising and flowing over the river ice.

The solutional load of the Mellte varies considerably with stage. In dry weather, the water at the Porth yr Ogof resurgence has been found to contain 100 mg/l of $CaCO_3$, while in flood conditions the content is reduced to 35 mg/l by dilution from surface water (Williams, 1963; Groom & Williams, 1965). Nevertheless, in the year monitored by Williams (1960–1), 38 d of flood flow accounted for substantially more erosion of the limestone than the 152 d of dry conditions (Table 14.1). A spring just upstream of the main cave has a mean solute load of 150 mg/l, due to its higher content of percolation water flowing first through soil and then constricted bedrock fissures.

Geology and geomorphology of the cave

The headwaters of the Afon Mellte, the Afon Llia and the Afon Dringarth drain off the Old Red Sandstone of the Fforest Fawr Mountains and converge near Ystradfellte. Not

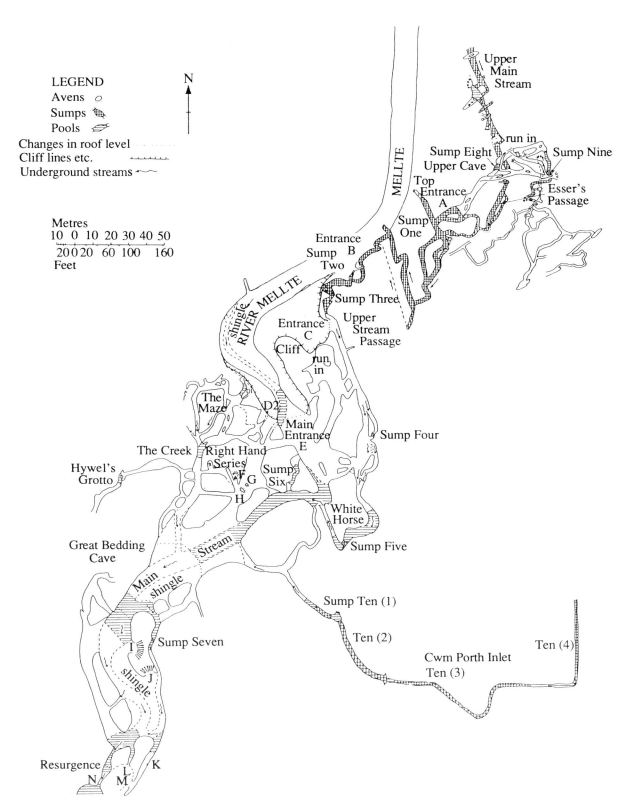

LEGEND
Avens   o
Sumps   🐚
Pools   🐟
Changes in roof level ....
Cliff lines etc.   ┴┴┴┴┴┴
Underground streams   ～～

Metres
10 0 10 20 30 40 50
20 0 20 60 100   160
Feet

Fig. 14.3. Plan of Porth yr Ogof (modified from surveys by the
University of Bristol Spelaeological Society and Cave Diving Group,
reproduced with permission).

Fig. 14.4. The Great Bedding Cave in Porth yr Ogof (photo by P. R. Deakin).

Table 14.1 *Solutional erosion by the River Mellte at Port yr Ogof* (from Williams, 1963)

	River flow conditions			
	Dry	Normal	Flood	
Number of days (1960–61)	152	175	38	
Mean river flow	0.37	0.63	3.65	m^3/s
Solute accretion	52	29	23	mg/l $CaCO_3$
Daily solute removal	1675	1568	7500	kg d^{-1}
Total solutional erosion	255	275	285	kg/y
Total river flow	4860	9520	12080	m^3/y

long after the confluence, the Afon Mellte crosses onto the outcrop of the lowest beds of the Carboniferous Limestone, here dipping southwards at between 4° and 10°. As soon as massive limestones are reached, the river sinks in dry weather at the Main Mellte (Church) Sink (SN 931 133) in the east bank. Repeated attempts to dig out the swallet have been frustrated by flood debris, and the course of the underground Mellte is unknown for some 600 m. In wet weather, the river remains on the surface and flows into a gorge incised some 50 m into the massive limestones, here patchily covered by boulder clay. At Porth yr Ogof the river goes underground for a distance of some 300 m beneath the floor of an earlier valley, resurging some 700 m upstream of where the limestone outcrops are terminated by a fault crossing the valley (IGS, 1980).

Upstream of the entrance, the gorge is apparently at least in part a collapsed former extension of the cave, as shown by the large joint blocks still littering the riverbed. Doubtless, as the cave is enlarged, further collapses will take place until the river is entirely flowing in a gorge. Porth yr Ogof may thus be seen as an example of incomplete incision of a gorge by cave collapse, in the classic Cvíjič style.

The cave itself is largely controlled by the occurrence of four strong bedding planes rather less than 1 m apart, as can be seen in the entrance and adjacent ledges. These beds span wide

areas of roof, as in the Great Bedding Cave. They are, however, crossed by two strong joint sets oriented NNW–SSE and NNE–SSW, and many passages are oriented along joint–bedding intersections. The development of the cave system was thus by phreatic solution opening up routes through the limestone, with vadose enlargement by the Afon Mellte. As it is in a valley floor it is a young cave system, effectively new in transition from the phreatic to the vadose zone. Earlier stages must have occurred, but, apart from the isolated Hywel's Grotto and some stalagmite-cemented gravel in an oxbow off the Great Bedding Cave, there is little evidence. The high-level passages of the Maze are probably part of the earlier stages of cave development.

The potholes of Gwaen Hepste

The upland moor between the valleys of the Mellte and Hepste is largely composed of a wide expanse of peat-covered basal conglomerate of the Millstone Grit, overlying the limestone. A number of small potholes occur close to the western boundary of the conglomerate outcrop, and the highly acidic, peaty waters which sink have been shown to resurge from various springs close to the Afon Mellte downstream of Porth yr Ogof. Low- and high-level resurgences are active according to weather conditions. It comes as something of a surprise to find that some of the potholes retain a partial roof of conglomerate, but an even bigger surprise was the discovery of Ogof Coeden Prop (SN 932 124), Ogof Ganol (SN 933 121) and Ogof Ffynnon (SN 934 120), well within the conglomerate outcrop (Lloyd, 1979). These caves consist of a series of shafts up to 20 m deep, joined by bedding caves along the conglomerate–limestone boundary, so they also are roofed by the insoluble conglomerate. It has been shown (Burke & Bird, 1966; Burke, 1967) that these potholes have been formed by the acidic waters seeping through the conglomerate, etching out and widening the jointed limestone beneath, and furthermore, that this process is still operating by increasing the widths of the vertical fluting in the shaft walls. These shaft caves can thus be regarded as originating within the vadose zone. The seepage waters drain away via inaccessible bedding planes and lower joints in what can probably be classified as epiphreatic systems.

To the north of the Hepste Valley lies the moorland of Gwaen Cefn y Gareg. Like the comparable Gwaen Hepste Moor which overlooks the valley, this has many potholes along the margin of the Basal Grit and also a number of sinks within the Grit outcrop. The major sinkhole of Pwll Derw (SN 941 123) has so far defied all attempts to dig into a cave beneath, but its water has been dye-traced to the Hepste Resurgence. Pulpit Hole (SN 941 138) (Fig. 14.1) lies beneath a cap of Basal Grit and is a complex of several parallel shafts up to 11 m deep with a bedding cave linking them directly beneath the grit cap (Fig. 14.5). These caves have a direct bearing on the hypothesis of wholesale collapse of large areas of the Millstone Grit into presumed caverns in the underlying limestone (Thomas, 1959, 1974) in that they demonstrate the existence of at least small potholes beneath the Millstone Grit. It has yet to be discovered whether the vast chambers proposed by Thomas actually exist.

The shaft caves of Gwaen Hepste are also unusual in that

Fig. 14.5. The vertical shaft entrance of Pulpit Hole under the roof of Millstone Grit (photo by Ian Davinson).

Fig. 14.6. Peat stalagmite in cavern M/N, Cwm Porth, east of Porth yr Ogof (photo by A. Burke).

they contain peat stalactites and stalagmites (Burke, 1970) (Fig. 14.6). Suspended peaty material is brought in by water percolating through the conglomerate cap, and deposited as low-density pendant stalactites. Gradual dehydration by compaction, creep and flow has led to the formation of stratified peaty stalagmites and flowstone. Bacterial oxidation of iron compounds results in the co-precipitation of limonite and goethite within the peaty flowstone during compaction. The

iron oxide stalactites in Pulpit Hole are hard and brittle.

There are many more examples of large collapse sinkholes within the Grit outcrops, both on Cefn Cadlan, south of the Hepste, and on Mynydd y Glog, south of Pant Sychbant. It is significant that all these moorlands are geographically close to the intervening Hepste Valley with its cave systems beneath the grit of the valley floor. Further study of the area could yield useful data on the early stages of cave development, and attendant collapses, beneath an insoluble cover rock.

The Hepste River Caves

The caves of the Hepste River stand comparison with Porth yr Ogof, on the adjacent Mellte, as they may represent an early stage in the evolution of a similar valley floor system. The River Hepste drains sandstone uplands to the northeast and, shortly after crossing onto the limestone, the river sinks in a series of fissures, the last of which is Ogof Glan Hepste (SN 952 118) also known as Tuck's Rift. This contains 100 m of tight joint and bedding tubes explored by divers, but the continuation of the underground course is choked (Fig. 14.7).

For something over a kilometre the riverbed is normally dry, until, shortly below Hepste Bridge, the tributary Nant y Mawr enters just as the river course leaves the limestone. From there on, a stream is maintained on the surface for 1500 m, to the Middle Hepste Main Sink situated in a shallow gritstone ravine. In the next few hundred metres there are several entrances to the complex Hepste River Caves System (Figs. 14.7, 14.8). These are mostly sited in riverbed limestone inliers too small to appear on published geological maps, and, like the potholes of the adjacent moors, the River Caves lie in the limestone immediately beneath the conglomeratic Basal Grit.

The Hepste Resurgence (SN 936 097) is a fissure in sandstone which pours water from the limestone about a metre beneath the riverbed (Fig. 14.9). Ogof Tarddiant Hepste (SN 937 097) has an obvious entrance beneath a waterfall, and has been known for a long time, but the main cave, Ogof Afon Hepste (SN 938 097), was found only in 1969 by digging away boulders. Most of its passages are only accessible to divers.

The lower end of the cave system doubles back and forth beneath the riverbed, and sections may be entered at the Hepste Resurgence, at Ogof Tram Trucks, Ogof Tarddiant Hepste and downstream from Ogof Afon Hepste. These passages are essentially a series of bedding–joint intersections trending roughly parallel to the strike of the Coed Hir Fault, lying about 100 m to the southeast. This throws down shales against both the Basal Grit and the underlying limestone forming a hydrological barrier. Ogof Afon Hepste is a similar series of passages oriented more or less parallel to the fault and trend of the surface gorge, with three parallel feeder streamways entering from the north. The Western Streamway carries an active distributary from the main Hepste Sinks and is too tight for progress. The Main Streamway has been explored for approximately 500 m through a series of sumps along a strong NNE-oriented joint system. It appears to carry most of the Afon Hepste from its sinks 2 km upstream. The Eastern Streamway is of similar length and orientation, but tends to be a series of muddy rifts and canals with little active flow. Both Main and Eastern Streamway passages almost reach where they pass beneath the course of the surface Afon Hepste a

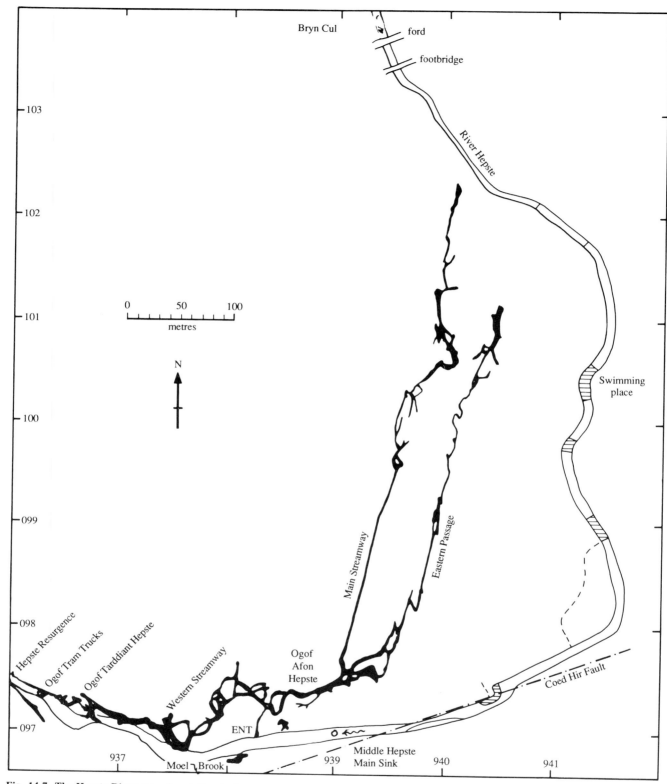

Fig. 14.7. The Hepste River caves in relation to the great bend in the Afon Hepste (based on surveys by the Cave Diving Group, 1969–79, and by O. Lloyd, 1979).

short way downstream of the Bryn Cul Ford footbridge (Lloyd, 1979).

Much of the Afon Hepste System can only be reached in low water conditions, for the flow in some of the sumps is too great

for a diver to make progress. In the more easily accessible parts water rises and flows through open passages to join the Western Stream, but in low conditions the Main Stream sinks in Sump One just inside Ogof Afon Hepste, and is not seen

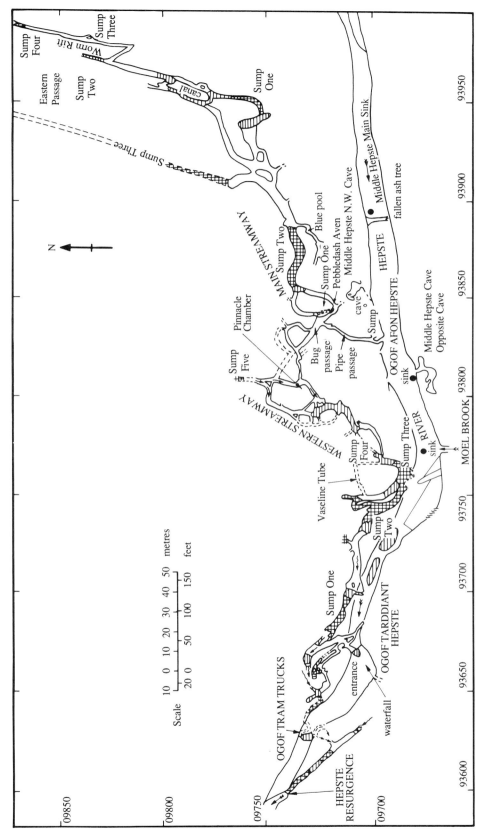

Fig. 14.8. Detailed plan of the Hepste River caves (based on Cave Diving Group surveys, 1969–76, reproduced with permission).

Fig. 14.9. The Hepste Main Resurgence: water rises from the limestone through a fissure in the Millstone Grit cover (photo by A. C. Waltham).

again until the Resurgence. The route taken by the surface water sinking at the Middle Hepste Main Sink is unknown.

While the greater part of the known system is effectively at the water table, there is a greater thickness of limestone at the northern extremities and more high-level old passages may be expected there, but considerable digging through gritstone chokes and cap rock will be needed to reach them. The total drop from Ogof Glan Hepste to the Resurgence is 68 m over a straight-line distance of 2640 m.

The caves of the Sychryd Gorge

The downstream end of the River Sychryd is deeply incised as a consequence of its recent capture by the River Neath. Its final section of narrow gorge, 500 m long, is cut along a major fault, separating the vertical slice of limestone forming Craig y Dinas from a tight anticline of limestone just to the south (Fig. 14.10). This anomalous geological structure is due to its position within the Neath Disturbance.

The main caves are on the north side of the Sychryd, beneath Craig y Dinas. At the upper end of the system, Ogof Coed y Ffrynau, now blocked, contains 90 m of narrow rifts. Wills Hole is the most important cave, with nearly 400 m of known passages (SN 915 080). An 11 m shaft drops into an almost level stream passage which is 3 m wide and up to 7 m high. Upstream it is choked, and its downstream sump is nearly at the level of the resurgence. Ogof Pont Sychryd (SN 911 079) is a fossil phreatic maze which may once have been a distributary of the phreatic ancestral Wills Hole, although no connection is now open.

On the south side of the gorge lie the remnant Bwa Maen Arch and the short abandoned phreatic passages of Ogof Bwa Maen (SN 916 080). It appears that the River Sychryd is a recent invader which has incised its gorge through a complex of far older caves. It has therefore fossilized all the caves at higher levels, but it has in turn lost some of its water to the lower levels of Wills Hole.

The caves of Penderyn

Sections of surprisingly large fossil cave passages have repeatedly been revealed and then totally removed by the quarry workings which are progressively consuming the western end of Twyn y Glog. This is a rugged limestone ridge on the narrow outcrop just east of Penderyn village immediately south of the Neath Disturbance (Fig. 14.11).

Penderyn Quarry Cave (SN 959 090) was opened in 1939 on the northern flank of the workings. Calcite formations of every type adorned the walls, roof and floor of the passage, which according to contemporary descriptions was completely fossil and dry (Fig. 14.12). Its tubular form suggested a phreatic origin with vadose modification to North (1940, 1962) who visited the cave in 1939.

Ogof Twyn y Glog (c. SN 959 091) was exposed in 1966 on a quarry bench further east, and probably represented a continuation of the same trunk cave. It was only 60 m long, but had a passage up to 20 m high and 8 m wide. Then, in 1972, a quarry

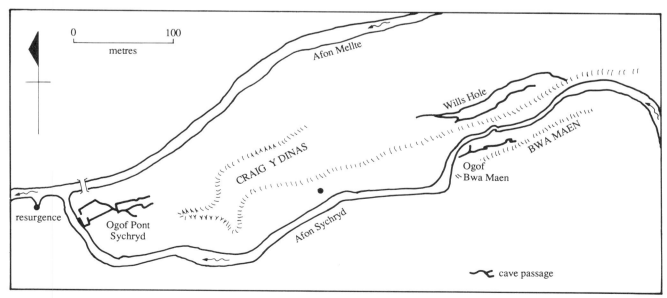

Fig. 14.10. Sketch map of caves near the confluence of the Rivers Mellte and Sychryd.

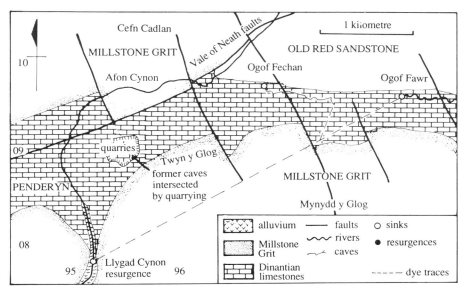

Fig. 14.11. Sketch map of the limestone area east of Penderyn showing the relationship of the caves to the Afon Cynon.

Fig. 14.12. Some of the magnificent speleothems in Penderyn Quarry Cave later destroyed by quarrying (photo by A. Coase).

drop-ball fell into a large hole, revealing about 150 m of passage which was 15 m in diameter and choked at both ends. This appears to have been another section of Ogof Twyn y Glog, and was also phreatic, fossil and dry.

All of these caves have now been destroyed (and the quarry company does not permit any access for cavers). They are the remains of a major ancient cave system which was feeding water, probably westwards, to an unknown resurgence, long since fossilized by the river captures of the area.

The caves of Pant Sychbant

The Pant Sychbant is an almost level, and poorly drained, high-altitude valley extending east–west (Fig. 14.1). Its southern flank is cut in the Carboniferous Limestones which are exposed almost to the crest of the scarp face of Mynydd y Glog, whose southerly dip-slope is formed by the Millstone Grits. The floor and northern flank of Pant Sychbant are cut in the Lower Carboniferous Shales and older sandstones. The valley therefore conforms to geological structure, but lies

across the regional direction of ice flow. In the Pleistocene, ice moved from the north before being deflected probably to both east and west by the resistant barrier of Mynydd y Glog.

The stream in the upper part of Pant Sychbant flows west from an area of marsh, before cascading through a trench incised in the broad glaciated valley floor, and then sinking into Ogof Fawr (SN 985 096). Lying at the foot of an imposing 20 m cliff, this sink consumes huge flood flows; the water drops through a narrow fissure before being lost in a massive boulder choke. The resurgence at Llygad Cynon is 3.5 km away and another 80 m lower (Fig. 14.11).

Beyond Ogof Fawr, the dry valley soon collects another small stream which in turn sinks below another limestone cliff. Adjacent to this, a shakehole has been dug out to yield access to Ogof Fechan (SN 970 097) (Fig. 14.13). This system, also known as Ogof Fach, is cut in the lowest beds of the limestone succession, and some passages have their floors cut into the underlying shales. The cave contains a complex of small phreatic passages, leading to a lower streamway of roomy

Fig. 14.13. The cramped diving base at Sump Two in Ogof Fechan (photo by T. Bennett).

proportions, which, although broken by sumps, can be followed for over 400 m. The limit of exploration in the constricted Sump Four is still 47 m above the presumed resurgence at Llygad Cynon. Ogof Fechan shows a close response to geological structure as it follows the impermeable base of the limestone; its major passages are strike-oriented, with the dip at 20–30° south. The Main Boulder Choke is located on a fault zone with conspicuous breccia and veining.

West of Ogof Fechan, Pant Sychbant collects another small stream which flows to join the Nant Cadlan clear of the limestone outcrop. The combined waters then cross the limestone by the village of Penderyn, where they do not sink but are added to from numerous springs. The largest of these is Llygad Cynon (SN 952 077) which appears to return most of

the drainage to daylight from the more distant parts of the limestone outcrop, including that sinking into Ogof Fawr. There is no accessible cave at the resurgence, where a pumping station takes 65 l/s for the local villages even in drought.

The known caves of Pant Sychbant appear to be mainly remnants of ancient systems. Ogof Fechan is largely phreatic; it is also renowned for its mud and thick sand deposits. It would seem that a complete Pleistocene clastic fill has only been partially removed from the cave. The two caves in the valley may represent successive stages of development by water sinking from essentially the same catchment. Even older are the remnant phreatic tunnels of Penderyn, now bypassed as the water takes a lower, down-dip route direct to Llygad Cynon.

References

Burke, A. R. (1967). Geomorphology and spelaeogenesis of vertical shafts in Carboniferous Limestone at Ystradfellte, Breconshire. *Proceedings of the British Speleological Association*, **5**, 17–46.

Burke, A. R. (1970). Deposition of stalactitic and related forms of peat; genesis and bacterial oxidation. *Transactions of the Cave Research Group, GB*, **12(4)**, 247–58.

Burke, A. R. & Bird, P. F. (1966). A new mechanism for the formation of vertical shafts in Carboniferous Limestone. *Nature, London*, **210**, 831–2.

Groom, G. E. & Williams, V. H. (1965). The solution of limestone in South Wales. *Geographical Journal*, **131**, 37–41.

Institute of Geological Sciences (IGS) (1980). 1:50000 Geological Map of the Merthyr Tydfil area. Sheet 231.

Lloyd, O. C. (1979). The Hepste River caves and a study of the Hepste–Mellte area. *Proceedings of the University of Bristol Spelaeological Society*, **15(2)**, 107–27.

Lloyd, O. C. (1980). Porth yr Ogof, Ystradfellte, Powys: a revised survey. *Proceedings of the University of Bristol Spelaeological Society*, **15**, 259.

North, F. J. (1940). A newly discovered South Wales cave. *The Penderyn District Review*, **43**.

North, F. J. (1962). *The River Scenery at the Head of the Vale of Neath*, 4th edn. Cardiff: National Museum of Wales.

Standing, P. & Lloyd, O. C. (1970). Porth yr Ogof, Breconshire. *Proceedings of the University of Bristol Spelaeological Society*, **12**, 213–29.

Thomas, T. M. (1959). The geomorphology of Brecknock. *Brycheiniog*, **5**, 55–156.

Thomas, T.M. (1974). The South Wales interstratal karst. *Transactions of the British Cave Research Association*, **1(3)**, 131–57.

Williams, V. H. (1963). A Study of the Solutional Processes and Phenomena in Limestone with particular reference to the North Avonian Outcrop of South Wales. Unpublished Ph.D. thesis. Swansea: University College of Wales.

15

The Afon Nedd Fechan caves

SAM MOORE

Topography and geology

The Nedd Fechan valley extends for about 12 km from the southern slopes of the Fforest Fawr mountains to the village of Pontneddfechan, where the Nedd Fechan combines with the Mellte to become the River Neath. The lower Nedd Fechan flows for 5 km entirely on Millstone Grit and speleological interest is confined to the middle reaches, where the river crosses the south-dipping outcrop of the Carboniferous Limestone over a distance of some 2.5 km.

The headwaters of the Nedd Fechan drain the Old Red Sandstone slopes of Fan Fraith (640 m) and Fan Nedd (660 m). The river flows southwards to meet the limestone outcrop near Blaen Nedd Isaf Farm and here the whole of the dry weather flow sinks into the east bank over a distance of 200 m. Apart from a small volume of surface tributary and some resurgence water from the west, the riverbed is then virtually dry for over 2 km.

In wet weather the sinks south of Blaen Nedd Isaf are incapable of taking the increased flow and the river flows past Bridge Cave before cascading spectacularly into Pwll y Rhyd. The whole of the section from Bridge Cave to the flood resurgence at Pwll Du forms an impressive gorge cut in the main Dowlais Limestone (S_2 or Holkerian). It is within these beds that the underground drainage of the area has developed.

Details of the stratigraphy are given in George (1970) and Owen (1971), the general succession of the area being outlined in Chapter 2. An important feature in the lower reaches of the valley is the outcrop of quartzitic sandstone (Millstone Grit) in the Pwll Du area. From here, 500 m upstream of Pont Rhyd y Cnau, the Nedd Fechan flows as a continuous surface stream. The position of this impermeable sandstone outcrop is strongly influenced by the effects of the Neath Disturbance. This is a complex feature, the main structural elements of which include impersistent WSW–ENE faults, NNW–SSE-trending shear faults and normal NW–SE faults. Several of these NNW–SSE faults affect the Nedd Fechan Valley, each fault displacing the limestone outcrop further to the southeast (Fig. 15.1). One of these faults runs the length of the limestone gorge, while adjacent faults on the east bank are highlighted by a linear

zone of dolines. Many cave passages are also aligned in a NNW–SSE direction.

To the northwest of the river sinks lies the broad gently sloping plateau of Pant Mawr, a large proportion of which also drains underground to the Nedd Fechan valley. The small stream now associated with Pant Mawr Pot sinks in the Lower Limestone Shales in the north. To the south there is no significant surface drainage for over 2 km, although the southern half of Pant Mawr is capped with substantial thicknesses of Millstone Grit. Pant Mawr Pot itself is formed in the lower beds of the Dowlais Limestone, at a similar stratigraphic level to the Nedd Fechan caves.

Weaver (1973), in a study of joints and cave passages in South Wales, found that four dominant joint sets are developed in the Tawe–Neath interfluve region. These trend NE–SW, NW–SE, N–S and W–E, approximately. Collapse dolines in the Millstone Grit have been described by Thomas (1974) on both sides of the Nedd Fechan valley, although they tend to be better developed to the west. On Pant Mawr, many of these dolines are aligned NE–SW, demonstrating the significance of this joint set in the underlying limestone.

History of exploration

The limestone gorge of Cwm Pwll y Rhyd attracted early attention and several prominent nineteenth-century travellers visited the spectacular entrance of Pwll y Rhyd. North (1928) published an important paper on the Nedd Fechan and neighbouring valleys but serious cave exploration did not take place until the mid-1930s. In 1938, Braithwaite and others visited the valley and explored the accessible regions of Pwll y Rhyd, White Lady Cave and Town Drain. They also followed a low passage from a large shakehole near the main river sink for a distance of 45 m to a boulder choke. In 1947, members of the newly formed South Wales Caving Club passed the choke and were able to follow a sizeable streamway for 150 m to a sump. They named the discovery Bridge Cave after a fine rock bridge over the stream.

To the west of the valley, Braithwaite also descended Pant Mawr Pot, but was unable to pass the Second Boulder Choke.

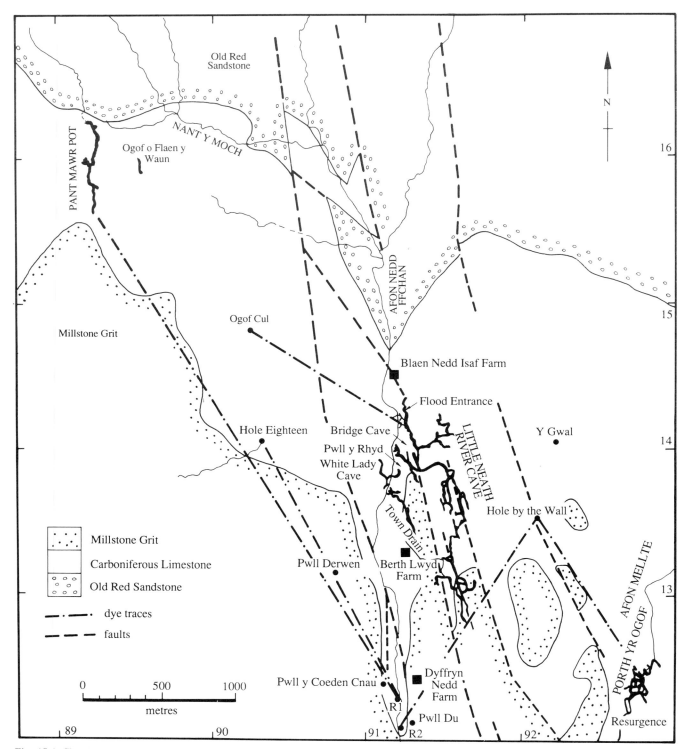

Fig. 15.1. Sketch map to show the relationship of Little Neath River Cave and Pant Mawr Pot to the geology and drainage pattern.

This remained impenetrable until 1953 when it was passed by the SWCC (South Wales Caving Club), who were able to explore essentially all of the present system. To the east of the valley, the Mendip Nature Research Committee found a number of small caves in the early 1960s. In 1960, the Cave Diving Group connected Pwll y Rhyd to White Lady Cave.

Access to the area for Mendip-based cavers became much easier with the opening of the Severn Bridge in 1965. Divers from the University of Bristol Speleological Society passed the sump at the end of Bridge Cave in January 1967. A large streamway (860 m) was explored to Sump Two, before the party followed a tributary passage back to daylight. This point was later located on the surface and easily made passable. The cave was named Little Neath River Cave and the new entrance

'Flood Entrance', since it was situated in the bed of the Nedd Fechan. Sumps Two, Three and Four were soon passed and led to almost 3 km of streamway and spectacular high-level passages, which are still accessible only to divers. Sump Eight, the present limit of exploration, is at a similar level to the resurgence and further dry extensions seem unlikely.

Little Neath River Cave (LNRC)

Main Stream Passage

The Bridge Cave section of the system is entered from a large shakehole close to the bridge below Blaen Nedd Isaf (Fig. 15.2). A crawl of 60 m leads to a boulder choke, which is followed by a short, meandering streamway to the impressive Main Stream Passage (Fig. 15.3). A slender rock bridge crosses the stream before the passage closes down towards the beginning of Sump One, a shallow sump normally 18 m long (Norton, Savage & Standing 1967; Standing, Newson & Wilkins, 1971; Mullen, 1987).

Beyond Sump One, water from the upstream sinks enters from the north as Tributary Passage. The combined flow turns east and enters the Canal, a wide, low, strike-orientated bedding plane, the upper end of which sumps in wet weather. At the end of the Canal, water enters from North East Inlet Series at Junction Chamber. The stream then turns SSE and enters Main Stream Passage, a fault-guided passage over 600 m long (Fig. 15.4). The water initially cascades over collapsed blocks before passing a bank of fill 15 m high near the entrance to Genesis Gallery. Downstream, further evidence of roof collapse is seen at the fallen slab and in Bouncing Boulder Hall. Here, the stream leaves the main passage in a joint-guided canyon, before rejoining it via the Wet Loop. From the lower end of Bouncing Boulder Hall, where the passage is 10 m wide and 20 m high, Main Stream Passage diminishes in size to the Duck, while maintaining its SSE trend. Beyond, the passage continues low, past Cairn Passage and Ubbs Aven, before turning west along the strike once more. This section terminates in a second Duck at the head of a substantial meander, before the passage swings SSE again, enlarges briefly and then closes down to Sump Two.

Side passages in the Upstream Cave

The 'dry' entrance to LNRC is a sink in the east bank of the Nedd Fechan, some 250 m north of the bridge below Blaen Nedd Isaf. After even moderate rain, the entrance becomes submerged and impassable. Inside, a low and constricted passage is followed for 30 m to a duck. Beyond, Blaen Nedd Isaf Passage, a 600 m long inlet, joins from the north and the enlarged stream flows SSE down Tributary Passage. After 200 m the entrance to the Canal Bypass is passed on the left and Tributary Passage continues as a low bedding plane leading to Sand Chamber, the Canal and the water entering from Bridge Cave. From here a large passage, Mud Hall, leads back north but soon chokes.

The Canal Bypass (Fig. 15.5) is the alternative route to the main streamway and is generally a crawl until the water from North East Inlet Series crosses it to reappear as a waterfall a little further on. The water then flows south to Junction Chamber and the downstream end of the Canal. North East Inlet Series is a mixture of small inlets and some larger dry canyon passages. The Canal Bypass, Tributary Passage and North East Inlet Series all have extensively developed solution networks, which suggest a phreatic origin.

Between Junction Chamber and Sump Two there are a number of high-level passages. Genesis Gallery is entered by a bedding-plane crawl above a large bank of fill in Main Stream Passage. A fault-controlled rift passage leads to a small bedding-plane chamber and the entrance to the 3D Maze. Keeping to the left-hand wall, the fault line is regained as a series of well-decorated avens. After further crawling, the second part of Genesis Gallery is reached. Water from Genesis Inlet, a well-decorated series at higher level, enters down an aven, but soon sinks in the floor. This water is presumed to be that flowing out of Exodus Crawl, near Sump Two.

Old World Series, discovered in 1970, is entered through the 3D Maze. Emerging from this, a large bedding chamber is entered and a 3 m square passage is followed for 100 m to Gooseberry Pot. A further 100 m of smaller passage can be entered via an oxbow half-way down this.

High Level Series is entered by climbing Straw Aven or through a tight vertical squeeze in the roof of Main Stream Passage. The series contains a small stream which sinks in an impenetrable bedding plane.

The Downstream Cave

Sump Two is 40 m long and emerges in a single chamber with a short, choked high-level passage. Sump Three is 70 m long and leads to 45 m of stream passage with a short side passage just before Sump Four, which is 50 m long.

The LNRC 5 streamway is 200 m in length, with two distinct passage forms. The first is a semi-circular tunnel up to about 2 m wide and high. The second is much larger, with the eastern wall formed on a NNW–SSE-trending fault. Sump Five can be bypassed via New World Passage. The LNRC 6 streamway is only 120 m long, but is initially an impressive canyon passage with numerous calcite veins in the roof, walls and floor. Water from Seventeenth Street enters from the east, the entrance to Sump Passage being high in the wall some 75 m further on. The stream then flows through Lake Chamber (Fig. 15.6), joining the continuation of New World Series, before the roof suddenly lowers to Sump Six.

Sump Passage is the abandoned main stream passage and ends after 135 m at Sump Six(A). Sumps Six and Six(A) lead to the same airspace, LNRC 7. The route to LNRC 8 leads off half-way through Sump Six(A) (Sump Seven) and was passed in 1972. LNRC 8 contains 200 m of streamway leading to Sump Eight, which lies on the Grit–Limestone boundary on a fault. Sump Eight remains the limit of exploration. Since its level is similar to that of the resurgences, 650 m away, the prospects for further open streamway seem remote.

Side passages in the Downstream Cave

Some 2 km of side passages have been explored in LNRC 5 and 6. They include the largest and most spectacular high-level passage in the cave – New World Passage – and two

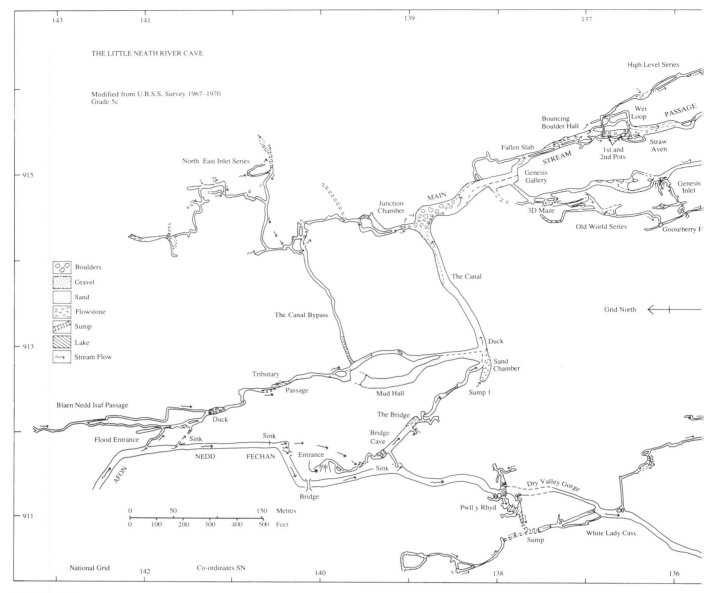

Fig. 15.2. Plan of Little Neath River Cave (modified from University of Bristol Spelaeological Society surveys 1967–70, reproduced with permission).

important tributaries, Seventeenth Street and Foot and Mouth Passage.

New World Passage can be entered at three points upstream of Sump Five and two points downstream. The most obvious route is up a steep boulder and gravel slope on the east side of the passage, 30 m below Sump Four. Eventually, a wide bedding plane is reached which opens out after 50 m to reveal the 20 m New World Aven. The floor of New World Passage drops away steeply and the second route up from the LNRC 5 streamway enters low down on the western wall as a steep climb up through boulders. Further on, the passage becomes 25 m wide and 15 m high and New World Oxbow, itself of impressive proportions, leaves on the west. The third point of access to LNRC 5 is from New World Oxbow, but requires a ladder for a 6 m pitch to the stream. After New World Oxbow

rejoins, the main passage continues for 50 m to a boulder choke on the east wall, down through which Lake Chamber is reached. Continuing past the choke, however, the roof soon descends and the passage leads into the roof of Sump Passage.

Foot and Mouth Passage enters the streamway about 60 m below Sump Four. There is an extensive series of high-level passages, oxbows and avens around its start and the tributary itself is approximately 350 m long. Several sections are dry, with chert ledges, but the final 100 m is in the streamway, which eventually becomes too tight. This point is about 150 m away from and 15 m lower than the end of Town Drain with which it may connect.

Seventeenth Street can be reached by two routes leading off from the eastern wall of New World Passage, of which the most obvious is a tortuous canyon passage. Upstream Seventeenth

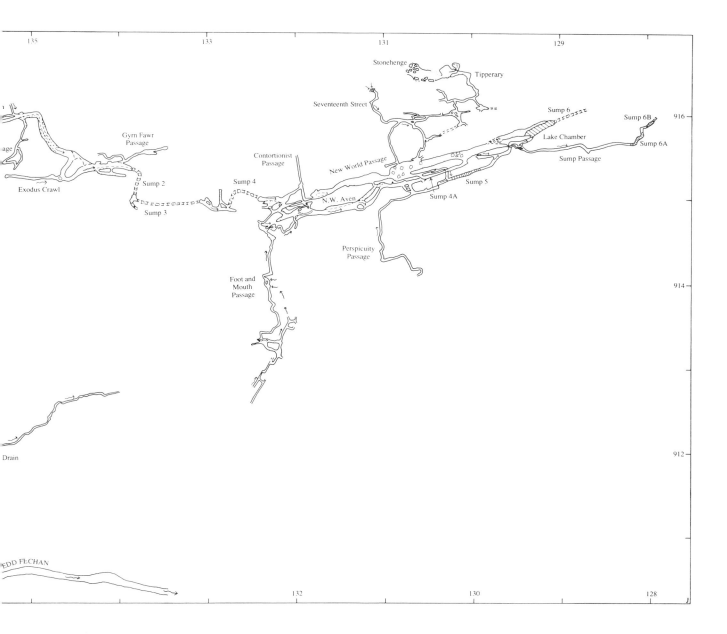

Street is a pleasant walking-sized passage leading to a short canal and a boulder choke. Downstream leads to a sump, the water reappearing briefly in Tipperary before entering another sump *en route* to LNRC 6. Tipperary is entered by a rift in the southern wall, 40 m upstream of the sump. It consists of a series of awkward rifts and large, dry chambers ending at a breakdown chamber 45 m long and 13 m wide blocked by Stonehenge, a choke of colossal boulders.

Cave morphology

The passages immediately below Flood Entrance are clearly of recent origin, formed by water leaking from the present riverbed. As a result, there are many right-angled bends where the passage follows closely spaced joints. Shortly inside the entrance is the bedding-plane enlargement of the

Duck, however, and Tributary Passage also appears to be controlled by the bedding. Mud Hall, of which the lower end of Tributary Passage is the continuation, is much larger and older, and the stream is slowly clearing it of fill.

In contrast, parts of the Canal Bypass and North East Inlet Series contain solutional networks and are clearly of phreatic origin. The Canal itself is a puzzling feature which is difficult to explain by reference to currently known passages. Through it, the large passage of Mud Hall leaves the dip and crosses the strike to Junction Chamber and Main Stream Passage. It is possible that the Canal is indeed as recent as it appears and that the continuation of Mud Hall lies through the fill of Sand Chamber.

The Main Stream Passage is again generally down-dip, and several bedding planes can be seen to have influenced its development. Here, the stream is re-excavating older fluvial

Fig. 15.3. The main passage of Bridge Cave (photo by
P. R. Deakin).

Fig. 15.4. Slab collapse in the main streamway of Little Neath River
Cave (photo by C. Westlake).

Fig. 15.5. A fine group of speleothems near the Canal Bypass in
Little Neath River Cave (photo by C. Westlake).

Fig. 15.6. Lake Chamber, near the downstream limit of Little
Neath River Cave 6 (photo by C. Westlake).

material, such as that found near the entrance to Genesis
Gallery. More recent debris is also accumulating as a result of
the breakdown which gives the passage its rectangular
cross-section.

For large sections of the cave, passage orientations are
guided by the predominant NNW–SSE faulting. The dip in the
area is of similar orientation and the two together, with cross
passages developed across the strike, give the cave its rectangular plan. The best example of a fault-controlled passage is
undoubtedly New World Passage, which is of truly impressive
proportions. This fault, and its neighbours to the west, provide
a considerable obstacle to the return of the water to the Nedd
Fechan. A 400 m block between the end of the cave and the
resurgence has been downthrown by about 50 m interposing
grit (Standing *et al.*, 1971). Either a deep phreatic loop or
cross-faulting in the grit is required to account for the drainage,
but the question is unlikely to be resolved unless Sump Eight is
passed or greatly extended.

Avens are found in several areas of the cave. Those in
Genesis Gallery are typical. They are high, narrow and elliptical, tapering upwards to end in small solutional tubes. Substan-

tial calcite deposits are associated with some, in contrast to the
general paucity of formations in Porth yr Ogof, in the near-by
Mellte Valley. There, the lack of deposits has been attributed
to rapid percolation through open joints, the water therefore
still being unsaturated when it reaches the cave. This is clearly
not true of LNRC, which is generally very much further below
the surface.

Pant Mawr Pot

The entrance to the cave is in a steep-sided depression
some 100 m south of the stream sink (Alexander & Jones,
1959) (Fig. 15.7). A 13 m pitch, in a shaft of generous propor-

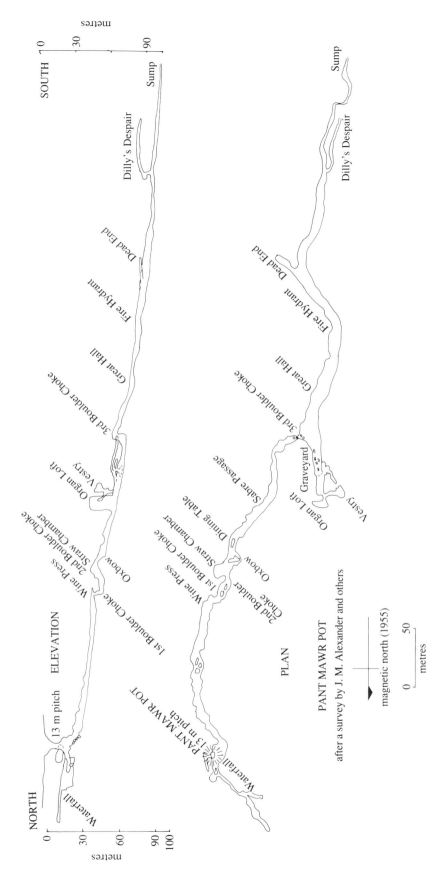

Fig. 15.7. Plan and elevation of Pant Mawr Pot (after a survey by Alexander & Jones, 1959).

Fig. 15.8. The main passage of Pant Mawr Pot with breakdown boulders widespread (photo by C. Westlake).

Fig. 15.9. Straw stalactites and helictites in Pant Mawr Pot (photo by B. A. Speyer).

tions, gives access to a large passage and the stream, flowing from north to south. Upstream, a smaller rift leads past a 7 m high waterfall and up a 12 m climb into a low bedding chamber. The top of the waterfall can be reached and the stream followed for a short distance before it becomes too tight.

Downstream, the passage is spacious, up to 8 m wide and 5 m high in places, with breakdown littering the floor (Fig. 15.8). After 120 m, the First Boulder Choke is reached, followed closely by the Second. The passage then regains its original size and turns southwest for a further 120 m to the Third Choke. A large, choked, high-level passage crosses this section at Sabre Junction, 25 m high.

Beyond the Third Choke a larger passage is entered. To the north is the Graveyard, leading to the Vestry and the Organ Loft, but the main route lies south, dipping steeply at first, through the Great Hall, 13 m wide and 5 m high. The passage gradually narrows but becomes higher until the Fire Hydrant, a percolation inlet issuing from the west wall, is reached. The Dead End, a large sand-filled passage, goes off to the southwest, but the main passage continues south, the passage size beginning to decrease rapidly after 70 m. Dilly's Despair, a higher level parallel passage goes off from a small aven but is choked with sand, while the stream passage, becoming smaller and lower, continues for a further 100 m to the sump. The position of this can vary considerably and in very low flow conditions it retreats to a position at which the passage is impassably small. The resurgence is at R1 in the Nedd Fechan Valley, some 100 m lower and 2.5 km distant.

Cave morphology

The Waterfall Series and the Sump Series appear to be the only parts of the cave related to the mean discharge of the present stream, which is a gross underfit in the larger passages. Speleothems are abundant in places (Fig. 15.9). It seems certain that an earlier main stream continued through what is now the Dead End and that the present stream is a recent invader. Surface features support this, a shallow valley continuing west beyond the sink towards the Nant y Moch, a tributary of the Nedd Fechan.

There is little evidence of a phreatic precursor to the cave. Half-tubes in the Organ Loft, the Vestry and in the roof at Sabre Junction have similar orientations but appear to have been intercepted by the cave rather than to have influenced its development. The Fire Hydrant is certainly supplied by percolation water, a statement supported by reports of bad air, but is again of recent origin. The main passages are orientated NNE–SSW and appear to be fault- or joint-guided. The overwhelming impression is of vadose development, with the passage later being modified by extensive breakdown.

Minor caves of the Nedd Fechan area

On Pant Mawr Moor

Ogof o Flaen y Waun is a juvenile cave associated with a small stream sink approximately 0.5 km southeast of Pant Mawr Pot. It is 150 m in length, generally of crawling dimensions and ends in a too tight bedding plane. Ogof Cul is also a juvenile cave, which runs close to the surface throughout its 200 m length. The entrance is in a dry valley a short distance north of Pant Mawr farmhouse. The cave stream is derived entirely from percolation water, since there is no surface drainage in the area.

In the Nedd Fechan Valley

Pwll y Rhyd (Fig. 15.10) is an unroofed phreatic rift crossing the riverbed about 250 m below Bridge Cave. The passage leads quickly to deep water and the sump which connects it to White Lady Cave. The resurgence of White Lady Cave (Fig. 15.11) is shortly downstream of Pwll y Rhyd. The large entrance leads in about 75 m to the sump which connects to Pwll y Rhyd. A second sump leads off this to an inlet series which is about 250 m in length. A dry entrance to this series has recently been opened.

Town Drain is in the opposite bank, a few metres downstream. The cave, although nearly 500 m in length, is essentially a single rift passage which follows the prevailing NNW–SSE jointing until it becomes choked with mud and gravel. It appears to be of much more recent origin than Pwll y Rhyd and

Fig. 15.10. The Pwll y Rhyd rift entrance (photo by P. R. Deakin).

Fig. 15.11. Gour pools, now smashed, in White Lady Cave (photo by A. C. Coase).

White Lady. The end of the cave is some 150 m from the end of Foot and Mouth Passage in LNRC 5, to which it could connect if its direction and inclination were maintained.

In the woods to the west of the river and the south of Pant Mawr are a number of small caves. The most extensive of these, about 200 m in length, is Pwll y Coeden Cnau, the entrance to which is just within the tree line and almost opposite Dyffryn Nedd Farm. There are few caves of any significant length to the east of the river. Worthy of note, however, is Y Gwal, a keyhole-shaped remnant 6 m high and 3 m wide on the high ground to the east of Blaen Nedd Isaf.

Hydrology

Little Neath River Cave and the east bank

The Old Red Sandstone catchment area of the Nedd Fechan is approximately 7 km^2. The river begins to sink into its bed south of Blaen Nedd Isaf Farm, although water normally flows as far as the bridge below the farm. The upstream sinks are clearly of recent origin and are unable to absorb the whole flow after anything more than moderate rainfall. Since the annual rainfall in the catchment averages 1900 mm, Flood Entrance can rapidly become impassable!

To the north and east of the cave is an area of limestone with no surface drainage. The LNRC share of this catchment area is poorly defined, but 1.5–2.0 km^2 is probably a reasonable estimate. There are, however, relatively few feeders to the system. Blaen Nedd Isaf Inlet and North East Inlet Series are presumed to derive from this area, but most of the drainage remains unaccounted for. Hole by the Wall (Fig. 15.1) seems to mark the watershed between the Nedd Fechan and Mellte valleys, since water sinking there has been traced to Porth yr Ogof and to the LNRC resurgence. The transit times were <24 h and <21 h, respectively (Young, 1986, pers. comm.).

In the lower part of the cave are the inlets of Foot and Mouth Passage and Seventeenth Street. The source of the water in Seventeenth Street is unknown. That in Foot and Mouth Passage is presumed to come either from Town Drain or from Berth Lwyd Swallet, a sink at the edge of the grit, almost directly above.

The resurgence for the cave is at R2, an unspectacular and impassable rising in the riverbed. Almost opposite is Pwll Du, a deep pool on the east bank, which acts as an additional resurgence in all but the driest weather. In dry weather, the flow-through time from the sinks by Bridge Cave is <36 h. As mentioned previously, the end of the LNRC streamway is at the same level as the resurgences and the mechanism by which it crosses the intervening block of grit is uncertain.

Pant Mawr Pot and the west bank

The catchment area of the stream sinking into Pant Mawr Pot is only about 0.6 km^2, but a further 3 km^2 to the south has no surface drainage. Much of this is capped with grit, although its thickness is <30 m. The large number of dolines in the area result from grit collapses and allow rainwater to penetrate to the underlying limestone. No resurgence is known for this area. Pant Mawr Pot itself resurges at R1, another impenetrable rising in the riverbed, about 200 m north of R2. There is no evidence of any connection between these risings. Ogof Cas, a sumped passage in the riverbed a little further north, may be a flood rising for the system behind R1.

To the east of Pant Mawr Pot, a low limestone ridge appears to have developed a localized drainage system, with a small stream resurging on its eastern flank. This flows across the drift for several hundred metres before sinking close to the upper edge of the Nedd Fechan Valley. It then resurges and sinks several times before entering the river as a surface stream. The shallow percolation stream of Ogof Cul, 750 m to the south, has been traced to the LNRC resurgence at R2 (Gascoine, 1985). Although some doubt remains over the point of entry,

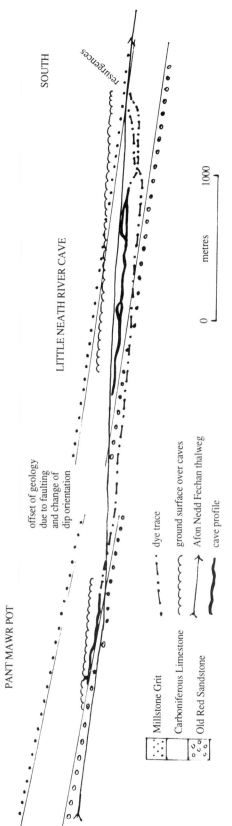

NORTH

PANT MAWR POT

LITTLE NEATH RIVER CAVE

SOUTH

resurgences

offset of geology
due to faulting
and change of
dip orientation

····· dye trace

∿∿∿ ground surface over caves

↗ Afon Nedd Fechan thalweg

━━ cave profile

Millstone Grit

Carboniferous Limestone

Old Red Sandstone

0   metres   1000

Fig. 15.12. Profiles of Pant Mawr Pot and Little Neath River Cave
to show the input of water at the base of the limestones and
resurgence at the top.

the balance of evidence suggests that the water flows under the Nedd Fechan, to reappear in Tributary Passage (Young, 1986, pers. comm.).

One kilometre west of Bridge Cave a significant stream flows off the grit and sinks at the end of a well-developed blind valley. This sink, known as Hole Eighteen after the numbering of the second UBSS paper on LNRC (Standing *et al.*, 1971), has been shown by Gascoine (1985) to resurge at R1. The flow-through time was <36 h in high water conditions. In the same study, Pwll Derwen, a sink 0.5 km to the northwest of Dyffryn Nedd Farm, was also traced to R1.

Solution studies

The surface above Little Neath River Cave has an extensive vegetation cover in an area which receives 1750–2250 mm of rainfall y^{-1}. There is, therefore, the potential for intensive solutional activity (Bridges, 1971). The ability of the soil water, charged with CO_2, to dissolve the limestone has been measured by Standing & Martell (unpublished notes) during the period November 1972–November 1973. The method suggested by Stenner (1969) was used to calculate the degree of saturation. Under varying conditions, the aggressiveness of soil water at sites above the cave was usually 15–90 ppm $CaCO_3$. At a depth of 12 cm, the range was 30–140 ppm. In winter, snow cover or frost gave rise to the highest values. As expected, soil activity associated with higher summer temperatures led to greater aggressiveness. $CaCO_3$ hardness values were <2 ppm.

Drip and percolation water within LNRC, at positions corresponding approximately to the surface soil water sampling sites, were always saturated with $CaCO_3$. These inputs may therefore be considered to be adding to the formations found within the higher parts of the cave. In contrast, however, the water sinking into the cave at Flood Entrance had calcium hardness values of 10–78 ppm $CaCO_3$. On almost every occasion on which samples were taken, this water was aggressive, the highest levels of aggressiveness usually coinciding with high flow conditions.

Mechanical erosion is also most effective at high flow conditions. Evans (1969) found that stream lengths on the Old Red Sandstone catchment of the near-by Hepste expanded three-fold in flood conditions, while suspended sediment loads increased by a factor of 20. He concluded that the transport of cobbles and boulders during flood was probably more rapid than in any other British limestone area.

Development of the drainage and caves

The present surface drainage bears little relationship to that existing before the last glaciation. North (1962) suggested that the Nedd Fechan, Mellte and Hepste originally flowed southeast with the dip, to form part of the headwaters of the Cynon. The formation of the main Neath Valley by glacial

action subsequently captured the three rivers, which now flow southwest. All three rivers appear to be juvenile, with no evidence of 'knick-points' and it seems probable that the course of the present Nedd Fechan, at least, is entirely post-glacial. The relationship of the present drainage to surface and geological features is shown in Figure 15.12.

To the northwest of Pant Mawr Pot is a wind gap, beyond which the Nant Byfre now drains to the Tawe Valley. The catchment for Pant Mawr Pot could therefore have extended some considerable distance to the northwest. A stream much larger than the present one may have sunk at the present open shaft, since there is no evidence to suggest that the main passage ever extended north of this. It now sinks a little to the north, reaching the base of the shaft through small passages which are clearly of relatively recent origin.

The major passages in LNRC drain to the southeast and the stream may have originally resurged in the Mellte Valley. Passages such as New World Passage may date from this period and may have originated considerably to the northwest of the present cave. Evidence of this early development in the area is provided by Y Gwal, a sizeable remnant passage on flat ground to the east of the cave and at an elevation comparable to that of Pant Mawr. The ground surface here has been lowered to such an extent that entry to the cave is through a surface collapse in the roof.

The Nedd Fechan has now cut a deeply incised valley. The earliest effect of this, on the argument above, would have been to behead the inlets which formed the major high-level passages in the cave. More recently, Bridge Cave has been intercepted and may once have taken all the flow of the Nedd Fechan. The Main Stream Passage may date from this period. The riverbed has now cut down still further and water enters Bridge Cave from the river by immature routes. The sinks by Flood Entrance are the most recent of all, but will eventually capture the whole river. Interestingly, the results of water-tracing tests from Ogof Cul, on the edge of Pant Mawr, suggest that older passages may still exist beneath the riverbed.

The down-cutting of the valley has captured the drainage of both LNRC and Pant Mawr Pot. Their resurgences (R2 and R1, respectively) are close together, but there is no evidence to suggest that they are linked hydrologically. About 0.5 km upstream of R1, up on the west bank, is Cwm Huw Bwb, a small natural amphitheatre which bears a marked resemblance to Pwll Du. It seems probable that this is an abandoned resurgence for Pant Mawr Pot, active when the riverbed was perhaps 30 m higher than at present. There is no comparable abandoned resurgence for LNRC, except perhaps (very speculatively) White Lady Cave.

Acknowledgements

The author has drawn extensively from an unpublished manuscript by P. Standing & N. Martell, and wishes to acknowledge their contribution.

References

Alexander, J. M. & Jones, J. C. (1959). *The Survey of Pant Mawr Pot, South Wales*. Cave Research Group, GB, Publ. No. 9.

Braithwaite, T. A. J. (1938). Porth yr Ogof and its Neighbours. *Caves and Caving (BSA)*, **1(3)**, 93–8.

Bridges, E. M. (1971). Pedology. In *Swansea and its Region*, ed.

W. G. V. Balchin, pp. 73–85. British Association Annual Meeting. Swansea: British Association.

Bull, G. (1972). On the Pant Mawr Master System. *Westminster Spelaeological Group Bulletin*, **7(9)**, 134–8.

Evans, M. (1969). An investigation into the relationship between morphometric and hydrological properties of the Upper Hepste. Undergraduate dissertation. University of Bristol, Geography Department.

Gascoine, W. (1985). Dye testing on Pant Mawr. *South Wales Caving Club Newsletter*, **100**, 25–7.

George, T. N. (1970). *British Regional Geology, South Wales*, 3rd edn. London: HMSO.

Mullen, G. (1987). Little Neath River Cave. *Caves and Caving*, **37**, 8–11.

North, F. J. (1928). The River scenery at the head of the Vale of Neath. *Transactions of the Cardiff Natural History Society*, **61**, 12–54.

North, F. J. (1962). *The River Scenery at the Head of the Vale of Neath*, 4th edn. Cardiff: National Museum of Wales.

Norton, M. G., Savage, D. A. & Standing, P. A. (1967). The Little Neath River Cave, South Wales. *Proceedings of the University of Bristol Spelaeological Society*, **11(2)**, 186–200.

Owen, T. R. (1971). The headwater region of the River Neath. In *Geological Excursions in South Wales and the Forest of Dean*, ed. T. R. Owen, pp. 74–85. Newton Abbot: David & Charles.

Standing, P. A., Newson, M. D. & Wilkins, A. G. (1971). Second Report on the Little Neath River Cave. *Proceedings of the University of Bristol Spelaeological Society*, **12(3)**, 303–25.

Stenner, R. D. (1969). The measurement of the aggressiveness of water towards calcium carbonate. *Transactions of the Cave Research Group, GB*, **11(3)**, 175–201.

Thomas, T. M. (1974). The South Wales interstratal karst. *Transactions of the British Cave Research Association*, **1(3)**, 131–53.

Weaver, J. D. (1973). The relationship between jointing and cave passage frequency at the head of the Tawe Valley, South Wales. *Transactions of the Cave Research Group, GB*, **15(3)**, 169–75.

16

Ogof Ffynnon Ddu

PETER L. SMART and NOEL S. J. CHRISTOPHER

Ogof Ffynnon Ddu (OFD) (SN 848 152) is the second longest cave in Britain with over 40 km of explored passage and, with a vertical range of 300 m, it is Britain's deepest cave. It is located 4 km northeast of Abercraf on the east side of the Upper Tawe Valley in a narrow belt of southward-dipping limestone. The cave is one of the most complex in Britain, both physically and morphologically. It extends from near its sink at Pwll Byfre (SN 875 166) to the resurgence of Ffynnon Ddu (the Black Spring) (SN 847 151) from which it derives its name. The cave and enclosing land were designated a National Nature Reserve by the Nature Conservancy Council due to their national significance and threats from quarrying.

Close to the resurgence is the Lower Entrance that gives access to OFD I (Fig. 16.1). This was the first part of the cave explored. The Entrance Series consists of a complex multi-levelled sequence of rectilinear passages, preferentially developed in a north–south direction due to fracture control, coincident with the generalized regional dip of 8–15° to the south. Beyond the Entrance Series a southwest-trending vadose stream derived from Pwll Byfre is encountered, which emerges from a sump after 400 m. Upper Flood Passage is an intermittently active continuation of the Streamway which ends at Boulder Chamber. Above the stream are abandoned series of passages, including Waterfall Series, Railton–Wilde (R–W) Series, Maypole Series and the Rawl Series. The Rawl Series is a major high-level fossil passage developed on a prominent shale band.

Beyond Boulder Chamber is Dip Sump, above which a restricted crawl gives access via Dip Series to OFD II in the Piccadilly–Smithy area. This area is also accessible via Cwm Dwr Quarry Cave whose entrance is close to Penwyllt (SN 856 156). The Main Stream is again encountered at the Confluence near Piccadilly and can be followed for 3130 m to Top Waterfall. The Streamway is varied and exhilarating, containing an extensive series of potholes whose origin is obscure (Fig. 16.2). An aesthetically appealing area, the Marble Bathroom, is thought to be associated with faulting and an extensive area of ledges and small waterfalls is associated with dolomite beds.

To the north of the Streamway and at greater altitude are the high-level fossil series of Marble Showers (Fig. 16.3), Fault Aven, Great Oxbow, Gnome Passage, Chasm and Nyth Bran, which are entered at Maypole Inlet and lead up to the Top Entrance (SN 864 159). Near by are the famous Columns (Fig. 16.4). The Top Waterfall can be bypassed via the fossil series following the route via the Chasm and Traverses to OFD III. This consists of a single stream passage (Fig. 16.5) and a coincident fossil series that terminate at Smith's Armoury and Grit Choke, respectively (Charity & Christopher, 1978).

History of exploration

Only the briefest outline of the long and fascinating history of exploration of Ogof Ffynnon Ddu can be given here; for further details the readers are referred to Railton (1953), Hill (1956), O'Reilly (1967) and O'Reilly, O'Reilly & Fairbairn (1969).

The resurgence of Ffynnon Ddu had attracted the attention of cavers for many years. The first serious attempt to enter the system thought to exist behind the spring was in 1942, when members of the Welsh Section of the Mendip Exploration Society dug into Pant Canol. However, shortly after the formation of the South Wales Caving Club, on 3 August 1946, Peter Harvey and Ian Nixon dug into Ogof Ffynnon Ddu I near the present entrance. The present Lower Entrance Series and OFD I Streamway were rapidly explored (Figs. 16.6, 16.7); in September the Waterfall Series was entered and explored and this was followed shortly after by the discovery of the Railton–Wilde Series (1947), the Maypole Series (1948) and the Rawl Series (1950), accessible via a high-level traverse (Fig. 16.8).

Progress towards the sink at Pwll Byfre was halted, due both to a blockage in the Boulder Chamber and by a sump. It was 1958 before Gordon Clissold and Bill Little digging in Boulder Chamber found Hush Sump, which was dived and connected to Dip Sump. Meanwhile, digging in Starlight Chamber, above Boulder Chamber, gave access in 1953 to Coronation Series, which was eventually pushed to an unstable boulder choke and abandoned.

Diving in Dip Sump began shortly after it was discovered, and although the way on was soon identified the problem of entering a roof tube, 5 m above the deep water with nowhere to climb out, defeated divers until 1966. In the interim, various

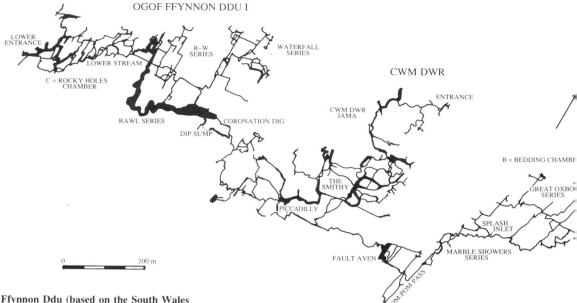

Fig. 16.1. Plan of Ogof Ffynnon Ddu (based on the South Wales Caving Club survey by O'Reilly *et al.*, 1969).

Fig. 16.2. Turbulent water in the Ogof Ffynnon Ddu streamway (photo by Ian Davinson).

Fig. 16.3. Heavy calcite veining in the limestone at Marble Showers (photo by P. R. Deakin).

people, particularly Clive Jones, Les Hawes and Bill Birchenough, were extending Cwm Dwr Quarry Cave, originally discovered by quarrying. In 1963 the Cwm Dwr Jama was entered, but progress was once again blocked by boulders. The

return of the divers in 1966 with a lightweight portable ladder allowed access to OFD II. The Streamway was rapidly explored (Fig. 16.9) and, despite a narrow escape with loose boulders and frustrated efforts in Coronation Dig, a dry route was found and made safe through the Cwm Dwr Boulder Chokes by mid-April of the following year (1967).

Some exploration was carried out in the next month, but it was May 1967 before Paddy O'Reilly and Colin Fairbairn maypoled one of the stream inlets and entered what is now known as Top Entrance or Clay Series; this latter name derives from the unusual slippery white deposit found. Exploration of

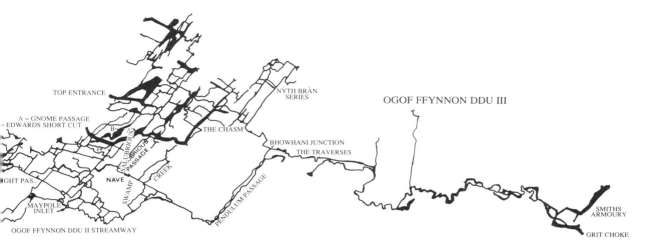

Fig. 16.4. 'The Columns' in OFD II (photo by J. Rowland).

this maze of passages, many kilometres long, occupied O'Reilly and his co-explorers for several months and 15 hour trips became common. During a camp in the cave a boulder pile with snails was found, and subsequent radio detection showed it to be close to the surface: following a smoke test a dig was started and within a few hours the Top Entrance was open. This shortened the trip duration considerably and allowed explorers to find OFD III in October 1967. Exploration and survey of this huge cave system occupied the next two years, culminating in the publication of the survey in 1969. The final safe link between the old cave and the new had to wait until the efforts of Radcliffe (1971): since then there have been numerous small finds but few big ones despite enormous effort. Many parts of the cave are superbly decorated (Figs. 16.4, 16.10).

Geology

The outcrop of Carboniferous Limestone in which Ogof

Fig. 16.5. A quiet stretch of the Ogof Ffynnon Ddu streamway (photo by P. R. Deakin).

Ffynnon Ddu is located is split by the NNW–SSE-trending Henrhyd Fault, which runs down the line of the River Tawe, and a NE–SW compression belt, the Cribarth Disturbance (Fig. 16.12). To the west of the Henrhyd Fault lies the cave of Dan yr Ogof (Coase & Judson, 1977), while to the east is Ogof Ffynnon Ddu. Several workers have given general accounts of the geology of the area. George (1927) concentrated on the

Fig. 16.6. The stream running roughly along the strike of inclined limestone beds in OFD I (photo by P. O'Reilly).

Fig. 16.8. The traverse route to the Rawl Series, high above the stream in OFD I (photo by J. Wooldridge).

Fig. 16.7. A shower falling onto a series of gour pools (photo by Ian Davinson).

Fig. 16.9. Vadose stream in a phreatic tube in the OFD II streamway (photo by C. Howes).

stratigraphy, while Weaver (1971, 1975) described the structure. Charity & Christopher (1977a,b) carried out a detailed stratigraphic and structural survey of the block of limestone encompassing the cave and it is from these works and other unpublished data that the following account is derived.

Stratigraphy

The oldest beds exposed in the area consist of sandstones and marls of the middle Old Red Sandstone Series.

These grade upwards into the Grey Grits which appear to be covered conformably by the Lower Limestone Shales of Courceyan (K) age (George, 1927; George et al., 1976). Lying unconformably above the Courceyan Stage are the beds of Holkerian (S_2) age, known locally as the Dowlais Limestone. A simplified section of the beds of this stage is presented in Figure 16.11. The basal beds, some 6 m thick and arenaceous towards the base, mark the unconformable Holkerian transgression; these are capped by a thick bed of the coral *Lithostrotion martini*. Above this are 10 m of pure limestones split after 7 m by a bed of *Composita ficoides*. These beds are overlain by 4.3 m of highly dolomitized limestones that outcrop over a wide area adjacent to Penwyllt and are extensively exposed in

Fig. 16.10. 'The Fingers' helictites in Ogof Ffynnon Ddu (photo by A. Coase).

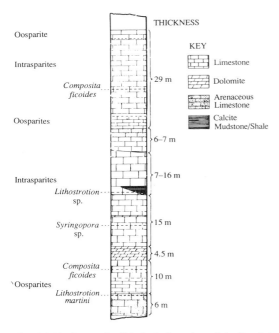

Fig. 16.11. Composite lithological section of the Dowlais Limestone (Holkerian stage) at Penwyllt, Breconshire (based on Charity & Christopher, 1977a,b).

the Main Streamway of OFD II. The next 15 m are generally of moderate purity yielding scattered *Syringopora* corals and thin dolomites. Above is a sequence of wedge-bedded bioclastic calcarenites 7 m thick at its maximum, which are seen in several parts of OFD I as small shale ledges. The upper 29 m of the Dowlais Limestone are characterized by massive bedded limestones of high purity containing many stylolites and split

centrally by a thick series of shell beds yielding a profusion of *Composita ficoides*, which is seen in the Top and Lower Entrance Series. At Top Entrance (SN 864 159) the uppermost beds of the Dowlais Limestone may be seen, overlain unconformably by the Honeycombed Sandstone of Asbian age (D_1). The Honeycombed Sandstone is locally an impersistent marker which may be replaced by up to 2 m of dark grey shales, seen particularly in the Rawl Series and Shale Chamber. The Asbian Stage is represented locally by some 10 m of oolitic limestone (the Penderyn Oolite), above which limestone beds of darker colour mark the transgression to the Brigantian Stage, some 35 m thick. The highest Lower Carboniferous Beds of the area comprise a poorly developed 2–3 m of Upper Limestone Shales (D_{2-3}), consisting of shales with impure limestones generally converted to rottenstone. The exposed rottenstone beds are overlain unconformably by massive quartzites with thin shale partings of the Millstone Grit (Homoceras Zone of the Namurian).

Ogof Ffynnon Ddu and its associated caves are developed exclusively within the Dowlais Limestone. However, the Asbian and Brigantian Limestones exposed in Brickworks Quarry (SN 856 154) and West of the Stump (SN 853 154) contain small caves that may have connected with OFD before truncation by surface erosion.

Structure

The area lies within a Hercynian compression belt influenced by structures in the underlying Lower Palaeozoic block (Weaver, 1975). A composite geological map based on Weaver (1975) and Charity & Christopher (1977a,b) is presented in Figure 16.12. The main features of the structure are the strong compression belt of the Cribarth Disturbance that trends NE–SW to the north of Penwyllt passing north of the cave. The Henrhyd Fault and its complementary splay faults pass NNW–SSE up the line of the Tawe Valley and marginally affect OFD I.

Moving east from this fault a series of N–S faults with throws varying from a fraction of a metre to 35 m have been mapped, while Charity & Christopher (1977a,b) identified for the first time a series of shallow asymmetric anticlines and synclines with a southerly plunge: the crests of these anticlines coincide with the principal high-level series of passages within OFD (Fig. 16.12). This series of anticlines and synclines is thought to result from strike–slip movement between the Henrhyd Fault and the fault at Pwll Byfre.

Near Pwll Byfre, Dinantian Limestone lies in close proximity to Namurian grits. The east wall of Smith's Armoury, in OFD III, shows considerable evidence of shattering and calcite infilling. This suggests the presence of a major fault that is probably responsible for the grit outlier close to Pwll Byfre, known locally as the Silica Pit, or Sand Hills. This contradicts the opinion of Thomas (1963) who suggested that it was due to collapse into a huge pre-existing cavern of mid-Tertiary age.

The joint pattern in the area has been examined by Glennie (1948, 1950), O'Reilly (1967) and Weaver (1973). The dominant joint set is N–S, with secondary E–W joints and some of NE–SW, NW–SE directions. These correspond to the longitudinal, transverse or shear fractures of the minor folds

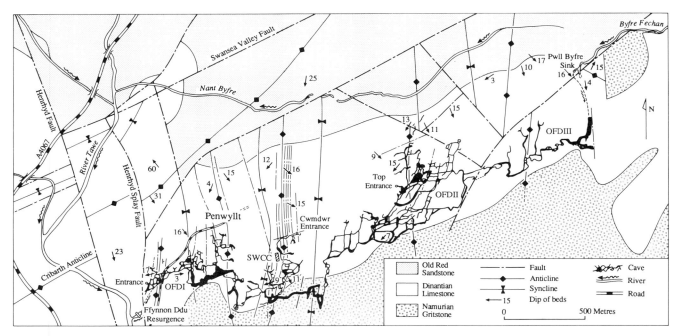

Fig. 16.12. Structural map of the Penwyllt area in relation to Ogof Ffynnon Ddu (based on Weaver, 1975; Charity & Christopher, 1977a,b).

described above, reflected in the joint–block structure of the walls of the Big Chamber near the Entrance (Fig. 16.13). The passage directions mirror the joint directions, and detailed examination of Top Entrance Series and OFD I reveals that the majority of the N–S passages are developed along calcite-infilled fractures with little or no apparent throw. However, some passages, notably Gnome Passage (Fig. 16.14), do not appear to be directly influenced by the N–S fractures. Gnome Passage passes NE–SW down the shallow limb of one anticline to the crest of the next tectonically influenced area. Such passages are possibly caused by water flowing in one bedding plane at a constant hydraulic gradient around the nose of an anticline.

Fig. 16.14. Gnome Passage in OFD II (photo by C. Howes).

Fig. 16.13. The 'Big Chamber near the Entrance' in OFD II, showing joint–block collapse (photo by C. Howes).

The structural control of the cave is well illustrated by Figure 16.15. Based on a structural survey of OFD I it can be seen that the main passages conform closely to the direction of the N–S fracture set. Glennie (1948) described the structure as limestone beds dipping at 16° south, divided by a single fault. Charity & Christopher (unpublished data) have found that the dip varies from 6 to 16° over a total directional range of 245° either side of south (Fig. 16.16), as a consequence of the folding now recognized. It can therefore be seen that Glennie's views are a considerable oversimplification.

Finally, collapse features dominate the scenery in certain areas, particularly in Top Entrance Series. Here a large area is roofed by the highly competent upper *Composita ficoides* shell bed, which, due to an absence of natural joints, presents a solid

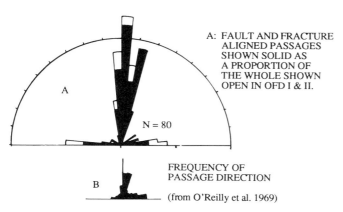

A: FAULT AND FRACTURE ALIGNED PASSAGES SHOWN SOLID AS A PROPORTION OF THE WHOLE SHOWN OPEN IN OFD I & II.

N = 80

FREQUENCY OF PASSAGE DIRECTION

(from O'Reilly et al. 1969)

Fig. 16.15. (A) Fault and fracture-aligned passages shown solid as a proportion of the whole shown open, in OFD I & II. (B) Frequency of passage direction (from O'Reilly et al., 1969).

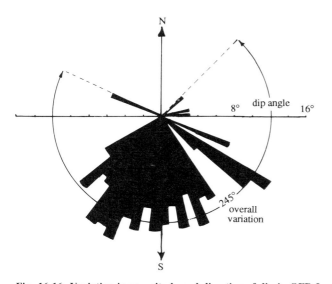

Fig. 16.16. Variation in magnitude and direction of dip in OFD I Entrance Series (from unpublished data by Charity & Christopher).

roof feature up to which lower less competent beds collapse. The cause of this extensive collapse is uncertain, but extensive vadose trenching, probably by invasive glacial streams, is likely to be a major factor. Structural relaxation following deglaciation and frost wedging in glacial episodes may also have contributed.

Geomorphology

A number of factors combine to make the geomorphology of OFD relatively difficult to describe. First, the cave is physically very complex, making detailed observation mentally taxing and laborious. Second, considerable fossil passage has been blocked by sediment or collapse and removed by surface erosion. Third, while an adequate plan of the whole system is available, elevation data is limited to OFD I. The previous geomorphological work of Glennie (1950) and Railton (1953) was limited to OFD I, prior to the discovery of OFD II and III. Subsequent work has concentrated on understanding geological controls on cavern development (Charity & Christopher, 1977a,b), although both O'Reilly (1973) and Eldridge (1977) have presented general accounts, the latter

emphasizing the surface denudation chronology for the area proposed by Brown (1960) and other earlier workers.

In this account we aim to identify the major passage types present in OFD, their mode of formation and their controls. Reducing the complex system to a limited number of simple components may permit progress in the explanation of the development sequence of the complete system. This possibility is demonstrated using examples from OFD I, where the surveyed elevation data essential in geomorphic studies is available.

Phreatic development

The most important control on the development of a phreatic cave system is the relative location of inputs and outputs in the limestone aquifer unit. The effective aquifer unit for OFD has been described above as the Dowlais Limestone, underlain by the Lower Limestone Shales and capped by the Honeycombed Sandstone or associated shales. Inputs to this unit have been at the base of the limestone from the underlying Old Red Sandstone, which has formed topographically higher terrain than the more soluble limestone during the long denudation history of the region. Drainage up-dip from the Millstone Grit, a mechanically strong and resistant rock which forms a prominent scarp and dip-slope at present, seems unlikely and is not suggested by any of the proposed denudation chronologies.

Output from the limestone has been controlled by the now deeply incised Tawe Valley since the development of this feature. Prior to this, Eldridge (1977), following Brown (1960), placed considerable emphasis on the significance of the Penwyllt Bench, a topographic flat developed in the vicinity of Penwyllt. This is proposed to be the floor of a proto–Tawe Valley, which drained to the Banwyn Gap, some 5 km to the south. It is in fact debatable whether the remaining surface evidence allows any realistic interpretation of the early topography and drainage of the area. Rather, the palaeo-hydrological conditions indicated by the cave passages are probably more reliable indicators of the position of the output.

The highest passages in the system discharge in a westerly direction, but the downstream portions of many of these have been destroyed by surface erosion. For instance, Gnome Passage (OFD II) is a major high-level canyon which terminates in a boulder choke 9 m below the surface. Thus, exact discharge points are not known, although minimum elevations could be deduced for each particular phase.

The current outlet is at the top of the limestone unit, and it seems probable that this has also been the case during earlier developments; certainly, in the upper levels of OFD II such as Bedding Chamber, the contiguous phreatic passages gain stratigraphic elevation in a downstream direction. Thus dip development was initially limited to the width of the exposed limestone (800–900 m at present), but as incision of the Tawe Valley occurred, this has increased to about 1.5 km as the limestone outlet has migrated down-dip to the south.

The detailed network geometry and passage location in a cave system are controlled primarily by the lithology and structure of the limestone unit, the regional topography and drainage governing only the level and direction of the system.

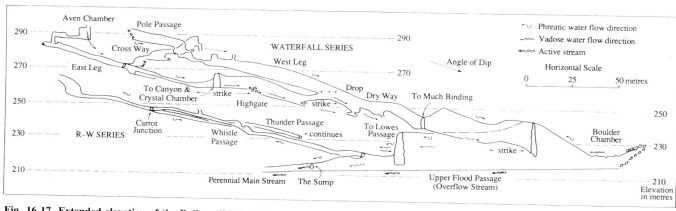

Fig. 16.17. Extended elevation of the Railton–Wilde Series and Waterfall Series (redrawn after Railton, 1953).

For OFD the influence of strike–slip flexures and bed lithology have been discussed above. Input to the dipping limestones is dominated by dip tubes formed at favourable points on the more open bedding planes, frequently at joint–bedding intersections. There may be many tiers of such dip tubes oriented approximately down-dip on the N–S joint or fracture set. Glennie (1950), for instance, identified five horizons of dip-tube development in a limited area in OFD I, and there may be many more in the complete Dowlais Limestone sequence. These dip tubes are integrated vertically by joint risers up which water can pass between different bedding planes. This permits water to gain stratigraphic height between input and output. Finally, strike integration can occur along E–W transverse fractures and NE–SW shear fractures in response to the head differences caused in the dip tubes by the regional hydraulic gradient. Not all dip tubes and strike integrator tubes are developed along joints; anastomosing tube networks formed along bedding planes can often be observed in OFD. The validity of this simple three-component network in dipping limestones has been demonstrated by the gypsum and salt block models of Ewers (1978).

The three basic units can readily be recognized in OFD; Figure 16.17, re-drawn after Railton (1953), is an extended section of the Railton–Wilde and Waterfall Series in OFD I. Dry Way–West Leg and East Leg are dip tubes developed on different bedding surfaces. Cross Way, developed on a strong transverse fracture, permits water to lift from East Leg to West Leg gaining stratigraphic elevation, a role also evident for the strike-oriented but joint-guided sections of Highgate. In fact these joint risers also serve as strike integrators, linking one segment of bedding dip tube to the other. Discharge from the Waterfall Series was by Upper Flood Passage, a strike integrator developed almost horizontally in the same bedding as the lower section of Dry Way. The adjacent Railton–Wilde Series, which is developed largely on a single bedding plane, shows several horizontal strike segments and also demonstrates that flow in dip tubes can equally be up-dip or down-dip under the pressure flow conditions present in phreatic systems.

The Railton–Wilde Series, which has been little affected by vadose erosion, also demonstrates the complexity of phreatic networks, with a series of cross-links short-circuiting the major

up-dip loops. Ford & Ewers (1978) have argued that such short-circuiting is associated with the progressive development of solutional openings in the limestone during prolonged phreatic circulation. However, they may also be a response to changing head distributions following a fall in the regional base level as the Tawe incised into the gradually rising Welsh massif.

Passage modification with falling base level

There are two possible responses of a phreatic cave system to base-level lowering: the cave may incise its floor giving rise to a vadose trench or it may develop a new phreatic passage to the lowered resurgence level (Fig. 16.18). The latter is possible because the head distribution on the bedding plane is altered, giving rise to a steeper hydraulic gradient. This causes enhanced diffuse flow along the bedding plane, which may be concentrated into the more open routes, some of which have been opened by solution under previous phreatic circulation. Once a mature passage is developed, the head difference across the bedding is reduced. For both these responses, a

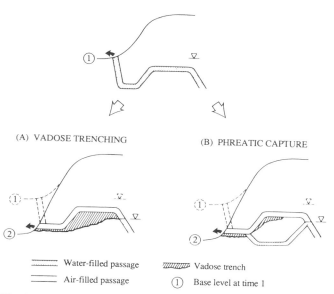

Fig. 16.18. Responses of the phreatic cave to base-level lowering. (A) Development of vadose trench to eliminate up-dip loop. (B) Development of phreatic capture.

decreased head at the resurgence is propagated upstream through the phreatic system to the crest of the next up-dip loop which lies above the new water level. A head difference thus again develops, and the loop can again be eliminated by trenching or capture.

In practice, because both responses are slow, many loops will show some trenching before capture by a lower phreatic passage; for instance, the Rocky Holes Chamber Passage has a limited trench formed prior to capture by the tube in the top of the Streamway Passage. Conversely, the trench may intersect a still immature capture as in Edwards Shortcut which captured water from the Gnome Passage trench. Furthermore, flood flows may still be accommodated by the abandoned high-level route for a considerable period until capacity in the lower levels has become adequate. This is, in fact, the present situation at the upstream end of OFD I, where Upper Flood Passage functions as a flood overflow for mainstream water from Boulder Chamber. At times of low flow, however, the capacity of the present phreatic passage discharging at the Sump is adequate. The Boulder Chamber–Upper Flood Passage route is clearly the most recently abandoned of the strike feeders in this part of the cave, but in the Waterfall Series, Much Binding and the Canyon form two higher systems, now much choked by sediment from Aven Chamber (Fig. 16.17). Unfortunately, elevations are not available for these passages, but it is apparent that, whereas the phreatic rift linking Crystal Chamber and East Leg was abandoned with relatively little modification, the Canyon was extensively trenched prior to the

development of the Much Binding cut-off. The compilation of such sequences allows identification of contemporaneous components of the active circulation. For example, discharge northwards of Carrot Junction must be contemporaneous with the Canyon or an earlier level, the vadose Much Binding input being too low to permit pressure flow in this phreatic passage (Fig. 16.17).

Vadose development

Examination of the network of currently active passages in OFD (Fig. 16.19) graphically demonstrates that the predominantly strike-oriented main streamway formed primarily by the vadose incision of phreatic loops, is only one of two major vadose components. The other comprises down-dip percolation water inlets. Indeed, the main stream follows a similar predominantly down-dip course from Pwll Byfre to Smith's Armoury, where it turns westward to form the OFD III Streamway.

Vadose flow is controlled by elevation difference, water moving under gravity from higher to lower elevations. Thus, where well-developed vertically continuous joint systems occur, vadose flows are predominantly vertical as at Splash Aven. More commonly, however, small dip tubes developed on bedding planes provide the most direct downhill routes, although these are often diverted from the regional dip direction by considerable irregularity of the bedding planes (Palmer, 1981). These tubes are rapidly modified to give narrow vertical canyons, which are developed in an upstream direction by progressive lowering of the floor below the forma-

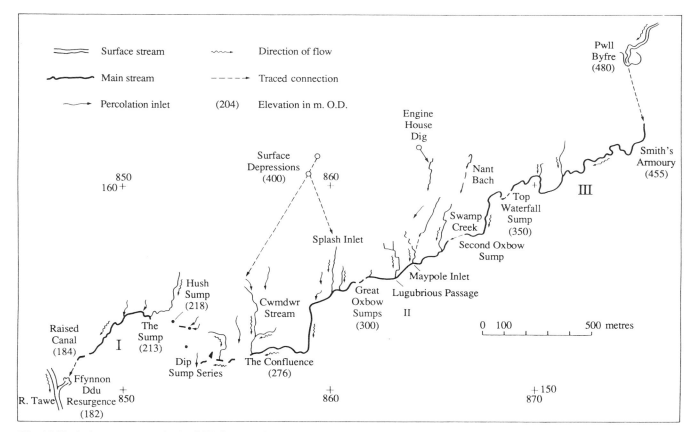

Fig. 16.19. Active streamways in the OFD System (elevation data from Eldridge, 1977, Fig. 14).

tive bedding plane. Thus, passages such as Swamp Creek or Maypole Inlet get progressively taller in a downstream direction as the floor incises towards the level of the main stream. Due to their much slower rate of incision, such passages hang above the trunk stream level, by 25 m in the case of Swamp Creek. A fossil example is the unnamed tributary canyon entering Gnome Passage from the north at the Terminal Boulder Choke.

Because the vadose zone contains a considerable number of abandoned phreatic and vadose passages, these routes are frequently used by active vadose flows; the Maypole Inlet–Salubrious Passage Stream provides a good example. The gravitational vadose flow may reverse the previous (pressure) flow direction, but is generally concordant with previous vadose flows. The width of a vadose passage is related in a general manner to the discharge of the formative stream, as is readily apparent on consideration of active streams in the cave at present; for example, Main Stream and Maypole Inlet. Thus, later vadose flows invading pre-existing passages can often be recognized by the change in passage width (for instance, Lower Dry Way OFD I or Lower Cwm Dwr Stream), even where both developments are now fossil.

The size of the current Cwm Dwr Stream gives an indication of the dimensions expected for an integrated vadose percolation stream in the OFD area. However, in OFD II there are numerous down-dip canyons whose dimensions are rather greater. These may have been caused by diversion and leakage of the main Pwll Byfre Stream by temporary blockage of the sink. However, the complexity of the area east of the Upper Entrance suggests that these inputs must have persisted during several shifts in base level, giving rise to multi-phase down-dip canyons, which truncate earlier vadose and phreatic passages. A persistent source, probably from the Old Red Sandstone, is therefore indicated.

Vadose processes are not simply erosional; roof and wall collapse has already been mentioned, but secondary sediments and speleothem deposition are both significant. Of particular interest are slippery white deposits found extensively in Ogof Ffynnon Ddu II, although an isolated patch occurs in the Waterfall Series of OFD I. They were originally thought to be montmorillonite clay (O'Reilly et al., 1969) but subsequent examination revealed that they were very light with a relative density of 0.3, and that they consisted of calcite with minor inclusions of quartz, clay minerals and sulphate (Picknett, 1969). This preliminary examination also showed the presence of many bacteria. A subsequent examination (Heathcote, 1977) showed that the calcite was present in the rare mineral form of lublinite, which strongly suggests a bacterial origin for the deposits, as this mineral is only found associated with moonmilk, a by-product of bacterial activity. There is, however, a great deal of work remaining before these deposits are satisfactorily explained.

The development of the Main Stream Passage

Three features of OFD are particularly puzzling and require specific consideration. First, the network geometry and morphology of OFD III is remarkably simple compared to OFD I and II. A single high-level passage links the Grit Choke upstream with the Nyth Bran Series and the major Chasm trench beyond Bohwani Junction. This tube shows no evidence of development up- or down-dip, although minor down-dip inlets are known. It has been trenched by some 10 m at its upstream end, increasing to 50 m in the vicinity of the Traverses. Second, the Main Stream Passage has a very limited active phreatic circulation at present (Fig. 16.19), the longest section being the sump complex between OFD I and II. This contrasts with the record of abandoned passages which suggests periods of extensive phreatic circulation have occurred. Third, the Main Streamway generally occupies the most southerly (down-dip) portion of the cave system. The only exceptions to this are Pom-Pom Passage in OFD I and the Rawl Series near Roundabout Chamber. The Streamway position correlates approximately with the location of the overlying Limestone–Millstone Grit boundary. There are, however, considerable departures and the coincidence is probably fortuitous. The additional overburden pressure of the grit cover will have only a marginal effect on penetration of joints by water, and solutional development of fractures by mixing corrosion can still occur as leakage through the Millstone Grit is known.

Much of the OFD II and III streamways are low in the Dowlais Limestone, displaying arenaceous and dolomitized beds, particularly downstream of Smith's Armoury and in the OFD II Streamway as far downstream as Maypole Inlet. A fold or fault south of the streamway, with the SW–NE orientation characteristic of the Cribarth Disturbance structure, could therefore bring up the underlying impermeable beds, preventing down-dip development of phreatic passages. In fact, there is little evidence for such a structure (Weaver, 1971, 1975), which would need to be relatively large to affect strike sections of streamway separated by predominantly down-dip passages, as in Pendulum Passage.

It is known that the arenaceous and dolomitized limestones (Fig. 16.20) are less soluble than the purer limestones higher in the succession (Christopher, 1967; Charity & Christopher, 1977a,b). They are also more thinly bedded. In combination this may retard the development of solutional voids, the bedding planes and joints remaining tightly closed, preventing phreatic circulation and the development of down-dip captures. This suggestion is supported by the occurrence of the most complex phreatic networks in the upper parts of the succession, for instance, near Top Entrance and in much of OFD I. Even if this lithological control was not important, a vadose trench incising through 50 m of limestone would have five times the probability of intersecting a potential phreatic capture point than one cutting through only 10 m. Trenches starting lower in the succession are therefore inherently less likely to be captured, a probability decreased further by the lower rate of incision in these less soluble lithologies.

However, if the above arguments are correct, it is surprising that the OFD III Streamway has not been significantly diverted down-dip from its original course, despite cutting through over half the limestone succession. One possible explanation is that the original streamway was guided by a series of strong fracture lines (which coincided with the direction of steepest hydraulic gradient) and these fractures have continued to guide passage development. The Traverse passage certainly conforms to this

Fig. 16.20. The OFD II streamway showing ledges developed as a result of alternations of dolomitic and non-dolomitic limestone (photo by C. D. Westlake).

suggestion, and although much of OFD III appears to be freely meandering, close examination shows strong fracture control at roof level.

There is an impression in OFD that captures from vadose passages are less frequent than in phreatic passages. In a vadose passage only the floor is water covered, compared to a phreatic passage, where roof and walls are also subject to solutional attack. Development of capture passages is therefore more probable in a phreatic system. A second possible explanation is that incision of a vadose trench is rapid in a stream fed by aggressive water and carrying a substantial load of abrasive silica sand during floods. Thus, where phreatic capture and vadose trenching are competing to drain an up-dip phreatic loop the latter may be dominant, providing the loop crest is relatively short. This argument is particularly compelling in the upstream parts of the system, where sediment load and aggressiveness are at a maximum, and could explain the remarkable OFD III stream trench. However, unlike canyon passages in County Clare, Ireland, which decrease in size downstream, as predicted by the above theory, the OFD III canyon increases in height downstream. This is related to the important capture, which caused abandonment of the Bohwani Junction route for Pendulum Passage. Upstream retreat of knick-points in the OFD II Streamway has permitted continued incision of the OFD III trench. This process continues today, with the knick-point at Top Waterfall eliminating the bedding-plane sump between OFD II and III, and the First

and Second Waterfalls (OFD III) relating to the abandonment of the Traverses passage.

It has been stressed previously that development of phreatic captures is more probable when head differences are large. Under laminar flow, discharge is directly proportional to head, but once turbulent flow develops, discharge increases with the third or fourth power of head. The sump complex between OFD I and II appears to be associated with a considerable downward step in the level of the main stream, supporting the view that maximum head differences are produced across phreatic loops being drained from downstream (a detailed survey is needed to confirm this observation). Thus, once reverse gradient phreatic risers have been eliminated from the system, the less steep hydraulic gradients associated with vadose streams greatly reduce the possibility of capture, while maintaining the ability to incise the passage floor. As phreatic capture occurs preferentially along the line of steepest hydraulic gradient, there is a considerable probability that the up-dip loop may be replaced by down-dip passage segment. This will be occupied by a vadose stream with the associated reduction in capture probability. Through time, there may therefore be a tendency to replace phreatic loops by a vadose streamway which will occupy the topographically lowest passage in the cave (Fig. 16.21).

This change may be assisted by the progressive development of solutional porosity in the limestones, as suggested by Ford & Ewers (1978). Thus, the early Pendulum Passage capture appears to involve very large head differences, and a remarkable sparsity of alternative routes compared to later captures in the history of the cave. This suggestion is confirmed by an analysis of the loop amplitudes in OFD I, compared to elevation of the loop crest (directly proportional to age for a uniformly falling base level) (Fig. 16.22).

To summarize, the lower units of the Dowlais Limestone are more resistant to solutional development of fracture porosity than the upper units. Vadose passages are inherently less likely to be the subject of phreatic capture than contemporaneous phreatic passages. The increase in solutional porosity of the limestone with time reduces the amplitude of phreatic loops, and therefore the head available for capture. Equally significant, however, is the uniform gradient which will result from the gradual replacement of up-dip phreatic loops with down-gradient (vadose) passage segments. Finally, where vadose entrenchment rates are high, this process will dominate the elimination of up-dip loop segments, not capture. It is therefore argued that in OFD there has been a gradual elimination of phreatic segments, with a corresponding increase in vadose streamway. This mechanism, recognized for the first time in OFD, gives rise to a hybrid cave, intermediate between the water table and vadose drawdown cave of Ford & Ewers (1978).

The effect of glaciation

That the hills of South Wales have been affected by glacial action during the Pleistocene has long been recognized. However, the number and frequency of glacial events is as yet uncertain. Deep-sea core data suggest that the northern hemisphere was glaciated as many as 10 times during the last million

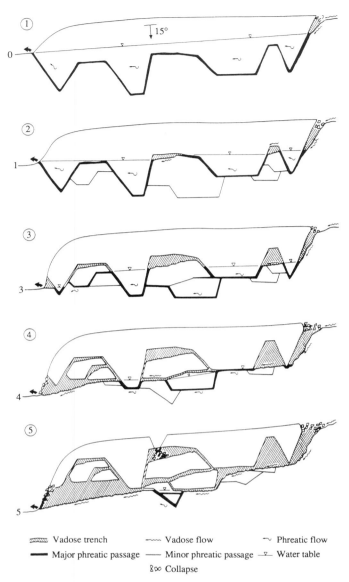

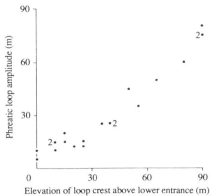

Fig. 16.22. Relationship between amplitude of phreatic loops and elevation of loop crests above the Lower Entrance to OFD I.

Vadose trench   ⌒⌒ Vadose flow   ∿ Phreatic flow

Major phreatic passage — Minor phreatic passage ᴤ Water table

ᴕ∞ Collapse

Fig. 16.21. Schematic vertical section to demonstrate development through time (1–5) with falling base level of predominantly vadose cave system by elimination of up-dip (reverse gradient) phreatic loops. Note that captures occur predominantly in phreatic segments, the amplitude of which reduces through time due to increased solution on fractures. Some deepening of the phreatic circulation does, however, occur. Vadose trenching becomes progressively more important than capture with time. The water table is defined by the level of the active passage floor in vadose segments, and by sump levels in phreatic segments.

years (Shackleton, 1977). Thus, Ogof Ffynnon Ddu has undoubtedly been affected by more glacial–interglacial cycles than the two so far recognized for South Wales. George (1970) identified two types of drift in South Wales: the newer is correlated with the late Devensian Glacial (ice maximum 18 kilo-annees (ka) BP), and the older drift with the previous (Pencoed) glacial thought to be locally equivalent to the Wolstonian (130–150 ka). No evidence of earlier deposits are recognized due presumably to their having been completely removed by subsequent glaciations.

The effects of glaciation on the landforms is not as spectacu-

lar locally as in the more northern parts of Britain, notably Scotland; however, they are none the less recognizable. There are areas of bare rock with glacial scour marks showing generally southerly ice flow, and the northern faces of the Fans have the characteristic features of glacial corries. The Tawe Valley below Craig y Nos has the characteristic shape of a glacial trough, the bottom of which is overdeepened and back-filled with moraine deposits. The Haffes and Nant Byfre valleys hang above this glacial trough. Further, former large areas of Ogof Ffynnon Ddu have been removed by erosion, probably as a direct result of glaciation. Finally, the direct erosional effect of a glacial–interglacial cycle, estimated to be 6–24 m by Gascoyne, Ford & Schwarcz (1983), would result in drainage of phreatic areas as a result of base-level lowering.

The secondary effects of glaciation are probably just as important. These include vadose erosion and subsequent infilling of the cave by fluvio-glacial deposits of sand and gravel, and fine-grained, laminated, ponding deposits formed sub-glacially, frost wedging in the near surface passages (Eldridge, 1977) and a contribution to collapse due to structural relaxation following de-glaciation. Further, the diversion of the present Byfre Stream from its former course down the Nant Byfre into the cave has been attributed to glacial ponding at the close of the Devensian Glaciation. However, the size of the present stream passage in relation to the active stream makes it obvious that this diversion has only restored the previous condition of a sink in the area of Pwll Byfre, before capture of the Upper Byfre by the rejuvenated Nant Byfre. Finally, there has been the sealing of all the former up-dip resurgence points by glacial moraine material which preserved OFD from premature discovery.

Chronology

Previous writers including Glennie (1948), O'Reilly (1967), Eldridge (1977) and Christopher (1983) have attempted to present speculative time-scales for the development of the cave. These have ranged from 10 to 15 million y (O'Reilly, 1967; Eldridge, 1977) to Glennie's (1948) 500 000 y. With our present imperfect knowledge of the landscape morphology it would be unwise to be too dogmatic about the time span involved. However, certain maxima and minima can be suggested. The work of Brown (1960) has coloured many subsequent interpretations of landform development. He maintained that the Welsh massif was shaped into a series of peneplains by a long period of sub-aerial denudation. The area around Penwyllt is loosely attributed to his middle Peneplain period assigned to Neogene times by Eldridge (1977). George (1970) maintained that the nature of the Welsh massif is caused by a pulsed sequential uplift and that the benches are wave-cut platforms. Subsequently he reported dolerite dykes of mean isotopic age 52 ka, that are now planed flush with the present land surface from sea level to over 780 m O.D. The surfaces also post-date movement on several major mid-Tertiary faults (George, 1974). Therefore, the Penwyllt area may have still been submerged in Miocene times.

Recent U/Th dating of flowstone from Top Entrance Series

has identified a sample which capped a previous clastic fill, now removed, with an age of 267 ka. Top Entrance Series cannot therefore be younger than this age.

The Traverses is a major high-level vadose canyon, some 50 m high, the floor of which is about 28 m above that of the OFD III Streamway. From a study of stalagmite ages of the walls of vadose trenches and their elevation relative to the outside valley, Gascoyne et al. (1983) have deduced vadose entrenchment rates of 5–20 cm/1000 y. Therefore, at 10 cm/1000 y it would require 780 000 y to develop the Traverses and OFD III Streamway by vadose incision. Gascoyne et al. (1983) suggested that the upper beds of the limestone of northwest Yorkshire were incised to form the Yorkshire Dales some 1–2 million years ago. By analogy the present authors favour a similar age range for Ogof Ffynnon Ddu.

References

Brown, E. H. (1960). *The Relief and Drainage of Wales*. Cardiff: University of Wales Press.

Charity, R. A. P. & Christopher, N. S. J. (1977a). The stratigraphy and structure of the Ogof Ffynnon Ddu area. *Transactions of the British Cave Research Association*, **4(3)**, 403–16.

Charity, R. A. P. & Christopher, N. S. J. (1977b). The Ogof Ffynnon Ddu Cave System related to geological structure. *Proceedings of the 7th International Congress of Speleology*, pp. 108–10. Sheffield.

Charity, R. A. P. & Christopher, N. S. J. (1978). Smith's fractured limbs. *South Wales Caving Club Newsletter*, **87**, 3–7.

Christopher, N. S. J. (1967). The differential solubility of limestone. *Cave Research Group, GB, Newsletter*, 109–11.

Christopher, N. S. J. (1983). The age of Ogof Ffynnon Ddu. *South Wales Caving Club Newsletter*, **97**, 2–8.

Coase, A. C. & Judson, D. M. (1977). Dan yr Ogof and associated caves. *Transactions of the British Cave Research Association*, **4(1–2)**, 245–344.

Eldridge, G. (1977). The geomorphological development of Ogof Ffynnon Ddu. Unpublished M.Sc. thesis. University of Swansea.

Ewers, R. (1978). A model for the development of broad scale networks of ground water flow in steeply dipping carbonate aquifers. *Transactions of the British Cave Research Association*, **5(2)**, 121–5.

Ford, D. C. & Ewers, R. O. (1978). The development of limestone cave systems in length and depth. *Canadian Journal of Earth Sciences*, **15**, 1783–98.

Gascoyne, M., Ford, D. C. & Schwarcz, H. P. (1983). Rates of cave and landform development in the Yorkshire Dales from speleothem age data. *Earth Surface Processes and Landforms*, **8**, 557–68.

George, T. N. (1927). The Carboniferous Limestone (Avonian) succession of a portion of the north crop of the S. Wales Coalfield. *Quarterly Journal of the Geological Society of London*, **83**, 38–95.

George, T. N. (1970). *British Regional Geology: South Wales*, 3rd edn. London: Institute of Geological Sciences.

George, T. N. (1974). The Cenozoic Evolution of Wales. In *The Upper Palaeozoic and Post-Palaeozoic Rocks of Wales*, ed. T. R. Owen, pp. 341–72. Cardiff: University of Wales Press.

George, T. N., Johnson, G. A. L., Mitchell, M., Prentice, J. E., Ramsbottom, W. H. C., Sevastopulo, G. D. & Wilson, R. B. (1976). *A Correlation of Dinantian Rocks in the British Isles*. The Geological Society of London Special Report, **7**, 1–87.

Glennie, E. A. (1948). Some points relating to Ogof Ffynnon Ddu. *Transactions of the Cave Research Group, GB*, **1(1)**, 13–25.

Glennie, E. A. (1950). Further notes on Ogof Ffynnon Ddu. *Transactions of the Cave Research Group, GB*, **1(3)**, 1–47.

Heathcote, J. (1977). A preliminary investigation of the white deposit in Salubrious Passage, Ogof Ffynnon Ddu. *South Wales Caving Club Newsletter*, **87**, 2–7.

Hill, A. (1956). *The First Ten Years*. South Wales Caving Club 10th Anniversary Publication. Penwyllt: SWCC. 54 pp.

O'Reilly, P. M. (1967). *The South Wales Caving Club's 21st Anniversary Publication*. Penwyllt: SWCC. 260 pp.

O'Reilly, P. M. (1973). The morphology and hydrology of the Ogof Ffynnon Ddu System. *Proceedings of the 6th International Congress of Speleology*, Olomouc. Reprinted in *South Wales Caving Club Newsletter* (1974), **76**, 6–17.

O'Reilly, P. M., O'Reilly, S. E. & Fairbairn, C. M. (1969). *Ogof Ffynnon Ddu*. Penwyllt: South Wales Caving Club. 52 pp.

Palmer, A. N. (1981). *A Geological Guide to Mammoth Cave National Park*. Teaneck, USA: Zephyrus Press. 196 pp.

Picknett, R. G. (1969). Bacteria in an unusual deposit in Ogof Ffynnon Ddu II. *Cave Research Group, GB, Newsletter*, **117**, 26–7.

Radcliffe, R. (1971). The complete cave. *South Wales Caving Club Newsletter*, **69**, 30–2.

Railton, C. L. (1953). The Ogof Ffynnon Ddu System. *Cave Research Group, GB*. Publication **No. 6**. 49 pp.

Shackleton, N. J. (1977). Oxygen isotope stratigraphy of the Middle Pleistocene. In *British Quaternary Studies: Recent Advances*, ed. F. W. Shotton, pp. 1–16. Oxford: Clarendon Press.

Thomas, T. M. (1963). Solution subsidence in southeast Carmarthenshire and southwest Breconshire. *Transactions of the Institute of British Geographers*, **33**, 45–60.

Weaver, J. D. (1971). The Swansea Valley Disturbance. Unpublished Ph.D. thesis. Swansea: University of Wales.

Weaver, J. D. (1973). The relationship between jointing and cave passage frequency at the head of the Tawe Valley, South Wales. *Transactions of the Cave Research Group, GB*, **15(3)**, 169–73.

Weaver, J. D. (1975). The structure of the Swansea Valley Disturbance between Clydach and Hay-on-Wye, South Wales. *Geological Journal*, **10(1)**, 75–86.

17

Dan yr Ogof

ALAN C. COASE

The caves considered in this chapter lie within the limestones of the Northern Outcrop of the South Wales Coalfield, on the dip-slope of the Carmarthen Fan, known as Mynydd Ddu or the Black Mountain. The principal cave is Dan yr Ogof (DYO) which measures about 15 km in explored length, while the adjacent Cathedral Cave (formerly Tunnel Cave) is over 2 km long. The relationship between these, the other caves in the area and the principal sinks, at Waen Fignen Felen (438 m O.D.) and Sink y Giedd (448 m O.D.), are shown in Figure 17.1. The resurgence cave at Dan yr Ogof is at 220 m O.D., very near the base of the limestone. The present catchment area totals over 10 km², although capture by the River Haffes has substantially reduced this from its original extent. The area is bounded by the Cribarth ridge to the south and the valley of the River Haffes to the northeast.

Brief history of exploration

While the resurgence cave has always been open (Fig. 17.2), the passages inside are not large and the discovery of Dan yr Ogof really dates from 1912 when the Morgan brothers climbed into what is now the Show Cave. Soon afterwards the Lakes were partly explored by coracle, but it was not until 1937 that they were crossed by rubber boat and the DYO I was explored as far as the start of the Long Crawl (Figs. 17.3, 17.4, 17.5). In 1939 the Show Cave opened to the public but closed soon afterwards to become a government wartime store and it was not until 1964 that exploration resumed. Diving in the Lakes brought limited results and it was not until Easter 1966 that the Squeeze in the Long Crawl was forced by Eileen Davies and the author who descended the pitch into a vast passage, the beginning of DYO II. Rapid exploration followed with delights such as the Crystal Pool (Fig. 17.6), Green Canal, the Abyss (Fig. 17.7) and Flabbergasm (Fig. 17.8) revealed. The Lower Series of active stream caves provided a round trip in DYO II (Figs. 17.9–17.14) and led off to Mazeways and the southwest sumps. Exploration was brought to a halt at an upstream sump, but a climb up a stalactite-coated rift, a traverse and a pitch of 14 m regained the streamway late in 1966, in what became known as DYO III. The whole cave turned north here and steadily grew to vast proportions in the

Great North Road (Figs. 17.15–17.18), eventually to close in a huge boulder choke. Numerous side passages were explored later and climbs reached Dali's Delight (Fig. 17.19) with its corroded Honeycombed Sandstone chokes and Rottenstone Avens with fallen blocks of what appeared to be greatly weathered limestone known as rottenstone. The main stream enters the System from the southwest, and diving has revealed complex extensions to Mazeways, with many boulder chokes, but the hypothetical Giedd Series has yet to be reached.

Meanwhile, explorations had revealed other caves: the pothole Pwll Dwfn was entered in 1946 but, although 97 m deep, it had little lateral extent. Sink y Giedd was dug out intermittently, and refilled by floods almost as often; Waen Fignen Felen Sink had much the same story, and no major cave has been penetrated in either. The short 'Tunnel' Cave (Railton, 1958), however, led through a boulder choke in 1953 into a fine series of dry tributary passages with magnificent stalactite formations (Fig. 17.20). Later the large Davy Price's Hall (Fig. 17.21) was developed as a Show Cave and renamed Cathedral Cave. Its course trends parallel to DYO III and there is little doubt that it is genetically part of the same story, although now a separate cave.

Dan yr Ogof now totals some 15 km of passages (Fig. 17.22) and is one of Britain's major cave systems. With the 2 km of Cathedral Cave they have been the subject of intensive study, and a full account was given in Alan Coase and Dave Judson's accounts in the *Transactions of the British Cave Research Association* (1977), **4**, to which the reader is referred for a full description, with large-scale plans, geological and geomorphological maps and sections.

Geology and structure

The geological structure of the area is relatively simple, the rocks generally dipping gently southwards. The area is crossed by Old Red Sandstone outcrops, Dinantian Limestones and basal Namurian sandstones which strike NW–SE as far as the Swansea Valley or Cribarth Disturbance (Fig. 17.1). The southeast margin of the caves is crossed by the faulted and folded limestones of the Disturbance, described by Coase (1967, 1975, 1977) and Weaver (1972, 1975), who both con-

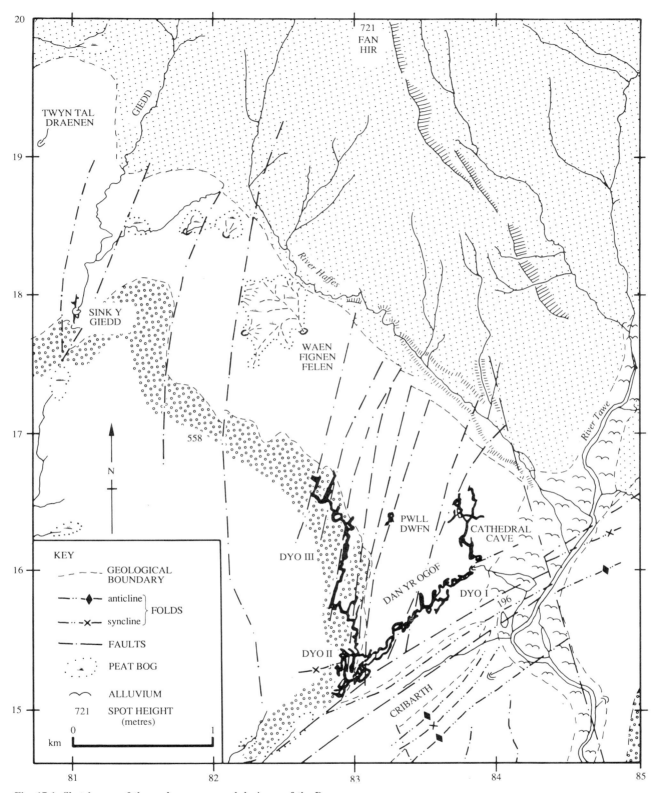

Fig. 17.1. Sketch map of the geology, caves and drainage of the Dan yr Ogof catchment area.

sidered these structures to have had an important effect on the geomorphology of the area in general and on the caves in particular. The Disturbance is composed of a series of sub-parallel faults and folds trending in a NE–SW direction. The fault zone is composed of two principal sinistral wrench faults. Of the accompanying folds, the most important to the caves is a shallow asymmetric syncline lying a little to the north of the Garth–Gwared Splay Fault, only 200 m north of the main

Fig. 17.2. The river entrance to Dan yr Ogof (photo by P. R. Deakin).

Fig. 17.3. 'The Three Nuns' – stalagmites in the Dan yr Ogof Show Cave (photo by T. D. Ford).

Fig. 17.4. One of the Cascades between the lakes in Dan yr Ogof (photo by P. R. Deakin).

Fig. 17.5. The Fourth Lake, Dan yr Ogof (photo by J. Rowland).

Fig. 17.6. The Crystal Pool in DYO II (photo by C. Howes).

Garth–Gwared Fault. Also associated with the Disturbance are a series of N–S and NNW-oriented faults which have influenced the development of Dan yr Ogof and Ogof Ffynnon Ddu (Charity & Christopher, 1977). One of the NNW-trend-ing faults, the Henrhyd Fault, is a wrench fault and displaces the main NE–SW zone. It lies a little to the east of the present Dan yr Ogof resurgence and may have influenced an earlier, more easterly resurgence point.

The Old Red Sandstone outcrops to the north of the area of the rising to 700 m O.D. in the escarpment of Fan Hir. It is subdivided into a Brownstones group below and the quartz conglomerates of the Plateau Beds; both are exposed in the River Haffes.

Fig. 17.7. The Abyss in DYO II.

Fig. 17.8. Flabbergasm Chasm in DYO II (photo by C. Howes).

Fig. 17.9. View from Flabbergasm into the Grand Canyon.

The Carboniferous Limestone ranges between 208 and 520 m above O.D. and presents a more varied landscape with an outstanding range of karstic features. These include some rather distinctive pavements, often with sandstone erratics, dry valleys, an extensive range of solutional and collapse features, disappearing streams, caves and potholes. The widespread drift cover is largely of local sandstone material, although soliflucted and cambered sandstone are extensive near the Millstone Grit escarpments; peat bogs, formed either on drift or collapsed grits, are widespread. The more rugged

limestone ridge of Cribarth has many bare rock exposures and extensive scree deposits.

Within the Dan yr Ogof area the limestone varies from a total thickness of approximately 170 m near the syncline to >200 m in the further reaches of the cave. Table 17.1 provides a summary of this lithology and Figure 17.23 relates it to the cave and the surface. It is divided into three main zones, but as the cave is formed mainly within the Dowlais Limestones of Holkerian age greater consideration has been given to this group of beds. The lowest subdivision is usually termed the Lower Limestone Shales of Courceyan age. This group name, however, is misleading in this area at least, for the limestone beds interbedded within the shales are often quite massive. While drawing exact boundaries is difficult the boundary with the Old Red Sandstone is taken to be a little below the Turbine House while that with the Dowlais Limestone is at or just above the head of the Llynfell Falls.

The Dowlais Limestone has a well-defined horizon of the coral *Lithostrotion martini* near its base and this band is found just inside the resurgence cave, at the bottom of Pwll Dwfn and in the Right Hand Series of Dan yr Ogof III. The lowest beds are generally dark and very pure, and formed of fine-grained crystalline limestone which is well bedded with thin shale partings. The lower 65–80 m are fairly consistent but colour lightens to grey and the rock becomes more oolitic in the 25–35 m of the upper beds. An important bed of slightly dolomitized limestone occurs about 10 m above the coral band

Fig. 17.10. Monk Hall in DYO II, named after the shape of one of its stalagmites.

Fig. 17.12. The multitude of straw stalactites in Cloud Chamber (photo by J. Rowland).

Fig. 17.13. Fault breccia wall and long straw stalactites near Flabbergasm.

Fig. 17.14. A muddy phreatic tube in Mazeways, DYO II (photo by C. D. Westlake).

Fig. 17.11. The Pool of Reflections near the Grand Canyon in DYO II.

Fig. 17.16. The gravel-floored canyon passage in the Far North of DYO III (photo by C. D. Westlake).

Fig. 17.15. The fault-controlled Great North Road passage in DYO III.

Fig. 17.17. The Pinnacle in Pinnacle Chamber, DYO III.

Fig. 17.18. The Mostest meander passage with its floor of gour pools (photo by C. D. Westlake).

Fig. 17.19. The deeply corroded limestone of Dali's Delight.

Fig. 17.20. Christmas Grotto in Cathedral Cave (photo by T. D. Ford).

and this acts as an important marker-horizon as well as having had a significant influence on the evolution of the cave passages. It is evident in both the Show Cave and the Lakes, where it is much faulted.

The Penderyn Oolite (Asbian) averages between 25 and 30 m thick and is marked at its base by a discontinuous bed of arenaceous limestone known as the Honeycombed Sandstone. This is easily recognized on both the surface and underground despite its thickness of only about 1 m, but it effectively marks the upper level reached in exploring Dan yr Ogof. It has been cut through in a number of places, notably in Dali's Delight. However, it is mostly seen as rubble in major boulder chokes or beneath high avens. The Penderyn Oolite seems structurally incapable of supporting major passage development.

The overlying Penwyllt Limestone (Brigantian) differs in that at least one significant cave – Sink y Giedd – is located within it. This group of dark crinoidal limestones totals 25–30 m thick and varies from dark grey oolite at the base to a

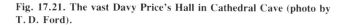

Fig. 17.21. The vast Davy Price's Hall in Cathedral Cave (photo by T. D. Ford).

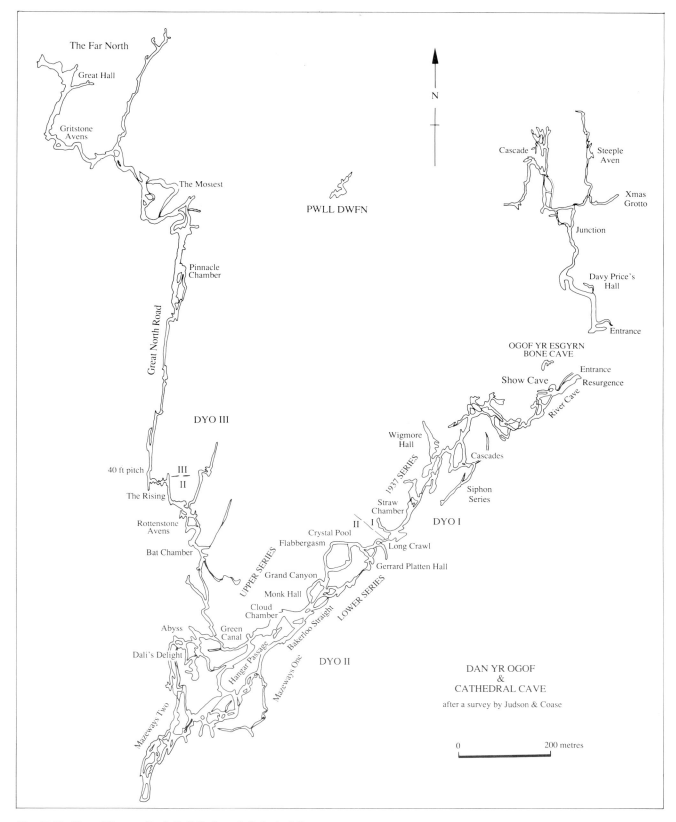

Fig. 17.22. Plan of Dan yr Ogof, Pwll Dwfn and Cathedral Cave (after a survey by D. Judson and A. C. Coase).

Table 17.1 *Table showing lithology of the limestone in the Dan yr Ogof area*
(Based on Section in Great North Road (829 160)
(Partly after George et al. 1976)

Stage	Rock units	Approx. thickness in m		Characteristics
	Namurian			Basal Grit. Massive exposures, coarse quartz conglomerates in highly jointed beds.
				——————————— U N C O N F O R M I T Y ———————
	D3	13		'Upper Limestone Shales'. Alternating dark grey shales and sandy limestones with chert. Top beds weather into fine siliceous 'Rottenstones'.
BRIGANTIAN	Penwyllt Limestones D2	26	7	Top: Medium grey crinoidal limestones with black fine-grained cherty limestones.
			9	Middle: Dark grey and black limestones with banded and nodular chert. Bituminous.
			10	Bottom: Dark grey oolitic and fine grained cherty limestones.
ASBIAN	Penderyn Oolite D1	20	18	Light grey, coarsely oolitic limestone. Massive beds, closely jointed with some false bedding and fragmented limestone.
			2	Honeycombed Sandstone, discontinuous, arenaceous, decalcified limestone
HOLKERIAN	Dowlais Limestone S2	106	35	Top: Lighter grey oolitic limestones with some pisolites and calcareous mudstones. Stylolites common.
			71	Middle and Lower: Very dark, black or blue, fine crystalline limestone. Well bedded with thin shale partings. Some oolite. Dolomitized beds near base. Coarse grained grey-brown. Coral bed at or near base with many colonies of *Lithostrotion martini*.
				——————————— U N C O N F O R M I T Y ———————
COURCEYAN	K	26		' Lower Limestone Shales'. Massive grey-brown limestones interbedded with impure arenaceous and carbonaceous limestones.
				——————————— U N C O N F O R M I T Y ———————
	Devonian			Old Red Sandstone. Grey Grits containing coarse quartz conglomerates are uppermost.

central section of dark siliceous limestone and a medium-grey upper crinoidal limestone. All contain chert and this is well represented in the form of outstanding ledges in Sink y Giedd. This 30 m deep pothole is developed in a series of distinctive joints but at its base the water disappears into what have so far proved impenetrable bedding planes, which presumably form the uppermost beds of the Penderyn Oolite. If this is so the implications are exciting to cave explorers for only a few metres below should lie the top beds of the cavernous Dowlais Limestone.

The 'Upper Limestone Shales' (Upper Brigantian) are usually regarded as only a few metres thick, although this seems a considerable underestimate here. The beds consist of thinly bedded siliceous shales alternating with arenaceous lime-

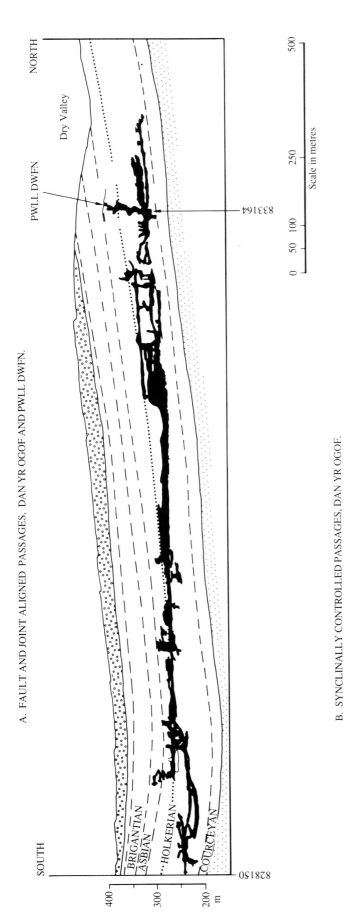

A. FAULT AND JOINT ALIGNED PASSAGES, DAN YR OGOF AND PWLL DWFN.

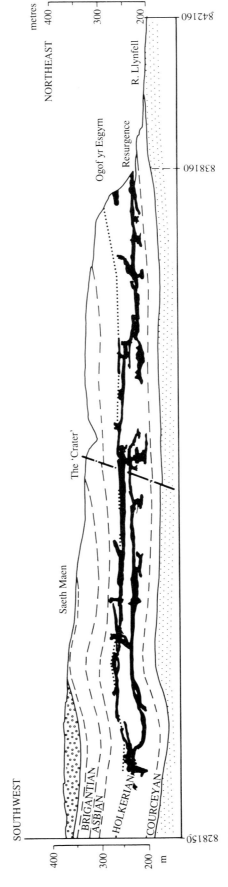

B. SYNCLINALLY CONTROLLED PASSAGES, DAN YR OGOF.

Fig. 17.23. Longitudinal sections of Dan yr Ogof related to stratigraphic levels.

Table 17.2 *Tentative correlations between the development of the Dan yr Ogof Cave System and Middle – Late Pleistocene stages*

Period		Stage Period in y BP	Local name or character	Climatic conditions	Main developments in area	Possible effects on cave development
Postglacial		PRESENT		Fluctuating but broadly cool, temperate and wet	Continued downcutting. Peat deposits eroding.	Mainly misfit streams reworking fluvioglacial deposits. Increased organic content from peat–growth in aggressiveness.
		FLANDRIAN 5000–8000		Sub-boreal: fluctuating temperatures	Slight marine transgression. Peat and lake deposits–possibly drainage deviations–Alluvium.	Waen Fignen Felen depression occupied by peat bog. Possible effects on resurgence location. Considerable introduction of and reworking of deposits.
				Moderate to warm temperature	Final localized drift and periglacial deposits.	Increased run-off from corrasive melt water. Fill stage?
Late	DEVENSIAN	10 000–12 000 Late Zone iii	Corrie Glaciations	iii. Sub-arctic: final cold phase	iii. Final period of surface run-off on frozen sub-soil.	iii. Periglacial deposits (solifluction). Little solution. Possible capture of upper Haffes? Backcutting of Llynfell?
		Zone ii		ii. Boreal–temperate: warmer Interstadial	ii. Aggressive water–increased vegetation.	ii. Increased run-off from aggressive streams. Considerable cave development.
		Zone i		i. Arctic–sub-arctic: fluctuating, possibly Interstadial	i. Periglacial conditions.	i. Fluctuations between fill and aggresive water stages and surface stream flow.
		26 000 Middle	Paviland Interstadial	Boreal–possibly sub-arctic	Sea level fluctuations.	Base level fluctuations and some rejuvenation.
		50 000 Early	Margam Glaciation	Arctic: initially fluctuating with Interstadials	Newer drift. Extensive glaciation from Beacons with considerable drift and morainic material. Downcutting of Tawe Valley.	Glacial and periglacial conditions. Some plugging of caves and potholes and solifluction deposits. Little solution.
Pleistocene (Late)	IPSWICHIAN 75 000		Minchin Hole Interglacial	Generally warm temperate but fluctuations–cool temperate–boreal	Fall in sea level–major river rejuvenation especially including Haffes.	Cathedral Cave loses headwater? Rejuvenation–Lower Haffes captured? Aggressive water and corrasion. Caves well developed in basic form. Considerable run-off, corrasive melt water and solifluction deposits. Introduction of early cave fills.
	WOLSTONIAN 130 000		Pencoed Glaciation	Arctic–sub-arctic	Older drift. Extensive glaciation, sea level low. Downcutting of Tawe Valley.	Periglacial conditions, surface drainage. drift deposits and solifluction active. Caves relatively quiescent, possibly plugged.
	HOXNIAN 150 000		Gower Beach Interglacial	Temperate	Rejuvenation of main rivers, possible Haffes begins to cut back more significantly.	Considerable cave development with downcutting commencing.
Middle	ANGLIAN 240 000		Ante-Penultimate Glaciation on Continent	Cold–sub-arctic	No evidence of glaciation.	Abrasive sediments. Little cave development–solifluction deposits etc.
	CROMERIAN 310 000		Interglacial	Temperate	Possible fluctuations in base level and rejuvenation.	Presumed period of cave development.
	BEESTONIAN 340 000		Early glaciation on Continent	Cold	No evidence of glaciation.	

stones, muddy sandstones and massive dark limestones with chert. The upper few metres have weathered to form rottenstone which has been quarried in the past. The resultant pockmarked appearance of the ground provides a fine indicator of the upper limit of the limestone, although in some cases the zone has been totally obscured by collapsed escarpments of Millstone Grit quartzite.

The overlying Basal Grit is highly quartzitic and contains several beds of coarse quartz conglomerate. These create marked scarplets which contribute to a distinctive profile, which is remarkable for the openness of its joints. This characteristic has often led to a process of outward collapse, and cambering frequently occurs with a resultant pile of quartzite scree. This, coupled with slumped and soliflucted boulders, often masks the exact grit–limestone boundary. The bare rock surface of the quartzite shows evidence of considerable ice-scouring and striations, orientated in a S- or SSW-direction, are fairly common. These, together with the sandstone drift

deposits on the limestone and the extensive morainic deposits in the valley, provide considerable evidence of comparatively late ice movement. The chronology of this and its relationship with the caves' evolution is discussed below and summarized in Table 17.2.

These glacial scree and solifluction materials, together with the deltaic and alluvial deposits of the Tawe and Haffes valleys and the peat bogs already mentioned, combine to create a veneer which conceals much of the solid geology and which has contributed sediments to the caves beneath.

Folds and faults

Attention has already been drawn to the relatively shallow syncline which influences the direction and drainage of much of the outer part of Dan yr Ogof and the orientation of Cathedral Cave. This syncline, hitherto unrecognized on the surface, is so important to the geomorphology of the cave that it is subsequently referred to as the Dan yr Ogof Syncline. It is asymmetric in form with a relatively gentle dip, normally between 2 and 6°, from the north, but angles of 12–20° are more common from the south and southeast. Its overall trend is approximately 050° but it is much interrupted by north–south faulting. Underground, in the eastern half of the catchment area, over 100 separate instances of faulting have been recorded, 90% of which have been observed within the northernmost sectors (350–019°). The most outstanding is that which runs for much of the length along the Great North Road in Dan yr Ogof III. This passage displays spectacular slickensides and extensive brecciation although, because of its size, the accumulation of sediments and calcite deposits in the area, the displacement along the fault has been impossible to establish. It is, however, clear that the orientation of this important part of the cave is closely related to the fault. Elsewhere there is considerable evidence of similar influences on the caves' morphology.

Joints and passage orientation

Jointing in the area is well defined and is relatively consistent across the major rock types. Surface jointing has been analysed by Weaver (1973) and underground jointing has been analysed by the writer (Fig. 17.24). The general picture

that emerges from these studies is of one major point set slightly east of north and another almost E–W on the surface. Underground, however, the N–S set is dominant and a third set trending 300–310° is developed at the expense of the E–W set. The sets are illustrated on the rose diagram (Fig. 17.24a).

A close relationship between faults and the main joint set is also apparent and this has had a significant influence on passage orientations. The latter have been plotted for all the surveyed and known passages of Dan yr Ogof (Fig. 17.24b). The most obvious characteristic is the dominance of the northerly sectors but, although still evident, this was found to be more pronounced in passages away from the Dan yr Ogof Syncline. Indeed, this observation, together with the recognition that passages in the syncline conformed fairly closely to 'classic' phreatic forms while those outside it frequently showed far more vadose characteristics, led to a recognition of two main subdivisions within the cave. These were the passages and sectors which were synclinally controlled or guided and those which were closely aligned with faults and joint sets. Further analysis of these divisions in terms of passage orientations proved to be significant (Fig. 17.24c,d). In the syncline there is a relatively broad spread of orientations although still with a northern maximum; however, within the joint-aligned

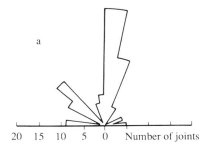

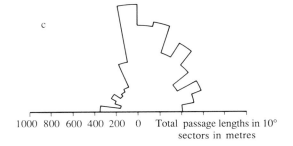

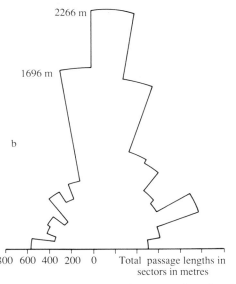

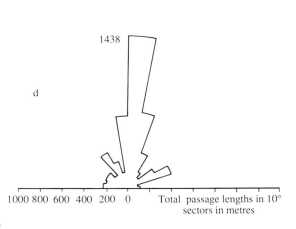

Fig. 17.24. Rose diagrams of passage orientations in Dan yr Ogof.
(a) A random sample of 110 joints in all parts of Dan yr Ogof.
(b) Orientation in 10° sectors of all passages in Dan yr Ogof. (c)
Orientation of synclinally-controlled passages. (d) Orientation of
joint and fault-aligned passages.

passages this sector is far more dominant. This subdivision is reinforced by Cathedral and other caves, most of which 'belong' in the joint-aligned category.

Thus passage orientations are strongly influenced by the principal north–south fracturing of the limestone, especially as this is also the general direction of dip. Other joint directions also influence passage trends in certain areas, especially adjacent to those where faulting has been significant, but the overall trend of the Dan yr Ogof Syncline is clearly the second most important influence.

The caves

Detailed investigation shows the known caves to be mainly located within the Dowlais Limestone. In Dan yr Ogof the lowest passages appear to be closely related to the Lower Limestone Shales while the highest occasionally reach and once or twice breach the lowest beds of the Penderyn Oolite. However, the majority of passages are located within the middle and lower beds of the Dowlais Limestone, although Cathedral Cave is limited mainly to the middle beds of that formation. Only Ogof Sink y Giedd is known to show significant cave passage development within beds of the Penderyn Oolite, although recent work shows some small and apparently old and abandoned caves in similar horizons further west.

Dan yr Ogof

The location of passages in Dan yr Ogof within the Dowlais Limestones is variable. Other than in the present active series, which is developed near the base of these beds, there is little evidence of a more than localized adoption of specific beds. Elsewhere the cave transgresses through the beds, in a manner dictated more by gradient than by adaptation to 'preferred' beds. Close examination of the cave gradient shows it to be developed along a comparatively regular slope which is disturbed only occasionally by faulting and folding. The main exceptions to this generalization occur in the Lower Series although, even there, fluctuations are limited to a relatively short distance. Both series appear to 'grade' towards a common level at the resurgence, thus suggesting that this, or a closely related level, has been dominant throughout the history of the cave.

The joint-aligned passages include the further parts of DYO II, all of DYO III, Cathedral Cave, Pwll Dwfn and Sink y Giedd. All of DYO I and the Upper and Lower Series of DYO II are regarded as conforming to the synclinally controlled group. In the undiscovered links, notably between Sink y Giedd and DYO II, and between Waen Fignen Felen and DYO III, the passages are expected to be primarily aligned in a north–south direction, although there is likely to be a certain degree of deflection towards the southeast. Part, at least, of the hypothetical Giedd Series is likely to occupy the Dan yr Ogof Syncline and although geological evidence suggests that this could continue to the southwest, hydrological evidence as indicated by the throughput of water in flood conditions, suggests that the present route, at least, may be more direct.

The passages in the southwest of the System, notably those of Mazeways I and II and the Hangar Passage Extensions, (which are collectively termed the Cribarth Passages) have some individual characteristics. They lie to the south of the synclinal axis, dip more steeply than other passages and do so in a northerly direction. Initially there was a temptation to regard them as the first manifestation of a new category of passages draining northwards from Cribarth. This may still be the case, although proof is likely to depend upon the nature and trend of further extensions of the Dan yr Ogof Syncline. Until further explorations have provided fuller information in this direction the area is regarded as a part, albeit a steeply dipping one, of the synclinal unit.

The adjacent series, known as Dali's Delight, is another problem area in terms of classification for it lies across the synclinal axis and yet shows very few of the characteristics of that zone. Because of the close relationship between its passages and the dominant joint and fault patterns this series is regarded as part of the joint-oriented group.

Cathedral Cave

Cathedral (Tunnel) Cave (Railton, 1958) appears to be fairly consistently located in the middle beds of the Dowlais Limestone, and the overall gradient closely coincides with this horizon except in the highest and furthest passages of the System. Its main chamber, Davy Price's Hall, has recently been opened to the public and while modifications to the floor have considerably altered the sediments exposed and created several new 'lakes', electric lighting has done much to reveal the true nature of the bedding and roof features. Particularly evident above the fluting of the 'Organ Pipes' is a bedding plane with numerous small phreatic tubes which occasionally provide quite aggressive water.

The smaller caves

Ogof yr Esgyrn (the Bone Cave) is located 45 m above the Dan yr Ogof resurgence and is important both archaeologically and geomorphologically. The former has implications for the latter, as Romano-British pottery dating back 1600 y was recovered from under a 75 mm thickness of stalagmite (Mason, 1968, 1977). The cave is also located almost exactly on the axis of the Dan yr Ogof Syncline and angles of dip and strike are minimal. The main feature is a wide roof which is generally flat apart from unusual decomposing linear calcite deposits which occur along almost every joint.

Pwll Dwfn or the 'Deep Pot' is significant as it extends from just below the Honeycombed Sandstone to the *Lithostrotion* beds at the base of the Dowlais Limestones. It is developed at the intersection of two faults on the floor of the dry valley to which it gives its name, although there is no indication either above or below ground of any major sink.

Sink y Giedd (Harvey, 1948) differs from all the other caves of the area in being located in the Penwyllt Limestones. The cave is developed mainly along a number of joints in very dark limestone characterized by well-defined chert ledges. Access is not always possible as the entrance is frequently blocked by winter floods and boulder movements. Thus, while it offers considerable hope for future extension, and perhaps a 'back door' into the whole of Dan yr Ogof, it is one with considerable attendant dangers. Investigation of related, but abandoned, sinks might be as profitable and perhaps less dangerous.

Hydrology

The past and present drainage patterns (Figs. 17.25–17.27) differ substantially as a result of river capture and truncation of supplies. The present pattern reflects the obvious

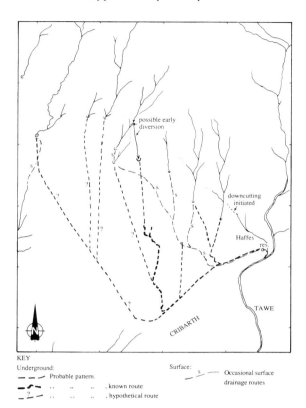

KEY
Underground: Surface:
– – – Probable pattern.   s —— Occasional surface
–––– „   „   „ , known route   drainage routes
–?–– „   „   „ , hypothetical route

Fig. 17.25. The probable drainage pattern in the early to middle Pleistocene. (The marginal ticks refer to National Grid Reference figures – see Fig. 17.1).

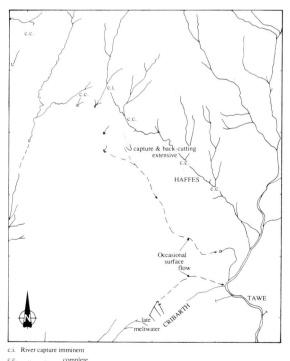

c.i. River capture imminent
c.c.   „   „   complete

Fig. 17.26. The surface drainage pattern in the late Pleistocene.

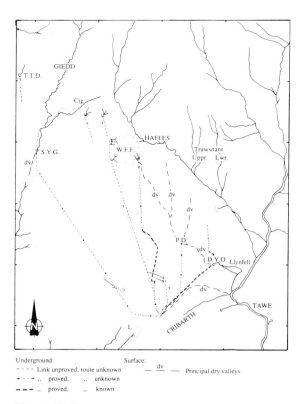

Underground: Surface:
– – – – Link unproved, route unknown   —— dv —— Principal dry valleys
–•–•– „   „ proved,   „ unknown
– – – „   „ proved,   „ known

Fig. 17.27. The present drainage pattern (initials are explained on Fig. 17.1).

input streams and sinks but many minor streams and peat bogs exist, especially on the Basal Grit, and percolation is derived from them. However, hydrological studies by Newson (1970), Ede (1972) and Gascoine (1983) suggest that because of the openness of the joints and complex microstructures the percolation water rapidly assumes the features of swallet water. Water entering the main swallets is also very aggressive, especially when derived from extensive peat bogs like that at Waen Fignen Felen. Such bogs also have a significant storage capacity and the release of large quantities of water in 'flash floods' must have had a significant effect on the caves' evolution.

The principal sink is Sink y Giedd, where the water is derived both from a large sandstone area and from a smaller peat bog. It is therefore slightly less aggressive and differs in other chemical characteristics. Studies of the chemistry of these cave waters by Bray (1969, 1972) showed that the constant mixing of differing waters (Bögli, 1971) within the System has also been highly significant. The structure of the cave, and notably its joints, faults and principal syncline have been influential in determining the actual routes followed by the water, the general pattern of which is shown in Figure 17.28 (Harvey, 1948; Hunt & Jones, 1963; Christopher & Bray, 1968; Gascoine, 1983).

Other sources of aggressiveness are important in the underground solutional processes and include both carbonic acid and sulphuric acid derived from peat, shale and drift sources (Burke, 1967; Stenner, 1969). The significance of abrasion is also of great importance, for the cave's gradient is relatively steep. There is also a large amount of 'aeration' and tur-

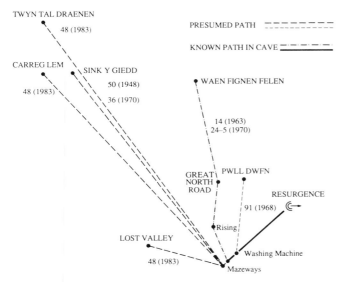

TWYN TAL DRAENEN
48 (1983)

CARREG LEM
48 (1983)

SINK Y GIEDD
50 (1948)
36 (1970)

PRESUMED PATH
KNOWN PATH IN CAVE

WAEN FIGNEN FELEN

14 (1963)
24–5 (1970)

GREAT
NORTH
ROAD

PWLL DWFN

RESURGENCE
91 (1968)

Rising

LOST VALLEY
48 (1983)

Washing Machine

Mazeways

Fig. 17.28. Summary of dye-tests in the Dan yr Ogof catchment; numbers refer to the shortest recorded flow times (h) with the date of the test (compiled from Harvey, 1948; Hunt & Jones, 1963; Christopher & Bray, 1968; Coase, 1975; Gascoine, 1983).

bulence arising from the passage of water through the frequent massive boulder chokes. The varied nature of the water from the very numerous minor input points is another factor which has strongly influenced the erosiveness of the cave streams.

Directions of flow are generally easily traceable from wall scalloping and with a few exceptions and anomalies, generally bear out the straightforward pattern of drainage outlined, that is, a series of down-dip, north–south consequents, relating closely to both dip and joint–fault orientations, until channelled into the main SW–NE-oriented synclinal 'drain'.

Cave sediments

This is another aspect of the caves' geomorphology that merits more attention, along the genetic classification approaches outlined by Ford (1974) and Bull (1975). Newson (1970) did carry out some studies on the balance between mechanical erosion in these, as well as other cave waters, and showed that abrasion was especially important in flood conditions. However, further study related to the abundance and variety of sediments in Dan yr Ogof should reveal much more about the cave's evolutionary chronology.

The sediments include widespread deposits of massive limestone blocks derived from undercutting and internal collapse as well as smaller pebbles, and gravels found in most of the streamways. Additionally there are extensive deposits of quartzitic sands and gravels especially in the Great North Road where they are often in thick well-bedded deposits. Considerable re-sorting of these deposits is occurring in the lower, active passages and large quantities of fine sand are deposited in the Show Cave after winter floods. They contribute marked turbidity to the waters. Derived peat deposits are also common in many active parts of the cave but to date no inter-bedded peats have been found. Other deposits include sticky red clays, coarse red sands and fine light brown sands and silts.

The relationship between surface and cave features

Some observations on these topics are essential for there appears to be a close relationship between surface morphology and cave features in the Dan yr Ogof Syncline. Here the cave's depth below the surface is at its shallowest and for much of the area is covered only by the highest limestones and relatively thin superficial deposits of drift and solifluction materials. Many of the boulder chokes in this section can be fairly closely related to numerous deep surface depressions, the most obvious example being the Crater, which has been proved to be little more than 30 m above the Long Crawl and the Platten Hall–Shower Aven chokes.

There is also considerable evidence of underground collapse contributing to surface depressions on the Basal Grit, as at Pwll y Wydden Fach. Indeed, Thomas (1954, 1959, 1963, 1974) showed that the greatest number of collapse dolines on the grit lie where it is <30 m thick above the limestone, although he also noted other examples where he estimated the grit to be >100 m in thickness.

Dry valleys are also a feature of the surface above Dan yr Ogof and its associated caves. The first of the two principal valleys passes close to Pwll Dwfn, and is believed to have been an intermittent surface route for the stream which would normally have disappeared underground in the vicinity of Waen Fignen Felen before its truncation by the Haffes. Below Pwll Dwfn it enters a deeper valley and then a gorge before plunging past Ogof yr Esgyrn and Cathedral Cave entrance, into the Dan yr Ogof resurgence gorge.

The second dry valley has been termed the Saeth Maen Valley from its proximity to the ancient stones, and this too has a shallow middle and upper section only to turn sharply away from the Dan yr Ogof Syncline near the Crater, and enter a gorge section ending at Nant y Gwared Farm. This is close to the present end of the Cribarth ridge and it is from higher up the same ridge that it begins. It is considered likely that this valley originated from surface flow from late snow or ice patches in periglacial or immediate post-glacial periods, a possibility which also has bearing on subterranean developments in the Cribarth Passages.

Chronology

Any attempt at reconstructing the chronology of speleogenetic evolution is bound to be speculative; none the less, a close relationship between it and the downcutting of the Tawe and the Haffes can be suggested and the following summary should be used in conjunction with Table 17.2. The basic assumption accepts Brown's (1960) view that the Tawe was initiated into its present course during one of the earlier interglacials, at a time when the general pattern was of a fairly simple series of sub-parallel streams. These flowed down-dip in a roughly N–S direction and are assumed to have entered the limestone at an early stage. Their waters are likely to have been highly aggressive and they also carried considerable quantities of abrasive debris (Williams, 1963). The early pattern underground would have been one of numerous small phreatic tubes and, although percolation would also have played an important part, it is likely that the water from the discrete stream sinks would have been instrumental in creating the early framework. These streams rapidly exploited the joint and fault fractures so well developed in the area and the initial pattern would have been of a series of sub-parallel passages trending from north to south until meeting the Cribarth folds. These, or more accurately the outermost syncline, led to an abrupt diversion of drainage to the NNE.

Subsequently, the area experienced a complex series of ice advances and retreats (Synge, 1970), during which the basic framework of the caves was considerably reinforced and increased in size, contemporaneously with the down-cutting of the Tawe Valley and the rejuvenation of the Haffes. The earliest stage of importance occurred during the Older, or Pencoed, Glaciation and the subsequent Minchin Hole Interglacial (Bowen, 1970). The latter, with fairly high rainfall and, at times, warm temperate conditions, was particularly influential. Subsequent changes in base levels of the Tawe, and river capture by the Haffes, led to the loss of the headwaters of first, Cathedral Cave and then of the Waen Fignen Felen–Great North Road System. These captures by the Haffes of what are now its north bank tributaries are believed to have occurred before the early Devensian Margam Ice Advance, but the capture of other headwaters may have occurred subsequently.

At the end of the Devensian (perhaps 10 000 y BP if based on Thomas's (1970) work on the limestone pavements of the area) came a resumption of 'normal' conditions. The aggressiveness of input waters would have increased with the milder temperatures and there is likely to have been an increase in the exogenetic sediments carried underground. However, by this time only one of the main limbs of the System was still active and the more easterly Cathedral Cave would not only have lost its principal stream but almost certainly its physical link with Dan yr Ogof.

A brief summary of the above suggests:

(1) Initiation as a complex network of small phreatic tubes resulting from the sinking of at least three separate groups of streams.

(2) Development of these streams, including enlargement by corrosion as well as by solution, basically on a down-dip pattern.

(3) Fluctuations in quantities and characteristics of input water and associated sediments leading to marked stages of growth and stand-still, to adaptations to lower base levels (see (4) below) and consequent vadose entrenchment, and to increasing amounts of bedding collapse.

(4) Downcutting and river capture as a result of external base-level changes and internal readaptations causing the loss of principal input waters, initially in the east, later in the centre, and potentially to the west in the future.

At least three main input limbs can therefore be identified as a result of the underground drainage pattern but their development and relationship to the Dan yr Ogof Syncline are unequal, thus:

(a) The Cathedral Cave limb, the easternmost and shortest of the principal feeder systems and the one first truncated, is relatively high in the Dowlais Limestone and is comparatively immature in form; it shows little but short-distance, fluctuations in gradient and relatively little sign of any distinctive levels associated with major changes in base level.

(b) The DYO limb consisting of DYO III and the Far Series of DYO II contains fairly extensive evidence of a large and complex series of high-level tubes with many similarities to Cathedral Cave as well as massive vadose and phreatic passages. These are often closely associated with faulting and corrosion as well as solution which has played an important role. Its relationship with the main passages of the Syncline is an ambivalent one for although the gradient coincides with the Upper Near Series of DYO II, there is little evidence to suggest that the actual link, via the Green Canal, is anything but a relatively late act of underground piracy. The more obvious link, via the Abyss, shows a marked lack of concordance with the pre-existing lower series.

(c) The third limb is the almost totally unknown Giedd Series, although this may consist of several, rather than just one, discrete series. These also enter from the north, possibly associated with such major fault lines as those adjacent to Disgwylfa and Carreg Goch. The one known characteristic of the input is that its principal sink at Sink y Giedd is markedly different from the other major input points. Sink y Giedd is near the southern margin of the limestone outcrops, in contrast to the other sinks. By the time that the Giedd stream appears in the Syncline, it has cut down virtually to the bottom beds of the Dowlais Limestone.

Conclusions

Geologically, the relationship of the various caves with neighbouring impermeable strata and the relatively gentle and consistent dip of the limestone, coupled with the intensive folding and faulting of the southwestern boundary of the catchment area all contribute to a favourable location for speleogenesis. On a more local scale the near coincidence of dip with the major directions of faulting and jointing and the controlling influence of the Dan yr Ogof Syncline on the main conduit passages are factors of major importance.

Lithologically, the almost complete containment of the known cave developments within the Dowlais Limestone has been emphasized. Even so, most passages appear to be adapted to hydrological rather than local lithological influences. There also appears to be a relationship between the complex network of early high-level phreatic tubes and the approximate boundary between the upper and lower parts of the Dowlais Limestone.

There is a marked variety in the form and development of the main north–south contributary passages compared to those of the Syncline and the lack of adjustment between them is also apparent. The contrast between the two known joint-aligned limbs is also considerable and when access is finally achieved to the third, a unique opportunity should be available to study a cave system with three parallel branches, each with many features in common, but each, for external reasons, at different stages of development.

The chemical characteristics of, and the organic and clastic materials transported by, the input waters have also played an important role in the development of the caves to their present character and size. Although comparatively young, the systems have developed a large and mature form because of the aggressiveness and abrasiveness of their streams, coupled with the variety of waters in both origin and flow rates. The latter, both in the short term, as floods, and in the long term, as major fluctuations in quantities of precipitation and run-off arising from the varying climatic stages in the Pleistocene, have been particularly significant. The importance of the introduction of the clastic materials has already been stressed but it is also worth noting that Gams showed that the largest Slovenian caves were found where input water contained 'pebbles and coarse debris' (Sweeting, 1972). This fact, coupled with high initial rates of aggressiveness and the subsequent undercutting and block breakdown, is held to be partly responsible for the massive size of the furthest passages in Dan yr Ogof.

The range, size and character of sediments passing through the System, coupled with fast throughput times, suggests that there is little impediment to free through-flow. It equally implies that there is no deep phreatic development, a characteristic which is also confirmed by the lithological reconstruc-

tions shown on the sections. Indeed, the relative flatness of dip and shallowness of the limestone combine to place the Dan yr Ogof caves very much into the 'shallow karst cave' category with primarily horizontal development. In this respect it could be regarded as more akin to many American caves than to its geographical neighbours on Mendip or in Yorkshire (Miotke & Palmer, 1974). However, it is unrealistic to attempt to classify the Dan yr Ogof System into one particular 'type' of cave, for like most major caves it combines a number of elements and is the result of a wide range of influences. Despite its comparative youth it displays a considerable maturity of form because of the size of its catchment area and because of the volume, variety, aggressiveness and abrasiveness of its influent streams. Its phreatic origins are clear in most areas but in the more developed north–south elements these are dominated by massive vadose developments (except where minor deflections along the strike have led to phreatic characteristics on a major scale). Faulting and jointing have been of great importance as an influence on the orientation of passages in all parts of the cave, although folding has guided the overall trend of the synclinal 'drain'. The relationship between cave and surface features is also remarkably close and it is appropriate to end this chapter with the view that, above and below ground, this and the immediately adjacent areas are among the most developed and interesting karst areas in the British Isles.

References

Bögli, A. (1971). Corrosion by mixing of karst waters. *Transactions of the Cave Research Group, GB*, **13(2)**, 109–14.

Bowen, D. Q. (1970). South-east and central South Wales. In *The Glaciation of Wales and Adjoining Regions*, ed. C. A. Lewis, pp. 197–227. London: Longman.

Bray, L. G. (1969). Some notes on the chemical investigation of cave waters. *Transactions of the Cave Research Group, GB*, **11(3)**, 165–73.

Bray, L. G. (1972). Preliminary oxidation studies of some cave waters from South Wales. *Transactions of the Cave Research Group, GB*, **14(2)**, 59–66.

Brown, E. H. (1960). *The Relief and Drainage of Wales*. Cardiff: University of Wales Press.

Bull, P. A. (1975). Sediment studies in Agen Allwedd. *Bulletin of the British Cave Research Association*, **9(29)** (abstract).

Burke, A. R. (1967). Geomorphology and speleogenesis of vertical shafts in Carboniferous Limestone at Ystradfellte, Brecon. *Proceedings of the British Speleological Association*, **5**, 17–46.

Charity, A. R. P. & Christopher, N. S. J. (1977). The stratigraphy and structure of the Ogof Ffynnon Ddu area. *Transactions of the British Cave Research Association*, **4**, 403–16.

Christopher, N. S. J. & Bray, L. G. (1968). Dan yr Ogof hydrology preliminary phase. *South Wales Caving Club Newsletter*, **60**.

Coase, A. C. (1967). Some preliminary observations on the geomorphology of the Dan yr Ogof system. *Proceedings of the British Speleological Association*, **5**, 53–67.

Coase, A. C. (1975). The structural geomorphology of Dan yr Ogof caves, Tawe Valley, South Wales. Unpublished Ph.D. thesis. University of Leicester.

Coase, A. C. (1977). The structure and evolution of the Dan yr Ogof caves. *Proceedings of the 7th International Speleological Congress*, pp. 116–20. Sheffield.

Coase, A. C. & Judson, D. M. (1977). Dan yr Ogof and its associated caves. *Transactions of the British Cave Research Association*, **4(1–2)**, 245–344.

Ede, D. P. (1972). An investigation into some factors influencing the variation in hardness of streams and springs on limestone with

particular reference to South Wales. Unpublished Ph.D. thesis. Cardiff: University of Wales.

Ford, T. D. (1974). Sediments in caves. *Transactions of the British Cave Research Association*, **2(1)**, 41–6.

Gascoine, W. (1983). A hydrological study of Dan yr Ogof and Ffrwd Las resurgences. *South Wales Caving Club Newsletter*, **97**, 9–12.

George, T. N., Johnson, G. A. L., Mitchell, M., Prentice, J. E., Ramsbottom, W. H. C., Sevastopulo, G. D. & Wilson, R. B. (1976). A correlation of Dinantian rocks in the British Isles. *The Geological Society of London Special Report*, **7**, 1–87.

Harvey, P. I. W. (1948). Dye tracing from Sink y Giedd. *Cave Research Group, GB, Newsletter*, **15**.

Hunt, D. & Jones, J. C. (1963). Waen Fignen Felen dye trace. *South Wales Caving Club Newsletter*, **42**.

Mason, E. J. (1968). Ogof yr Esgyrn, Dan yr Ogof cave excavations 1938–50. *Archaeologia Cantabrica*, **117**, 18–71.

Mason, E. J. (1972). *The Archaeology of Ogof yr Esgyrn. Cave Guide Book*. Abercrave, Wales: Dan yr Ogof Caves Ltd.

Miotke, F. D. & Palmer, A. M. (1974). *Genetic Relationship Between Caves and Landforms in the Mammoth Cave National Park Area* (private publication).

Newson, M. D. (1970). Hydrological investigations into the balance between chemical and mechanical erosion in the limestone stream system. Unpublished Ph.D. thesis. University of Bristol.

Railton, C. L. (1958). *The Survey of Tunnel Cave, South Wales*. Cave Research Group, GB. Publication **No. 7**.

Stenner, R. D. (1969). The measurement of the aggressiveness of water towards calcium carbonate. *Transactions of the Cave Research Group, GB*, **11(3)**, 175–200.

Sweeting, M. M. (1972). *Karst Landforms*. London: Macmillan.

Synge, F. M. (1970). The Pleistocene period in Wales. In *The Glaciation of Wales and Adjoining Regions*, ed. C. A. Lewis, pp. 315–50. London: Longman.

Thomas, T. M. (1954). Swallow holes on the Millstone Grit and Carboniferous Limestone of the South Wales Coalfield. *Geographical Journal*, **120**, 468–75.

Thomas, T. M. (1959). The geomorphology of Brecknock. *Brycheiniog*, **5**, 129–36.

Thomas, T. M. (1963). Solution subsidence in southeast Carmarthenshire and southwest Breconshire. *Transactions of the Institute of British Geographers*, **33**, 45–60.

Thomas, T. M. (1970). The limestone pavements of the North Crop of the South Wales Coalfield. *Transactions of the Institute of British Geographers*, **50**, 87–105.

Thomas, T. M. (1974). The South Wales interstratal karst. *Transactions of the British Cave Research Association*, **1(3)**, 131–52.

Weaver, J. D. (1972). The Swansea Valley Disturbance. Unpublished Ph.D. thesis. University of Wales.

Weaver, J. D. (1973). The relationship between jointing and cave passage frequency at the head of Tawe Valley, South Wales. *Transactions of the Cave Research Group, GB*, **15(3)**, 169–73.

Weaver, J. D. (1975). The structure of the Swansea Valley disturbance between Clydach and Hay-on-Wye, South Wales. *Geological Journal*, **10**, 75–86.

Williams, V. H. (1963). A study of the solutional processes and phenomena in limestone with particular reference to the North Avonian outcrop of South Wales. Unpublished Ph.D. thesis. University of Wales.

18

Llygad Llwchwr

MARTYN FARR

Llygad Llwchwr lies on the northwest crop of the South Wales Coalfield, 5.8 km from Ammanford and 1.4 km from the village of Trapp (SN 669 178). The area lies about 1 km south of the limestone outlier of Carreg Cennen, where the Carreg Cennen–Church Stretton Fault zone is exposed. It is a major resurgence system, the source of the River Loughor.

History of exploration

Llygad Llwchwr was unquestionably one of the first caves in South Wales to receive systematic exploration. As early as 1841, for example, Thomas Jenkins of Llandeilo

examined the site and in the course of several visits succeeded in reaching River Chamber Four – the current termination for all but cave divers. With the aid of a rope ladder and a collapsible coracle it is clear that he examined most of the complex system. In an extract from his diary (10 September 1843) Jenkins recalled:

Went together with B. Morgan, David Lewis, John Thomas, Walter Jones, Puddy Combe and Owen Jones to Llygad Llwchwr. Entered the cave at 8.30 am and after turning to the left at right angles to the main branch and getting down

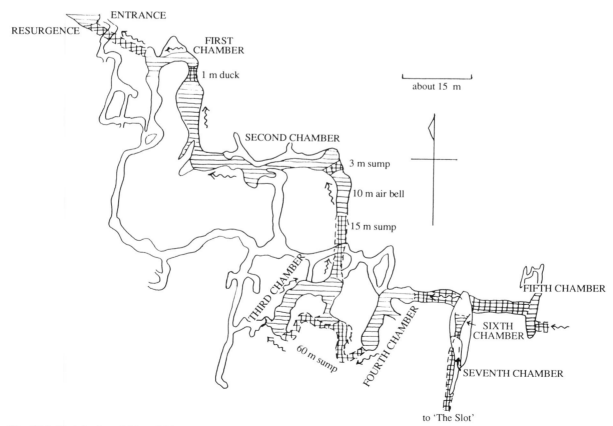

Fig. 18.1. Sketch plan of Llygad Llwchwr (not to scale).

over the rock by rope ladder over the stream, we made up the coracle and proceeded down the stream, over very deep pools (River Chamber 2) through several magnificent caverns where man had never dared to before. Came out to daylight at 1.30 pm ...

A sketch plan of the cave was made, together with other observations. The water temperature was measured at 49 °F and the total discharge estimated at 45 Hogsheads! Significantly, the neighbouring cave, Pal y Cwrt (according to Jenkins 'Palebryna') was also explored and little further progress has been made here to the present day, despite the obvious potential for open passage.

Following an early investigation by Weaver (1940) the next phase of exploration commenced with 'Operation Acheron' in May 1947. G. Balcombe, J. Sheppard and C. P. Weaver of the newly formed 'Cave Diving Group', visited the cave and explored upstream from River Chamber Four. Little actual progress was made at this time, but the early divers did prove that the way on was open, the visibility was good and that the route was quite practicable for an oxygen-rebreathing diver. The next dive took place in July 1960 and the first breakthrough, due to diving, was made. C. George and B. de Graaf followed the passage 320 ft (98.5 m) from base. They encountered an airbell after 150 ft (46 m) but unfortunately found that the river entered from a depth of more than 30 ft. Due to the impending danger of oxygen-poisoning below this depth alternative equipment was necessary. Moon and Fairbairn, supported by members of South Wales Caving Club, continued explorations here several years later, in May 1967. Using compressed air they succeeded in reaching the 'Slot'. About 350 ft (107 m) from base the river emerged from a constriction, at the bottom of a U-tube. When a diver approached the obstacle, he would be bombarded with gravel. Therefore, it was found impossible to continue, as the flow was simply too strong.

In May 1970 J. Parker made a determined effort on the terminal sumps and discovered 'Chamber Five'. By following a branch passage from the main route, a low and wide section led through a shallow sump to emerge in a substantial chamber, well adorned with formations. Beyond, 'the inlet' was found to be choked by sand and gravel. A duck, leading from Chamber Five, yielded about 30 m of small and awkward passage, but of a major breakthrough there was little possibility. Continuing his work Parker made limited progress into the 'Slot' itself. By excavating sand from the terminal constriction he was able to ascertain that the obstacle was of a localized nature and that every hope of open passage existed beyond. However, despite his work and that of the author, at a later date, the obstacle could not be passed. Even in drought the flow proved to be too great (Anon., 1973).

Late in 1977 a major project to extend the system was initiated by the author. It was considered that the major inlet entering Chamber Five was in fact part of the main stream. The objective was to remove a lip of rock, forming the floor of Chamber Five, in order to drain off the considerable quantity of gravel which prevented access to the continuation. To facilitate this removal large quantities of explosives were used.

Under the direction of R. Hall charges of 4.25 lb, 4 lb, 5 lb, 4.25 lb and 3.25 lb have already been detonated from a safe base in Chamber Four. The operations must of necessity be undertaken at least several days apart owing to the danger from toxic fumes in the confined chamber. The floor has been lowered about 1 m to date and it is hoped that some way can now be found to remove the gravel on the upstream side, as the actual flow of water here is incapable of flushing this away. Every hope of a major breakthrough exists and, should an unforeseen failure occur in Chamber Five, we shall undoubtedly renew the assault on the 'Slot' (Anon., 1979).

A strong draught can be detected at Pal y Gort, a feature which unfortunately is lost inside the cave, amid a large boulder ruckle. Indeed, the truncated portion of passage which is evident in Pal y Gort must certainly be associated with the main drainage system in Llygad Llwchwr. From all the evidence, it is clear that a major system remains to be entered. Just how far this is likely to extend is open to conjecture. Much of it may be in very fractured limestones owing to the considerable number of large north–south faults crossing the narrow limestone outcrop.

Geology
The stratigraphic succession is as follows:

Namurian		Basal Grit, sandstones and conglomerates	
Dinantian	Brigantian	Upper Limestone Shales	c. 15 m
		Mynydd y Garreg limestones – dark grey with chert bands and nodules	84 m
	Asbian	Penderyn Oolite – light grey limestones with oolitic bands	32 m
	Holkerian	Cil yr Ychen limestones (Dowlais Limestones) – dark grey with flaggy beds at the base	95 m
	Tournaisian	Lower Limestone Shales – with scattered beds of dark grey limestone	20 m
Devonian		Old Red Sandstone	

Beds in the vicinity of the cave dip 25–35° to the south and southeast and outcrops are frequently offset by north–south faults.

Cave description
Llygad Llwchwr is a complex network over 500 m in length. Although there is a pronounced dip (about 30°) evident in the cave, the effect of this is not readily apparent on the survey (Fig. 18.1). The entire system is developed eastwards along the strike. Access to the further reaches is made via the upper series, while many connections are possible between the abandoned network and the underlying streamway. The chambers, aptly named owing to their surprisingly large dimensions, are numbered in an upstream sequence, River Chamber One nearest the entrance and Chamber Four (the termination for

all but cave divers). A substantial proportion of the lower series is in fact occupied by sumps.

Phreatic passages are evident everywhere, particularly in the upper series. Below, evidence of vadose development can in fact be found only in limited locations, particularly in River Chamber Three.

The best example of the dip–strike pattern comes undoubtedly from the terminal section of the cave. From the air surface of the sump in Chamber Six, for example, the mainstream passage can be followed directly down-dip, until it reaches a depth of 9.2 m in the vicinity of the Slot. Over this distance the passage exceeds 1.6 m in diameter, until the final few metres. At the termination the direction deviates sharply to the left, along the strike. Should the cave upstream of this point follow the pattern of the preceding section of cave, it is quite likely that beyond the Slot the passage will tend gradually up-dip, to reach air in a comparatively short distance. The volume of resurging water is clearly undiminished at River Chamber Four.

Hydrology

Dye-tests (Chapter 4) have shown that some of the water is derived from sinks along the strike of the limestone to the east (see Fig. 4.15). Pwll Cwm Sych (SN 691 183), some 3 km to the east, and two sinks adjacent to the A4069 road at Sinc Ger y Ffordd (SN 732 189), some 7 km from the resurgence, have given positive traces. It is possible that there is some contribution of water from the limestone outcrop to the west of Llygad Llwchwr. An extensive cave system may be present as hinted at by the spacious mature passages of Pal y Gort, 0.69 km east of Llygad Llwchwr.

References

Anon. (1973). Llygad Llwchwr. *Cave Diving Group Newsletter*, **28**.
Anon. (1979). Llygad Llwchwr. *Cave Diving Group Newsletter*, **51**.

Weaver, C. P. (1940). Llygad Llwchwr – the source of the River Lougher. *British Caver*, **6**, 74–83.

19

Swallets and caves of the Gower Peninsula

DAI P. EDE and PETER A. BULL

The Gower Peninsula, extending some 20 km west of Swansea, has traditionally been associated with caves of archaeological or geological interest, usually found in the superb cliffs, particularly along the southern coast (see Chapter 7 for a discussion of the archaeological caves). Unlike the coastal caves of south Pembrokeshire these systems are of limited extent, rarely being longer than 60–70 m. The sites described in this chapter are inland drainage systems associated with sinking streams passing off the Namurian Shales underlying the common lands and the Old Red Sandstone of Cefn Bryn. Only two cave systems of significant length are known, both being

associated with the Llethrid–Parkmill Valley. Recent activity suggests, however, that a reasonably extensive system may be found beneath Bishopston Valley, and owing to the interesting form of the valley a full description is included.

Topography

The peninsula is formed of a broad plateau surface at or around 60 m altitude, rising in the east to around 100 m on Fairwood Common (Fig. 19.1). The only relief perceptible from the plateau surface is that created by the broad ridges of Cefn Bryn, Llandmadoc Down, Hardings Down and Rhossili

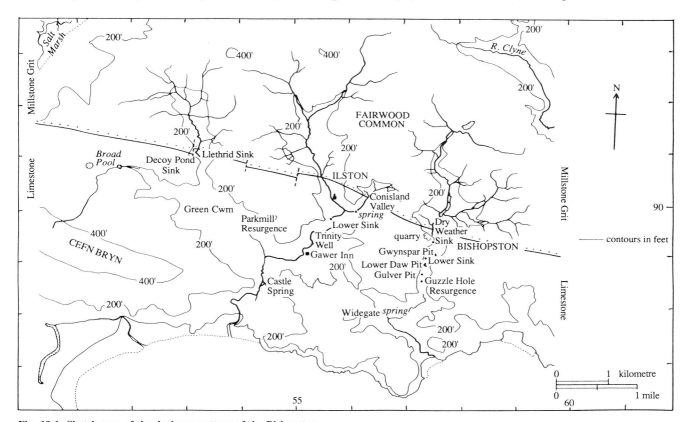

Fig. 19.1. Sketch map of the drainage pattern of the Bishopston, Ilston and Llethrid area, south Gower.

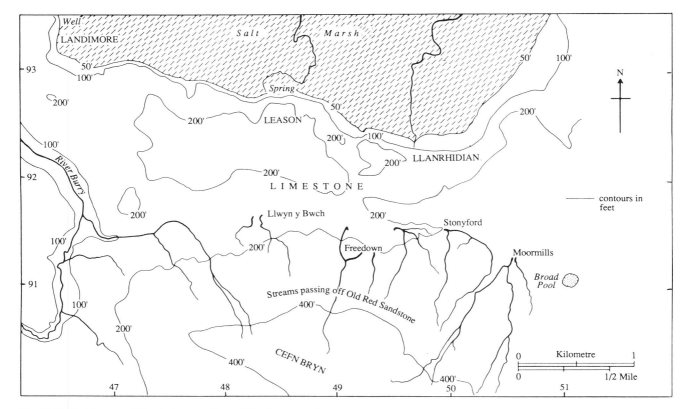

Fig. 19.2. Sketch map of the drainage pattern of the Leason and Llanrhidian area, north Gower.

Down. This general surface belies the complex structure of the limestone which controls the orientation of sections of the south Gower valleys. These valleys cut through the limestone and many of the distinctive features of the peninsula are associated with them. Valley cross-profiles within the limestone usually show high slope angles which contrast with the low-angled slopes on the less resistant Namurian Shales. Within the main valleys the drainage patterns demonstrate a wide range of variation both spatially and temporally but, nevertheless, the overall character is typified by flow patterns more closely associated with surface topography than is commonly considered normal in limestone terrain. In north Gower a small section of limestone receiving drainage off Cefn Bryn to a number of sinks and hence to resurgence at Leason and Llanrhidian has, at least, the outward appearance of form normally associated with classical limestone drainage patterns. Water passes off an impermeable rock, sinks into the limestone and re-emerges at the foot at a scarp-like slope, in this case represented by the old cliff line of the Burry estuary (Fig. 19.2).

Minor springs occur at many points on the peninsula and work carried out by Baynton (1968–9) and since then Chambers (1974) has done much to establish flow connections of many of the smaller systems. Apart from sinking streams, landscape features associated with karst terrain usually occur only as isolated examples and it is only in the Bishopston Valley that a range of forms can be seen in close association. Surface depressions so typical of the Northern Outcrop are generally absent, although the form of the sink complex at

Moormills (SS 505 913) suggests that sink retreat may be associated with depression formation or enlargement in an area where a number of other broad basin-like features are seen. The appearance of isolated collapse features is discussed below.

Geology

The Gower Peninsula is composed of two anticlines trending WNW–ESE, with the Carboniferous Limestone occupying more than half of the outcrop area. Each anticline has a linear outcrop of Old Red Sandstone in its core and the intervening syncline is occupied by sandstones and shales of the Millstone Grit. The northern anticline is terminated in part by a thrust fault relationship with Coal Measures. The limestone outcrops are cut by a large number of normal faults with small displacements and trending at right angles to the fold axes. The dip of the limestone beds is highly variable from <20° to vertical. The combination of steeply dipping bedding planes and the many faults and joints means that the limestone acts as a fractured aquifer with numerous pathways for the movement of underground drainage. This fact, coupled with the limited catchment areas on the Old Red Sandstone has meant limited opportunity for the formation of enterable caves. Glaciation, with the deposition of widespread boulder clay, has led to most potential cave entrances being choked with glacial sediments.

Perhaps the most interesting suite of limestone landforms are those associated with the coastal zone. These are ouside the scope of this chapter.

The Bishopston Valley

The source of the stream draining to Bishopston Valley is a broad area of indistinct drainage on Fairwood Common (Fig. 19.1). The headwater catchment consists of a mixture of agricultural and common land underlain by Namurian Shales and Coal Measures. Soil cover varies from peaty gley to brown earth and it is from these soils that much of the pre-limestone solute load is derived (Groom & Ede, 1972). As it nears the Namurian Shales–limestone junction the stream flows in a well-defined channel 2–3 m wide and passes onto the Oystermouth Beds (Upper Limestone Shales, D_3 zone) just north of the large quarry in the Oxwich Head limestone (D_2 zone) at Barlands (SS 577 896). Under low flow conditions the stream passes underground at the junction of the shales with the limestones. Barlands Quarry sink has a number of openings in an area of general collapse and is partially filled with flood debris and waste from the quarry above. Some 5 m above the sink is the opening of Barlands Quarry Cave (SS 576 896). This consists of a steep 10 m shaft leading to a constricted series of passages and small chambers 450 m in length. Down-valley from this sink the surface course is marked by a well-defined channel running through banks of alluvium over a bedload of Namurian Shales and limestone gravel and boulders. The valley side slopes steepen and evidence of collapse is seen at the Higher Daw Pit (SS 576 893). This is the local name for a pit 7 m deep on the down-slope side with a face 15 m high; up-slope it covers an area of $650 m^2$ and on the walls there is evidence of networks on undercut blocks. This suggests solution under phreatic conditions (Bretz, 1942), although not necessarily in a cavern-size passage. The steep sides of the pit and the absence of any signs of gradual settling suggest rapid downward movement. Thomas (1954) has described steep-sided depressions on the Millstone Grit of the Northern Outcrop and has attributed their formation to the collapse of the underlying limestone. The similarity of form of this and the other pits in Bishopston Valley suggests that they also have formed by collapse but that it has taken place wholly within the limestone. Water from the sub-surface section has never been seen in this pit and as far as the present drainage is concerned it appears to be a 'fossil' feature.

Downstream from the Higher Daw Pit the present course continues as a gravel-floored channel. Some 400 m below the first sink the channel is cut through a large deposit of alluvium. From the accumulated debris it appears to be partially derived from material produced during the construction of the Kittle–Bishopston loop road built to avoid the steep valley side and ford at SS 577 893. The deep channel cut by the stream through this material indicates a greater incidence of surface flows possibly resulting from the partial infilling of the first part of the underground course by debris, particularly quarry waste. At the southern end of this alluvial section is another sink, while in the eastern valley slope is another pit known as Gwynspar Pit. This, the deepest of the pits in the valley, is 30–35 m deep with a boulder-strewn floor and steep sides. Once again the pit appears to be the result of a dramatic collapse of a cave chamber.

The sink mentioned above consists of a number of openings in the streambed and has been observed to be capable of taking

relatively high flows, but no accessible passage has yet been found leading from it. To the south of the sink is the Lower Daw Pit (SS 575 891). This covers an area of $900 m^2$ and has collapsed below a general slope of 31°. The down-slope face (east) is 6.2 m high while the up-slope face is 24 m high. The west wall is formed of alternating massive limestones and shales of Arundian age. On the southern side, beneath an overhanging block, is a distinct spongework of the type described by Bretz (1942) as being due to solution under phreatic conditions. The overall form of the pit, being similar to the others described by cave chamber collapse, is again postulated as the origin. In flood it acts as a storage reservoir.

Below the Lower Daw Pit the normally dry stream course is partially vegetated and at a number of points there are minor seepage springs, while at SS 574 888 is Gulver Pit, which is partially roofed, 15–20 m deep, 28 m long and 8 m wide. The narrowness of this pit and its overall appearance suggest it was an aven in the former cave system, the partial roof surviving because its fissure-like form gave it a relatively greater mechanical strength in comparison to one covering a larger area.

Down-valley is Guzzle Hole (SS 574 886). This is a partially abandoned resurgence where the stream can be seen flowing underground. Just inside the arched entrance in the west wall is a small tube, accessible for approximately 25 m before a sump pool is reached. Under dry conditions the stream can be followed upstream from Guzzle Hole entrance until access to a small roof chamber can be gained. No further progress has been made. In flood conditions water resurges from Guzzle Hole and follows a surface course for some 140 m to the permanent resurgence. This resurgence has two openings 21.5 m apart, both being low impenetrable fissures. Beyond this point surface flow continues to the sea at Pwlldu Bay with only a limited number of additions from small seepage springs.

Llethrid Swallet

The main drainage to Llethrid Swallet (SS 531 912) is initiated on Welsh Moor, an area of rough pasture underlain by Namurian Shales with a soil cover of gleys and peat. This upper catchment has an area of about $5 km^2$ and incorporates part of the Pengwern Forest. At Llethrid Bridge (SS 531 913) the channel is 3.5 m wide and at this point the water passes onto shales of the Oystermouth Beds. This it crosses for approximately 40 m before passing onto the Oxwich Head limestone. At several points minor choked sinks can be seen on the western side of the channel, the main sink being at SS 531 912. This takes the form of a boulder ruckle and the first part of the cave beyond the sink is subject to movement of these blocks, making route-finding occasionally difficult. The Entrance Series proper is reached after some 15–20 m of crawling between large boulders.

Entrance Series

The Entrance Series, approximately 100 m long, consists of a number of constrictions and small chambers. Under low flow conditions the active stream sinks at the Rocking Boulder (see Fig. 19.3) and moving water is not seen again in the whole system. Increased discharge extends the length of flow through

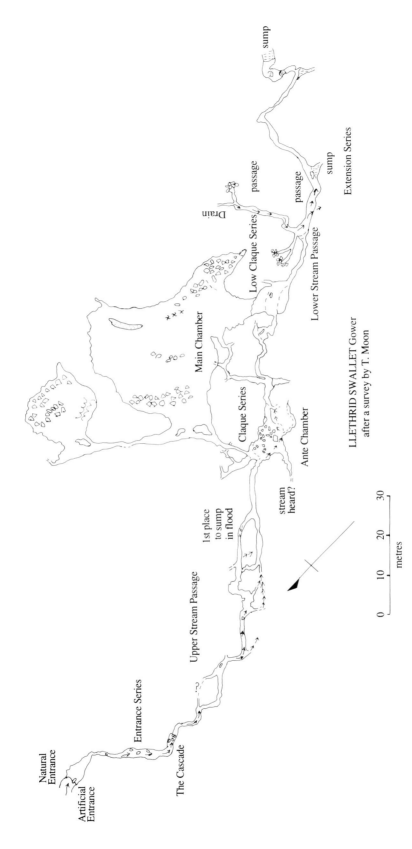

Fig. 19.3. Survey of Llethrid Swallet (after a survey by T. Moon, courtesy of South Wales Caving Club).

the system and can create a number of sumps in this constricted series. In a number of places boulder falls have taken place and it is only in the Rift Passages that a well-defined passage cross-section is seen. These are joint-controlled passages with well-developed scalloping indicating high-velocity flows. There are three such passages leading to the Ante-Chamber which is a high rift with a number of short choked passages leading off. From here a series of tight crawls below the level of the Main Chamber leads to the Extension Series.

Main Chamber

This large chamber is reached from the Ante-Chamber via a short slab at the top of which there is a noticeable increase in sediment. This upper level chamber, which is 40 m long, 30 m wide and up to 10 m high, is the most notable feature of the whole system. The scale of this chamber contrasts markedly with the dimensions of the passages up to that point. At the northern end is the Annex, containing high banks of silt- and clay-grade sediment and separated from the Main Chamber proper by a low arch. The upper part of the Annex appears to be close to the surface and what appears to be clay fill can be seen. In the Main Chamber itself there is evidence of roof breakdown with a large mound of boulders partially cemented with flowstone. The roof of the chamber is well decorated with stalactites (Fig. 19.4), while the floor beyond the boulder

Fig. 19.4. The Curtain stalactite, Llethrid Swallet (photo by J. Rowland).

mound consists of a silty clay fill. The rough stratification of this fill is difficult to interpret but the thickness, shown in a number of collapse pits to be up to 0.5 m, suggests an abundance of sediment. This sediment is of allogenic origin but it is impossible, at this stage, to judge whether it is reworked glacial drift or the result of primary erosion. At the southern end of the chamber is a long boulder slope with a number of flowstone deposits developed on it. Near the top a formation known as the 'church and steeple' shows the dislocation of a complete column caused by renewed movement of the boulder slope. The thickness of the deposits suggests that the boulder debris is of considerable age.

The total contrast between this chamber and the Entrance Series leads to the conclusion that the present course of the Entrance Series has little relevance to the upper cave system represented by the Main Chamber. The point of entry of drainage passing through the Main Chamber when active may have been near the top of the Annex. Although the general direction of water movement was probably the same the present course is a much later development.

Extension Series

This series is reached either by passing through the Low Claque Series from the Ante-Chamber or by climbing down a narrow rift on the western side of the Main Chamber. This rift roughly follows the bedding and because of an offset bedding block encountered approximately half-way down it is known as the Camel's Hump. At the bottom of the rift access is gained to a low series of crawls, often partially flooded. In this series roof scallops, fretting and partially developed spongeworks can be seen and it is clear that this section was, until recently, completely water-filled. From this, low access can be gained to Mud Chamber, a high-level chamber containing a great thickness of silty clay and possibly related to the Main Chamber development stage but now isolated, possibly owing to sediment fill.

The Extension Series leads to a short permanent duck (in wet conditions temporary sumps of varying length may be encountered both in this and the Entrance Series). On the other side is a small chamber and the first permanent sump some 5 m long. Beyond is another small chamber with the, as yet, impenetrable second sump. This forms the end of the explorable system, although it appears that this system may be linked, either directly via the second sump passage or indirectly as part of an overflow system to the Tooth Cave System.

Tooth Cave

The entrance to Tooth Cave is found in a small outcrop some 10 m east of the main track (SS 532 910). Almost immediately the first chamber is reached. The cave earth of this chamber has yielded some valuable archaeological remains, while a boulder fall within the chamber suggests the former existence of another entrance. From the chamber there is a low crawl of approximately 300 m to the Main Stream Passage. During the course of this crawl there is a junction with Razor Passage, which appears to contain a tributary stream flow during wet weather. As the cave is liable to flooding, Main

Stream Passage has to be dry for access to be gained to the rest of the system.

The passage itself is approximately 750 m long and has a low, arched cross-section. Re-solution of some of the minor flow-stone formations on the roof of this passage appears to be taking place, and at a number of points in the cave re-deposited peat is found. This is probably derived from Welsh Moor. Main Stream Passage is normally sumped at both ends, the up-valley sump, Top Sump, being at the bottom of a scalloped tube. The symmetry of the scalloping indicates up-slope flow under pressure. Big Sump at the down-valley end is a pool sump over 20 m long which rarely dries out. Beyond it is a further series of passages ending at Final Sump. It is likely that Tooth Cave is an extension of the Llethrid System, the absence of a permanent stream suggesting that its present role is as a flood bypass and storage system with permanent drainage taking place at a lower level.

North Gower drainage systems

In north Gower two sinking stream systems are found, namely the drainage passing to Llanrhidian and Leason. Both springs have variations in hardness with time typical of systems with well-developed conduit flows, and in at least one sink a short cave system has been explored (Chambers, 1974). In this area the development of sub-surface conduit flows may be a recent phenomenon as there are several points where there is evidence of former surface flow and, indeed, the present flow pattern is subject to seasonal variation. The relative youth of this pattern is also suggested by a continuing successional retreat of the sinking stream input point, particularly in the drainage feeding Llanrhidian Spring. At Moormills the actual point of stream disappearance can be seen to have retreated some 50 m resulting in the erosion of a broad trough forming a marked feature in the flat surrounding common land.

In the general area of both of these systems localized subsidence occurs. This is normally limited to the appearance of a minor shaft up to 5 m deep in the glacial drift cover. The diameter of these shafts varies from 0.5 to 1.5 m at the surface and may increase with depth, although bedrock has never been observed within them. The frequency of these subsidence effects is not known, as the local farmers are naturally keen to fill them in almost as soon as they occur. The appearance of the shafts may be linked to the active development of the sub-surface drainage system and the movement of overlying material via pre-existing solutional voids previously plugged with drift.

It is likely that at least one cave system exists as part of the Llanrhidian and Leason springs. Its dimensions and precise location are, however, matters of conjecture.

Karst landscape development

These concluding remarks will be largely restricted to the areas described in greater detail above.

The development of the marine-cut 60 m (altitude) surface in South Wales has been regarded as taking place in the early Pleistocene. As the plateau was exposed surface drainage took place across the limestone, while at depth phreatic cave systems were initiated. These systems were progressively invaded as a sinking stream drainage pattern developed with continued erosion. Parts of the systems grew to relatively large dimensions and may be considered to be represented today by the Main Chamber of Llethrid Swallet and the collapsed Chambers of Bishopston Valley. There may also have been some development of surface depressions during this period which were subsequently masked by the deposition of the older drift (Wolstonian?). Groom (1971) suggested that the main south Gower valleys were cut during the succeeding interglacial and if this is so the sediment in the Main Chamber at Llethrid could be the result of the reworking of glacial drift deposits at this stage. The cutting of these valleys by the erosion of material above these cave systems would make the roofs of large chambers more susceptible both to surface climatic effects and to the direct infiltration of aggressive water. It is likely that drainage along the north flank of Cefn Bryn developed, at this stage, as a tributary of the Burry River. Pre-existing drainage flowing from Cefn Bryn directly to Llanrhidian may have been incorporated.

The limits of the 'Newer Drift' (Devensian) glaciation given by Groom (1971) differ from those presented by Bowen (1970) who includes the upper part of the Llethrid Valley within his area of drift cover. The nature of the sediment at the northern end of the Annex in Llethrid Swallet does not confirm this limit but it is possible that the upper Swallet System may have been blocked either by primary or reworked drift. Intense periglacial activity in the area outside the glacial limit could increase boulder falls in shallow caves while, as the climate ameliorated, continued sub-soil or direct surface solution would further weaken the roofs of shallow cave chambers.

The collapse of the Bishopston Valley chambers is considered to be a relatively recent event as is the development of the present sub-surface flow system in Llethrid Swallet. In north Gower there may have been some reactivation of earlier sub-surface systems and the invasion of formerly blocked courses.

This broad chronology is extremely tentative and further detailed work is required, particularly on the sediments in the main caves and on the relationship between the minor caves of the peninsula and the Pleistocene history of this part of South Wales.

References

Baynton, R. (1968–9). The hydrology of Gower. *South Wales Caving Club Newsletters*, **58**, **59**, **60**, **62**.
Bowen, D. Q. (1970). Southeast and central South Wales. In *Glaciations of Wales*, ed. C. A. Lewis, pp. 197–227. London: Longman.
Bretz, J. H. (1942). Vadose and phreatic features of limestone caverns. *Journal of Geology*, **50**, 675–811.
Chambers, W. (1974). Limestone springs and individual flood events. *Transactions of the Cave Research Group, GB*, **15(2)**, 91–8.

Groom, G. E. (1971). Geomorphology. In *Swansea and its Region*, ed. W. G. Balchin, pp. 29–40. Swansea: British Association Handbook.
Groom, G. E. & Ede, D. P. (1972). Laboratory simulation of limestone solution. *Transactions of the Cave Research Group, GB*, **14**, 89–95.
Thomas, T. M. (1954). Swallow holes on the Millstone Grit and Carboniferous Limestone of the South Wales Coalfield. *Geographical Journal*, **120**, 468–75.

20

Limestones and caves of North Wales

PETER APPLETON

The limestones of North Wales outcrop from Anglesey eastwards through the county of Clwyd and south to the border country at Llanymynech. Rugged landscape features occur where thicker beds are well developed, notably in the sea cliffs of the Great Orme's Head at Llandudno and the impressive Eglwyseg escarpment above the Vale of Llangollen (Fig. 20.1). With so much limestone available, the lack of large open caves seems surprising, and this deficiency may be the reason why North Wales has never shared the romance of those parts of South Wales, where forbidding river caves and dark springs stir the imagination. The limestone in the north attracted a more commercial interest, and while travellers of former years marvelled at the peaks and waterfalls of the ancient mountains of Snowdonia, it seems that the less spectacular limestone hills had captured the attention of the lead miner. The mining of lead, and to a lesser extent copper ores, has been of importance at least since Roman times, metal mining ceasing on a significant scale only in 1958. In the heyday of the industry, in

Fig. 20.1. The Eglwyseg escarpment near Llangollen.

the eighteenth and nineteenth centuries, hundreds of mines were worked, some yielding large tonnages of ore. Working at depth was often severely hampered by influxes of water which entered through natural fissures, and there are numerous references in mining documents to caves being intersected. Such caves usually seem to have been difficult of access, even to the most inquiring of visitors, and descriptive accounts are relatively few.

Perhaps it was the archaeologist Sir William Boyd Dawkins who first created a popular interest in the caves of the region. Important discoveries were made in caves in the Elwy Valley as long ago as 1833 by Edward Lloyd and others. A description of these finds and accounts of his own excavations at Llandegla featured prominently in Dawkins's classic *Cave Hunting* (1874). At about the same time the geologist G. H. Morton was working on a classification of the limestone strata at Llangollen, based largely on the colour and lithology of the different beds exposed. His detailed study was extended in subsequent years to produce the first comprehensive stratigraphy of the Carboniferous Limestone in North Wales, forming a basis from which later workers drew heavily and which is only now being superseded by a new, formal stratigraphical nomenclature. With no large accessible entrances, interest in the exploration of caves was slight, although a visit by Ernest Baker merited several pages in his *Netherworld of Mendip* (Baker & Balch, 1907). At that time the most adventurous excursion appeared to be the tortuous 150 m of the Lower Ceiriog Cave.

The following account seeks to introduce a limestone area which is still relatively little known, and present a summary of progress in related studies. Since the evidence of mining is widespread, the mineral deposits and their exploitation are treated at some length. With improving exploration techniques the pace of cave discoveries has accelerated in recent years, with not inconsiderable help from the 'Old Man's' workings. Descriptions of the main cave systems are given along with discussion of their development in the context of local geology and hydrology.

Limestone geology

Of the numerous areas of Carboniferous Limestone or Dinantian in North Wales the outcrop flanking the eastern slopes of the Clwydian Hills is the most extensive, sometimes being described as forming the eastern segment of a dome. Twice stepped eastwards by faulting the outcrop continues southwards, finally ending in Shropshire near the town of Oswestry. Another nearly parallel outcrop extends from the coast along the western side of the Vale of Clwyd where it is dissected into a series of blocks in the south. In the northwest, scattered outcrops occur from Llandudno westwards into Anglesey (Fig. 20.2).

Classification of the Dinantian rocks

The North Wales limestones contain a wide variety of sediments, and at their fullest development reach a thickness approaching 900 m. The first comprehensive classification of the North Wales sequence was attempted by G. H. Morton (1876, 1879, 1883, 1897, 1901). In 1876–9 he described the Carboniferous Limestone of the Llangollen area, dividing the sequence into lithological divisions, based largely on exposures in the Eglwyseg escarpment. Subsequently, this work was extended to cover all the North Wales outcrops. Morton's (1898) divisions of the Carboniferous Limestone are as follows:

Sandy Limestone or Black Limestone of northern Clwyd
Upper Grey Limestone
White Limestone
Lower Brown Limestone

This classification was later to be broadly adopted by the Geological Survey (Wedd *et al.*, 1927).

Vaughan's zoning of exposures in the Avon Gorge, Bristol, was applied to northeast Clwyd by Hind & Stobbs (1906). They concluded that the North Wales limestones were represented only by the 'D' or Dibunophyllum zone at the top of the Avonian succession. There was some disagreement between workers over the position of the boundary between D_1 and D_2

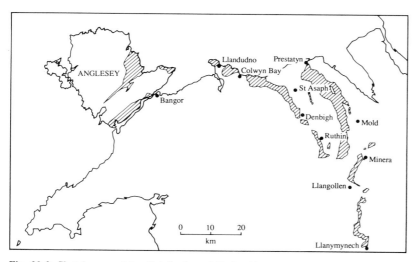

Fig. 20.2. Sketch map of the distribution of Carboniferous Limestone outcrops (shaded) in North Wales.

sub-zones. This was finally concluded by Neaverson (1946) as being at the transition from the White to the Upper Grey Limestones and George (1974) restated this classification with little change.

In keeping with the modern nomenclature of what is now recognized as a cyclic type of sedimentation involving transgressions and regressions of the Dinantian sea, Ramsbottom (1973), placed the D_1 and D_2 sub-zones into his fifth and sixth major cycles of deposition. These correspond to the Asbian and Brigantian stages of George et al. (1976), respectively.

In addition to the more important and widespread calcareous strata, 'Basement Beds' sometimes underlie the lowest limestones. Due to their generally red colour they were referred by Morton to the Old Red Sandstone, but they have been confirmed by fossil evidence as belonging to the Dinantian.

Lying above the Sandy or Black Limestone is the Cefn y Fedw Sandstone sequence, formerly included entirely in the Millstone Grit, or Namurian. The lower part of these beds, comprising a sandstone and a cherty shale at Llangollen, has been shown by its fossil assemblage to belong to the highest part of the Brigantian. The Cherts of northeast Clwyd, an apparently equivalent sequence of beds, are overlain by Holywell Shales containing Namurian (E_1) fossils, and are only by inference included in the Dinantian (George et al., 1976).

Somerville (1979a,b,c) proposed an alternative to Morton's nomenclature based on extensive fieldwork in the early and late Asbian Limestones north of Llangollen and in the early Brigantian Limestone east of the Clwydian Hills. Except for a slight difference in the division between early and late Asbian

Limestones, however, the vertical range of the respective formations corresponds exactly. More recent work in the area west of Mold by Somerville & Strank (1984a,b) has demonstrated that the lowest 200 m or so of limestones outcropping on the east side of the River Alyn, north of Llanarmon yn Ial, are of pre-Asbian age. Careful study of macro- and particularly of micro-fossils confirmed that the lower beds of the Lower Brown Limestone here belonged to the Holkerian and Arundian stages of deposition. Fossil assemblages in the Dyserth area have confirmed that the lowest limestones here are also of Arundian age (Somerville, Mitchell & Strank, 1986).

Morton's classification, in spite of some shortcomings, is still the only one which has been applied to the whole of North Wales, and for this reason it is retained in the following discussion. The relationship between the various classifications is indicated in Table 20.1 and includes Somerville and Strank's nomenclature for formations in the Mold district.

The Basement Beds are not always present but, where exposed, are seen to consist of sandstones, shales and conglomerates, sometimes with thin limestone bands. They are generally red or purple in colour and lacking in fossil remains. The best exposures in the western outcrops are along the coast and in the Vale of Clwyd. In the Llangollen Syncline at Eglwyseg the Basement Beds reach a thickness of about 100 m but are poorly exposed. At Lligwy Bay in Anglesey a basal conglomerate contains boulders up to 1 m in diameter. The nature of these deposits suggests that they were the first sediments laid down near the shoreline of the Dinantian sea, which gradually transgressed and submerged an eroded landscape of Lower Palaeozoic rocks. For this reason the different

Table 20.1 *Relationship between various geological classification systems*

Morton (1878)	Hind & Stobbs (1906)	Ramsbottom (1973)	George et al. (1976)	Somerville & Strank (1984 a, b)[a]	
Sandstones and shales–Holywell shales			(NAMURIAN)		
Cherty shale and lower Cefn y Fedw Sandstone–bedded cherts.	D₂ sub-zone	Sixth group of minor cycles	Brigantian		
Sandy Limestone– Black Limestone					
Upper Grey Limestone				Cefn Mawr Limestone	Early Brigantian
Middle White Limestone			Late Asbian	Loggerheads Limestone	Late Asbian
Lower Brown Limestone	D₁ sub-zone	Fifth group of minor cycles	Early Asbian	Leete Limestone	Early Asbian
				Llanarmon Limestone	Holkerian
				Llwyn y Frân Limestone	
				Spring Quarry Limestone	Arundian
Basement Beds	S₂?		Holkerian ?	Basement Beds	?

[a] Names of formations refer to Mold district only.

occurrences cannot be considered to be all of the same age, although work on micro-fossils from the Basement Beds of the Bangor district has indicated Holkerian age (Hibbert & Lacey, 1969).

The Lower Brown Limestone, as the name suggests, is usually dark in colour containing a significant proportion of clastic sediment. It is usually thinly bedded with frequent partings of shale which is sometimes carbonaceous and associated with a few millimetres of coal. Porcellanous limestones are generally present, particularly towards the top of the sequence. At the Great Orme, Llandudno, the beds have been extensively dolomitized, apparently long after their deposition since the horizon of dolomitization is irregular, and influenced by the presence of mineral veins.

At Eglwyseg the thickness of the Lower Brown Limestone is about 150 m but further north at Llandegla it reaches its maximum development of over 300 m.

The Middle White Limestone, often separated from the Lower Brown Limestone by a bed of shale, is a lighter coloured rock. It contains thick massive beds, especially in the lower part of the sequence and forms conspicuous outcrop features right across North Wales. At Eglwyseg the division is 100 m thick, but its maximum development is in northeast Clwyd where a thickness of 500 m is attained. In the south the individual beds are often separated by thin bands of shale, but to the north these are less obvious. Knoll-reefs occur in the north at Dyserth (Somerville et al., 1986). The Anglesey sequence contains two sandstones, each several metres thick with associated sandstone-filled 'pot-holes' (Walkden & Davies, 1983).

The Upper Grey Limestone is a series of dark, generally thin-bedded fossiliferous limestones, containing thick beds of shale especially in the upper part of the sequence. One persistent bed of shale can be traced from Minera northwards as far as Halkyn, its thickness varying from 3 to 7 m. The conspicuous Coral Bed marks the top of the Upper Grey Limestone between Llangollen and Llandegla. At, or just below this horizon, at Halkyn is a 2–4 m band consisting largely of crinoid debris and known as Halkyn marble. The thickness of the Upper Grey Limestone is 100 m at Eglwyseg, >250 m at Llandegla and up to 170 m in the Vale of Clwyd.

The Sandy and Black Limestone: in the south of the eastern outcrop, this group comprises a variable sequence of sandstones, sandy limestones and shales. Between Minera and Llangollen a bed of sandy oolite occurs at the base. In northeast Clwyd the sandy beds pass laterally into the Black Limestone of Halkyn, a thin-bedded, dark fine-grained rock with shale bands. In the Black Limestone at Teilia near Prestatyn many well-preserved remains of drifted plants have been obtained.

The Lower Cefn y Fedw Sandstone and Bedded Cherts: at Cefn y Fedw near Llangollen the sequence is represented by a hard white sandstone containing quartz pebbles which is succeeded by a cherty shale. In northeast Clwyd the beds appear to have passed laterally into the bedded cherts which reach a thickness of about 70 m.

Palaeogeography

The limestones and associated rocks present in North Wales and their distribution confirm that the Dinantian sea was transgressive in nature resulting in the gradual but incomplete submergence of the ancient landmass of 'St George's Land' during the latter part of the period. This old landmass is represented today in North Wales by a core of Lower Palaeozoic rocks – the Snowdonia Mountains. Deposition in Dinantian times was confined to its northern and northeastern margins. The first evidence of submergence is furnished by the 'nearshore' deposits of the Basement Beds. The limestones were formed as shallow water deposits, variation in depth of water being thought to give rise to corresponding variation in limestone types. Calcite mudstones or porcellanous limestones are considered to be the shallowest water type, followed by pale, thicker bedded limestones and finally dark limestones and shales with increasing depth (Ramsbottom, 1973). Towards the close of the Dinantian period a general shallowing of the sea occurred accompanied by a change from open water to the more deltaic conditions of the Namurian.

Large variations in the thickness of Dinantian rocks occur partly due to both the inclined surface of the submerged landscape and its undulating nature, giving a Dinantian coastline of bays and headlands. This situation has led in many instances to 'overlapping' of the earlier sediments by later sediments. Near Chirk, for example, both the Lower Brown and White Limestones are absent, the Upper Grey Limestone forming the lowest beds along the crest of the Berwyn Anticline. Only 7 km to the north, in the Llangollen Syncline, the whole succession is present, the Lower Brown Limestone and Basement Beds reaching their fullest development. In Anglesey the Dinantian rocks are eventually completely overlapped, Namurian sediments resting directly on lower Palaeozoic rocks in the west of the island.

Cyclic sedimentation

Superimposed on the generally transgressive nature of the Dinantian inundation, a cyclical sequence of sedimentation has been recognized, the result of repeated periods of transgression and regression due to variations in sea level. This situation has persisted into Upper Carboniferous times and is inferred from the repetition of beds and sequences of particular lithologies. Of six major sedimentary cycles identified by Ramsbottom throughout mainland Britain, cycles 5 and 6 occur widely in North Wales, each consisting of a number of minor cycles (Ramsbottom, 1973). Somerville has described minor cyclicity in the Lower Brown, the White and Upper Grey Limestones based on recent field work in the Llangollen and Mold areas (Somerville, 1979a,b,c). Due to the comparative shallowness of the sea, regressive phases frequently resulted in the emergence of the recently formed sediments exposing them to the processes of erosion. Palaeokarstic erosion surfaces thus formed can be readily identified and are important in establishing individual cycle boundaries. The limestone underlying these surfaces is sometimes brecciated, cracking probably being due to desiccation. Often the palaeokarst surfaces are hummocky and layers of orange or greenish clays may be present in thicknesses varying from a few millimetres to one metre. These deposits, known as 'K-bentonite' clays, are characterized by a comparatively high potassium content and are considered to be fossil soils or 'palaeosols', analysis inferring their probable source to be wind-blown volcanic dust (Somerville, 1979a). Sometimes a coal parting separates the clay from the succeeding limestone bed, indicating more prolonged emergence with establishment of vegetation cover.

Sandstone-filled pipes

Palaeokarstic surfaces occur in Anglesey where hundreds of sandstone-filled pipes occur at several points in beach and cliff exposures on the east coast. They occur on at least seven horizons and are most fully developed at Dwlban Point on Red Wharf Bay, where they reach a depth of 2.7 m with diameters up to 2.0 m (Figs. 20.3, 20.4). They are smooth-walled and circular in outline, rounded and closed at the bottom and with no honeycombing in the host limestone. The fill rock consists generally of a fine-grained quartz sandstone with occasional patches of angular quartz pebbles, the cementing medium being haematite in part (Baughen & Walsh, 1980). Further study of the Dwlban Point exposures by Walkden & Davies (1983), has inferred the occurrence of four separate erosive phases at the main pipe horizon, fill rock in the second phase being a poorly sorted conglomerate. A third phase involved the excavation of a channel, subsequently filled with limestone breccia, while in a fourth phase some new pipes have penetrated the breccia into the limestone, these pipes being filled by buff sandstones which unite to form a continuous bed above. An interesting feature of this multi-phase process has been erosion at the perimeters of the early 'plugs' with deposition of later sediments such that the resulting plugs are of complex lithology.

The present consensus favours a solutional origin for the pipes, although somewhat conflicting theories have been advanced for the precise mechanisms involved. Certain features are difficult to explain: for example, Walkden & Davies (1983) show a later perfectly circular pipe which has consumed part of an earlier conglomerate plug. Clearly, even if the conglomerate was susceptible to solutional attack, the

Fig. 20.4. Sandstone-filled pipes at Dwlban Point, Anglesey.

relative rates of removal of conglomerate and limestone would not have been so similar as to allow the later pit to assume an uninterrupted circular cross-section.

Whatever the mechanism, it must have been capable of repetition at successive horizons over a lateral range of at least 15 km. Although dismissed in the latest studies, it is tempting to consider a mechanical origin for the pipes, perhaps with subordinate surface solution contributing to crevice formation around the perimeter of earlier plugs. Since coarse conglomerate beds form part of the Old Red Sandstone sequence which unconformably underlies the limestone in this part of Anglesey, there would always have been ample abrasive material available at the contemporary shoreline. Abrasion by 'rock mill' action, perhaps of only partly lithified calcareous sediments, might then have taken place in a situation of marine erosion much as exists at the present day.

Pocket deposits

During the nineteenth century over 20 deposits of fine clays and sands were worked from pits in the surface of the limestone, mainly in the Halkyn and Mold areas. Known as 'pocket deposits' they occupied pipes and more irregular cavities formed by ancient solution, generally along mineral veins or other faults or joints. Typical dimensions were from 100 to 200 m in diameter by 50 to 100 m in depth.

The contents of one of the pits at Rhes y Cae on Halkyn Mountain were described by Strahan (1890), who found their contents to be very variable including shales, 'masses' of 'Millstone Grit', casts of small crinoids, white and variegated sands, and white, black and variegated clays. The sand was used in glassmaking, while cherty sandstone and the white clays were used in the manufacture of porcelain. The black clay was carbonaceous, rather like a lignite, which Strahan suggested might have been an organic soil which once covered the ground above.

Examination of the now poorly exposed sites and a review of the old literature has been made by Walsh & Brown (1971), who proposed that the contents of the pockets were introduced by subsidence of overlying alluvial sediments of Tertiary age. Unfortunately, although samples of lignite were recovered on re-examining the Rhes y Cae Pit, no datable organic remains

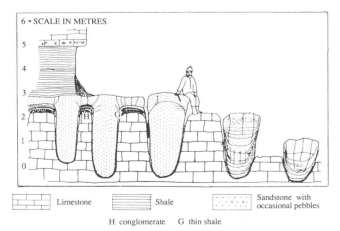

Fig. 20.3. Diagrammatic section through sandstone pipes in the palaeokarst at Dwlban Point, Anglesey (reproduced from Baughen & Walsh, 1980, with permission).

were found. Their suggestion that the infilling sediments were of Tertiary age is, therefore, based on comparison with similar pocket deposits occurring in Derbyshire, where a Miocene–Pliocene flora has been preserved.

While Strahan also concluded that the coarse material in the Rhes y Cae Pit could only have been introduced by subsidence of sandstone which once existed some distance above the present top of the pipe, he felt that such a mechanism was not necessary to explain the presence of the fine sediments which constituted the bulk of most other pocket deposits. These materials, he thought, were derived from the upper limestone strata and overlying cherty sandstones. Transported by underground streams, they would eventually settle in larger solutional cavities which particularly occurred at the intersections of E–W veins with N–S faults or 'cross-courses'. One such situation was at the west end of the Cathole Vein, 4 km west of Mold, where red, yellow and white clays and white and red sands were obtained from a large irregular hollow in the limestone outcrop. In 1982, exploration of underground lead workings in the Cathole Vein confirmed the presence of a number of smaller solution cavities within 200 m of this old pit. They occur in and adjacent to the vein fissure and contain what appears to be a similar range of sediments. Since some of the cavities were horizontal and none appear to have communicated directly with the surface, the evidence does tend in this case to support Strahan's view of underground transport of sediments.

The Pleistocene glaciations

North Wales was covered by ice sheets during several episodes of glaciation with the county of Clwyd being very much the scene of interplay between ice dispersing from Snowdonia and ice derived from Scotland, the Irish Sea area impinging on the North Welsh coast. However, no comprehensive glacial stratigraphy has been worked out for the areas with limestone outcrops and caves and thus it is at present impossible to relate episodes of karstic erosion and cave formation to the standard Pleistocene sequence. Comments are made in the ensuing paragraphs where appropriate but until a sequence of events in the glacial–interglacial cycles can be determined there seems to be little point in trying to fit episodes of speleogenesis to it. Suffice it to say that most of the glacial effects and deposits are probably of Devensian age. However, the sequence of deposits in Pontnewydd Cave indicates that a complex succession of climatic phases has affected at least part of the limestone outcrop (Green *et al.*, 1981).

The northern and western outcrops

Northeast Clwyd

The limestone outcrop in northeast Clwyd takes the form of an elevated plateau up to 6 km in width (Fig. 20.5). In the northwest the limestone is last seen at Dyserth in the bold hills of Graig Fawr and Moel Hiraddug, before being stepped down to the west by the Vale of Clwyd Fault. Further southeast the plateau is bounded firstly by the lower northern hills of the Clwydians, then by the Wheeler Valley with its small misfit

stream, a remnant of the former north-flowing Alyn River System (Embleton, 1957). At the eastern end of the area the ground rises to Halkyn Mountain, centre of the Holywell–Halkyn mining field, with its numerous quarries. The town of Holywell lies at the eastern edge of the plateau, its famous spring, St Winifride's Well, long being the object of pilgrimage for its healing properties.

Minerals and mining The discovery of several pigs of lead with inscriptions indicating a northeast Wales origin confirms that lead was mined in the area in Roman times. One of the pigs was found in 1950 at Carmel near Holywell and there is also evidence of lead smelting near the town of Flint dating from an ancient period. One site with Roman associations was the Talargoch Mine at Meliden in the northwest. Here narrow veins outcropping in the crags of Graig Fawr would have attracted early attention. The main Talargoch Vein was a fault downthrowing to the northwest. Over much of its length it was covered by a considerable depth of drift deposits in which large quantities of 'gravel' ore were found, sometimes in water-worn masses many tons in weight. In later periods of working such ore was obtained by driving tunnels in the gravel at its contact with the underlying limestone (Smith, 1921). The Talargoch Vein was a prolific producer yielding 58 000 tons of lead ore and 50 000 tons of zinc ore between 1845 and 1884. There was a strong zonation of ore minerals and it was the predominance of sphalerite in some of the deepest workings at 400 m below surface at a time of low zinc prices, which contributed to the closure of the mine in 1884. In some of the shallow workings smithsonite, locally known as 'coke' was obtained in significant quantities, one adit working being known as the 'Coke Hole'. The main producing area lay between Holywell and the Hendre Gorge, with Halkyn Mountain at its centre. Lead and zinc ores were won from fissure veins at horizons from the middle of the White Limestone into the Bedded Cherts.

The limestone is dissected by what are described as veins and cross-course faults, the former predominantly E–W, the latter N–S. In the Holywell–Halkyn area the relationship of these faults to the dip of the beds varies, since the dip veers from easterly at Hendre, to the northeast at Halkyn, to north at Holywell. In the area south of Hendre the cross-courses have almost without exception been barren. Their direction in that area coinciding with the strike of the beds. At Halkyn, along with the E–W veins, many of the N–S faults have been very productive, but here they cross the dip of the beds obliquely. The ground on the southwest side of Halkyn Mountain is stepped down by the major Nant–Figillt Fault, the throw of which increases along its SSE-trending course into the Coal Measures west of Mold.

Between Halkyn and Hendre the position of ore shoots in the vein fissures is strongly influenced by the presence of two thick shale beds in the Upper Grey Limestone sequence. The shale has the effect of clogging the vein fissures, preventing the upward circulation of mineralizing solutions and causing local enrichment by lead and zinc ores (Schnellmann, 1939, 1959). From Halkyn northwards, most ore has been found in the higher Black Limestone and cherts, vein fissures in these

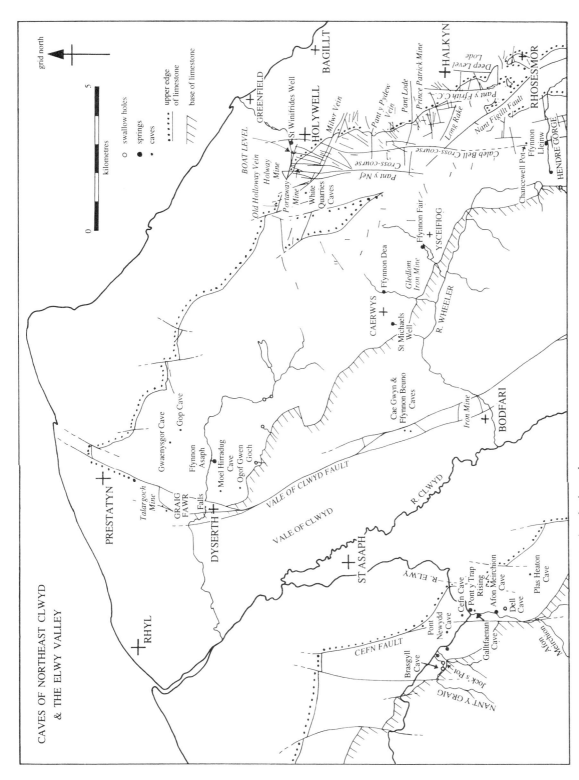

Fig. 20.5. Sketch map of the limestone outcrop, mineral veins and caves of northeast Clwyd and the Vale of Elwy.

situations being lost in the overlying thick sequence of Holywell Shales belonging to the Westphalian Lower Coal Measures.

Mining in the Holywell–Halkyn field progressed on a fairly considerable scale in the late seventeenth century, and in the early eighteenth century, the ore still being won cheaply from comparatively shallow workings. At that time there were also several lead smelting works on near-by Deeside. The ores mined were mainly galena, which in a few cases, as at Silver Rake Mine, contained silver at up to 18 ounces per ton. In some of the veins, large quantities of calamine or smithsonite were mined, but prior to about 1720 this ore was not identified and Pennant noted that it had been used to mend roads (Pennant, 1796).

Apart from the major veins, numerous small strings were worked at the outcrop and from shallow shafts, which today pock-mark the mountain top. More important was the occurrence of flats or pipe veins. One such deposit, discovered in the Prince Patrick Mine, proved to be extremely rich, the ore being found in a series of ancient solution cavities running down with the dip of the Black Limestone beds to the north for over 800 m (Davies, 1883).

Strahan visited the mine in 1879 shortly before its closure and described the deposit as, 'bounded on all sides by rounded walls of solid limestone, except in the roof and floor, where could be seen a joint, . . . the pipe was about 8 yards wide and was filled with beautiful calc–spar sometimes stalagmitic' (Strahan, 1890). The galena lay as a bed up to 'two feet' thick running down the centre of the cavity about 1 m above the floor.

An interesting occurrence in two N–S lodes at Halkyn, the Pant y Ffrith Cross-course and the near-by Old Rake, was the discovery of trees in unworked ground at depths of up to '45 yards'. In the Cathole Mine in the area to the south, near Mold, several trees with a girth of about '1½ feet' were found at a depth of '205 yards', at the junction of the vein with a N–S cross-course. The wood was dark and heavy and the centre of one of the trees was impregnated with pyrite (Strahan, 1890).

During the latter part of the eighteenth century some of the deeper mines were beginning to encounter water problems and this prompted the proprietors of the Holway Mine at Holywell to devise a radical solution to the problem. In 1774 the Holywell level was commenced at a height of 70 m above sea level or 3 m above the level of St Winifride's Well. It was driven a total of 1500 m and by 1795 was being used as a canal, ore being conveyed from the mine in narrow barges. The tunnel has been known since that time as the Holywell Boat Level and the mine worked profitably for many years. Several strong springs, associated with N–S faults, rose from the sides and bottom of the level. One of these, known as the 'Big Duke', flowed from the New Pant y Nef Cross-course, a fault which passed through the Portaway Mine some 600 m to the south and was reputed to contain a 'large stream of water running through a great subterraneous cavern' (Pennant, 1796).

In the Duke of Westminster's mines at Halkyn, major discoveries were made in the early years of the nineteenth century, but heavy influxes of water were met within a few years. In 1818 an attempt was made to unwater the mines by the driving of a tunnel, known as the Halkyn Deep Level, at about 55 m above sea level (Fig. 20.6). The tunnel ran into several areas of bad ground necessitating the use of cast-iron support segments. In 1822, after driving 1.5 km, the project was abandoned. It was resumed in 1838 and cut the productive N–S Crockford's or Deep Level Lode (Fig. 20.7) at a distance of 2 km from the portal. The Deep Level was continued southwards for a further kilometre thereby developing the lode and draining the mines in the immediate vicinity.

In the 1870s the state of the mines in the south of the area was desperate due to inability to cope with water. In response to the problem, the Halkyn District Mines Drainage Company was formed in 1875, the object being to continue the Halkyn Deep Level into that area allowing the affected mines to resume production. An act of Parliament empowered the Drainage Company to charge a levy on each ton of ore raised from the mines which benefited. The result was that the mining industry between Halkyn and Hendre was revitalized, an added bonus being the discovery of several new veins, of which the Great Halkyn Lode and Powell's Lode, both at Rhosesmor, were extremely productive. The tunnel was subsequently extended as far as the South Llyn y Pandy Vein in the Mold area, a total distance of over 8 km.

In 1896 a similar company, the Holywell–Halkyn Mining and Tunnel Company was formed to drain the mines to the north of the area covered by the 1875 company. In the following year it commenced driving the Milwr Sea Level Tunnel southwestwards from the Dee estuary at Bagillt. At the Herward Mine the tunnel turned south reaching the boundary of that company's area in 1919, a total length of driving of 5.5 km.

In 1928 the various mining and tunnel drainage companies were amalgamated as the Halkyn District United Mines Ltd, the new company extending the Sea Level Tunnel through the remainder of the area in the 1930s, with a major branch being driven to Powell's Lode at Rhosesmor. Production by this company, up to the cessation of large-scale operations in 1958, amounted to approximately 200 000 tons of lead and zinc concentrates, much of this being obtained from new veins located in driving through this area and that area immediately to the south of Hendre.

In addition to lead–zinc ores mining of high-grade limestone was carried out by Halkyn Mines in the 1950s and early 1960s, being extracted from chambers connected with the Sea Level Tunnel. Upper beds in the White Limestone were worked, just north of the Hendre Vein. The stone was raised and crushed at the Olwen Goch Shaft and sold for use in agriculture and as a raw material in glassmaking before the operation became uneconomical.

A number of comparatively small iron mines were formerly worked along the western margin of the main limestone outcrop from Dyserth to Ysceifiog, and also in a faulted-down block of limestone in the Vale of Clwyd at Bodfari. Haematite ores were associated with N–S jointing or occurred in more randomly orientated solution cavities, deposition arising from

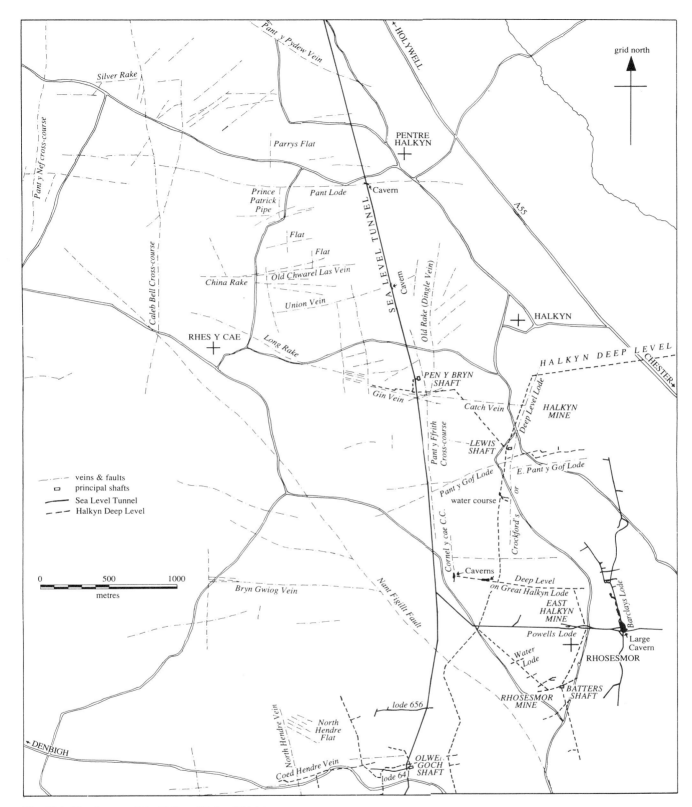

Fig. 20.6. Sketch map of the Halkyn Mining District.

the downward percolation of iron-bearing waters. Outputs were fairly minor compared with those in South Wales, Gledlom Mine near Caerwys, for example, producing 16 000 tons in nine years in the late nineteenth century. The Moel Hirradug Mine at Dyserth was worked in a series of steep pipes to a depth of 70 m and had the distinction of marketing 240 tons of earthy cobalt ore or asbolan, assaying approximately 30% of that metal (Davies, 1883).

Fig. 20.7. Stope in Crockford's Lode directly above the Halkyn Deep Level in Halkyn Mines.

Fig. 20.8. Solution cavern in the Great Halkyn Lode at Halkyn Deep Level.

Karst drainage and caves of the Holywell catchment
The largest rising for the area, prior to the driving of the Sea Level Tunnel, was St Winifride's Well from which the town of Holywell derives its name. The water appeared at 67 m above sea level from fissures in the chert beds which here overlay the limestone, at a point where a far-ranging N–S fault, the Caleb Bell Cross-course intersects E–W faulting associated with Old Holloway Vein (Fig. 20.5). Apart from its religious significance the flow from the spring, which averaged 6 million gallons per day, was an important source of power upon which the industries which grew up in the Greenfield Valley, between the town and the Dee estuary, were dependent. Although the flow was fairly constant, it did increase and become turbid in prolonged wet weather. The total hardness of the water is recorded as averaging 300 ppm $CaCO_3$. This high figure, and the absence of stream swallets on the neighbouring limestone uplands, suggests that the water may have been entirely derived from percolation. Evidence from mining records confirms that this percolation flow is concentrated by structural features, such as the network of N–S and E–W faults, and also by impervious strata in the upper limestones, into major deep-seated phreatic conduits. There is some evidence to show that caverns encountered during mining, and sometimes described as 'isolated', were, in fact, part of such an integrated system of karstic drainage.

Many solution caverns, known locally as lochs or vughs,

were intersected by the different drainage tunnels. One of a group of smaller openings in and adjacent to Crockford's Lode was entered from a 'rise' 40 m above the Halkyn Deep Level (Fig. 20.6). Described as a 'watercourse' it dipped at only a slight angle and was surveyed by the miners for 75 m. Also at the Halkyn Deep or 'Old Tunnel' level, larger openings were found some 600 m to the south in the unproductive western end of the Great Halkyn Lode in the White Limestone. One was associated with the minor Cornel y Cae Cross-course and contained large blocks of limestone. The other consisted of two openings: one in the vein fissure was 30 m long and extended about 15 m above the tunnel, but was filled with rubble below (Fig. 20.8). This chamber connected, via a short passage, to a 25 m high cavern in the south or hanging wall of the vein. The passage also communicated with steep irregular solution tubes and rifts formed in jointing adjacent to the vein which became choked with sandy debris and corroded limestone blocks about 20 m above the tunnel.

The main influx of water into the Deep Level north of Hendre was from the Rhosesmor Mine, where a branch tunnel struck the 'Water Lode' on passing from shales into limestone in an anticlinal structure. In a northward driving from this mine Powell's Lode was discovered in 1903, rich ore being followed below tunnel level at the expense of pumping 3500 gallons of water per minute.

It was in the driving of the Sea Level Tunnel that the main

waterflow in the Holywell System was encountered. In 1917, at a distance of 4 km from St Winifride's Well, a cavern was breached by the tunnel face on reaching the line of the Pant Lode. After firing, a massive burst of water occurred which swept loaded tubs a long way down the Tunnel. Eleven hours later the flow of water from the Well diminished and ceased. Great quantities of quartzose and cherty sand were washed in with the initial flood. Gaugings were taken 12 d later when a flow of 6000 gallons per minute was recorded. The flow subsided after a further 4 weeks to just over 4000 gallons per minute, the normal volume delivered by the Well (Smith, 1921). Presumably, by then, all natural conduits and old workings connected with the flow had drained to Tunnel level. A supply of water was subsequently reinstated to the Well from the Holway Mine, where, although the water level in part of the mine had also dropped, feeds of water into the Boat Level to the west of the Old Pant y Nef Cross-course were unaffected.

As the Tunnel progressed southwards, further heavy feeds were intersected, although others to the north of the Pant Lode dried up. At about 800 m south of the Pant Lode, on the line of the Union Vein, a winding cavern was intersected more than once by the Tunnel. The largest cavern of all, and one of the most impressive in British caves, was that found in 1932 when a branch of the Sea Level Tunnel was driven under the East Halkyn Mine (Richardson, 1936). The main chamber measured 70 m by 22 m, extending northwards along the N–S Barclay's Lode from its intersection with Powell's Lode (Fig. 20.9). The walls of the chamber are often formed into large rounded concavities by solution; in other places they are angular where large blocks have fallen away. The main features of the cavern were its two large lakes, although one of these has been completely filled with mine spoil. The remaining one, about 20 m in diameter, was plumbed at one time for 60 m without touching the bottom. The two main faults and several minor ones can clearly be seen in the walls and roof, beds of Upper Grey Limestone in places being faulted against White Limestone. The general roof level is about 20 m, although a large circular opening in the centre of the cavern rises through higher caverns, containing clays and sandy sediments to a height of 50 m, giving a total vertical development of at least 110 m, half of which is below sea level.

During the original inspection of the chamber, an improvised raft was constructed by which means one of the lakes was crossed. Disaster struck when the raft capsized leaving the two men struggling in the icy water. The only swimmer, Captain Jack Francis of Halkyn Mines was able to rescue the other unfortunate explorer (*Daily Sketch*, 5 December 1932). A drivage was continued northwards along Barclay's Lode, the tunnel being driven through a further series of solution chambers which extended continuously for 300 m where they dipped below the water level. To prevent the tramway sinking into the soft sediment floor, substantial amounts of brushwood were laid down under the tracks.

The present flow from Powell's Lode Cavern is about 5000 gal min⁻¹, and it is clear that this flow represents that which once issued at Holywell, some 8 km distant, plus an additional flow perhaps from the south. It is known that the

Fig. 20.9. Powell's Lode Cavern intersection by the Sea Level Tunnel in Halkyn Mines.

quantity of water entering the Hendre mines was affected by drainage operations further north inferring flow across the Nant Figillt Fault.

Although the picture of waterflow in the Holywell catchment is still incomplete, it seems that the area drained is of the order of 15 km² and that percolation is concentrated into large conduits associated with faulting in the higher parts of the White Limestone sequence. The flow in the part of the area remote from the spring passes deep into the phreatic zone, but although the major flow enters the mine system here, there is no reason to suppose that more direct lines of flow to the spring did not exist prior to mining operations. The problems of entering more than fragments of what is the largest and most unusual system of karst drainage in North Wales are great. With no surface openings the only possible routes are via mine workings which are steadily falling into disrepair.

The Ffynnon Asaph catchment: Ffynnon Asaph is the most important spring in the north of Clwyd, issuing at about 110 m O.D. and 1.5 km east of the village of Dyserth (Fig. 20.5). The reason for its discharge at such a high level when the limestone outcrops continuously to the west down to 50 m O.D. is unclear, but may be due to a barrier to westward flow of karst water imposed by a series of mineralized N–S faults. One of these has been worked near by for calc–spar where a thickness of >3 m of that mineral was exposed. The flow from the spring averages 3 million gal d⁻¹, but fluctuates more than did St Winifride's Well. Total hardness levels are normally about 250 ppm $CaCO_3$ inferring a mainly percolation source. Evaporation at the near-by Dyserth Falls has caused calcium carbonate to re-precipitate with the formation of a tufa curtain the height of the fall. Several small stream swallets are known at distances of from 2 to 3 km from the spring, their wet weather contribution to the system probably accounting for the rapid response of the spring to heavy rain.

Although no caves associated with the present drainage are known, there is a potential for cave discovery, particularly since the catchment lies along a former northward course of the River Alyn, by which the present pattern of karst drainage may have been initiated.

Several small fossil caves occur under Moel Hiraddug at

Dyserth, including the lost Ogof Gwen Goch. Some havoc has been caused by the now disused local quarry, but where intact, the caves are found to contain beautiful deep red calcite formations, their colour derived from haematite formerly mined in the same hill. The near-by Gwaenysgor and Gop caves are archaeological sites.

Drainage to the River Wheeler: the Wheeler, running along the southwestern margin of the limestone outcrop, is fed by several limestone springs, the principal ones being St Michael's Well and Ffynnon Dêa, west and east of the village of Caerwys, respectively, and also Ffynnon Fair at Ysceifiog. They have a combined flow of only about 2 million gal d^{-1}. A percolation source is suggested by the high hardness levels which, in each case, have led to extensive deposits of calcareous tufa. Formerly quarried, the tufa sometimes exceeded 6 m in thickness and in places had an interbedded layer of peat. Although a flood rising above Ffynnon Dêa was excavated a few years ago to 8 m depth exposing a stalagmite-coated rift, it seems that prospects of discovering caves here are slight. Water flows southwestwards up the dip of the beds to discharge at a 'sill' formed by the underlying Silurian Shales at a level lying above the floor of the main valley. In such a situation it is likely that any cave passages linked with the springs will be below the local water table.

Drainage to the Hendre Gorge: at the head of the Hendre Gorge there was formerly an 'ebbing and flowing' spring known as Ffynnon Lleinw, but by the mid-nineteenth century it had lost this property due to lead-mining operations near by. The spring occurred in the valley bottom on the line of the presumed Caleb Bell Cross-course (Fig. 20.5).

Although the catchment for Ffynnon Lleinw has not been established, it may be that a small stream encountered in a cave some few hundred metres to the north flows to it. The cave was explored in 1981 by the North Wales Caving Club and was found to have been developed along or close to the same N–S fault line. Entered from an old well, Chancewell Pot could be followed both north and south at a depth of 15 m below surface. The main passage, 160 m long, consisted of a narrow phreatically enlarged rift in a fault or joint fissure: 40 m from the well a further shaft appeared to have been used as a cesspit. Towards the south end of the rift passage a mine level had been driven along a spar-filled joint, while a circular pot in the floor with finely polished walls appeared to have contributed water to the rift above at one time. The cave probably acted as a resurgence prior to the incision of the Hendre Gorge, its water now finding its way to a lower outlet.

Anglesey, the north coast and the Vale of Clwyd

Anglesey

Although the several limestone outcrops in Anglesey make up an area of about 70 km^2 there is scarcely a cave to be found. The problem seems to be one of low relief, interbedded sandstones and a partial cover of superficial deposits. Below Castell Mawr on Red Wharf Bay are several choked phreatic caves, some exhibiting scalloping, while at Penmon in the extreme east a narrow ravine leads to a constricted swallet cave, the small stream reappearing from among quarry rubble on the sea cliffs below.

Llandudno

The Great Orme's Head overlooking Llandudno is a massive limestone headland separated from the main limestone outcrop by an isthmus of drift deposits upon which much of the town is built. Although containing a few small caves of archaeological importance, its main interest is in the copper mines which were worked by the Romans and at various periods up until the late nineteenth century. Stone hammers and querns have frequently been found along with several tools made from horn. In the mid-eighteenth century the mines were described as being 'formerly great' but now under water. Their most profitable period was between 1830 and 1850, and it was in 1834 that the half-mile long Pen Morfa adit level was commenced from just above the beach. The tunnel secured deep drainage and opened up rich new copper deposits.

Chalcopyrite and carbonate ores were extracted from irregular and ramifying deposits associated with four N–S veins and two oblique veins. Small quantities of galena and cerussite were also worked. Dolomite was the most common gangue mineral, the attractive variety known as satin spar, often tinted pale green, occurring extensively with the copper ores. Calcite was found much less frequently, and sometimes in an irregular dog-tooth form, grey in colour. Occurrences of "anthracite" were a curiosity.

The most productive horizons were in the Lower Brown and White Limestones, the former being generally dolomitic. Llandudno was the second highest producer of copper ores in Wales, production running into tens of thousands of tons.

Llandulas

In the prominent cliff overlooking the Dulas Valley are a group of entrances, which are possibly old sea caves, although they now lie 40 m above sea level. All are blocked with clay or consolidated masses of ill-sorted sandy clay with pebbles, thought by some to be Irish Sea drift.

A short distance along the Dulas Valley, on the east side, is a roomy cave entrance with remnants of a stalagmite floor. A phreatic tube, 3 m in diameter, has been partially excavated of silty sediments by miners and has been followed for 200 m eastwards. It runs parallel to an E–W mineral vein containing galena and barytes, access to which is gained by a short cross-cut from the cave. Otherwise Llandulas is probably better known for its enormous limestone quarries, a proportion of their products being loaded directly onto ships for export.

The Elwy Valley

The valleys of the Elwy and its tributaries, the Nant y Graig and the Afon Meirchion, contain several small active cave systems along with the large excavated passages of the Pontnewydd (Green *et al.*, 1981), Cefn and Plas Heaton archaeological caves (Fig. 20.5). Although the Elwy flows over limestone for over 5 km, no riverbed swallets are known, probably due to the low fall in the river and the presence of alluvial deposits in the valley floor. Near the cottage of

Brasgyll the Nant y Graig forms a narrow gorge and sinks into its bed in normal weather to rise in the main valley some 300 m distant. Several caves are known, the longest being Jock's Pot, discovered in 1965 by the Shropshire Mining Club. It consists of about 80 m of passage, developed mainly by vadose processes and containing an awkward stream pitch of 10 m.

In the valley of the northward-flowing Meirchion is a simple swallet–resurgence system (Wild, 1948). Dell Cave, the swallet, is blocked after 20 m or so, but the resurgence, the Afon Meirchion Cave, was explored in 1976 by G. Shone of Henllan, to a total length of about 300 m. The initial part of the cave is a streamway alternating between low sections going up-dip to the south and higher stretches following the strike of the beds towards the east. The extension is only accessible in dry weather, a constricted section leading eventually to a chamber with calcite formations.

The vigorous rising of Pont y Trap is found on the south bank of the Elwy, downstream from the point of entry of the Meirchion. One or two local swallets contribute insufficient water to account for the bulk of the flow. In 1977 a rift cave was entered by North Wales Caving Club members after excavating for 10 m through earth and limestone talus. The cave walls were of a roughly etched texture and some breakdown had occurred. After 30 m a waterfall was reached, the stream pouring from a constricted rift passage 5 m above, and too narrow for exploration. Rumours that this stream is fed from swallow holes up to 3.5 km distant have yet to be substantiated. Near by is the short fossil Galltfaenan Cave.

Vale of Clwyd

Of the number of faulted blocks and lenses of limestone on each side of the Vale, perhaps the site of most potential for cave discovery is a drainage system rising at Llanrhaiadr Church. There are two known swallow holes, the furthest being 2 km away. One stream sinks in the short cave called Ogof Rhewl which is blocked with stream debris.

The Ffynnon Beuno and Cae Gwyn caves at Tremeirchion near St Asaph are archaeological sites.

The upper Alyn Valley

Along the eastern flank of the Clwydian Hills, west of the town of Mold (Fig. 20.10), the upper Alyn Valley area is bounded in the north by the Hendre Gorge, an old meltwater channel extending across the divide between the Wheeler and lower Alyn valleys (Embleton, 1964). The southern limit is at Llandegla, where a major fracture, the Llanelidan Fault displaces the limestone 7 km eastwards to Minera. The limestone outcrop is about 2.5 km wide, the beds dipping to the east, typically at between 15 and 20°. In the neighbourhoods of Llanarmon and particularly Cilcain, the limestone extends for some distance up the rising ground of the 'Clwydians'. These lowest beds are generally a dark grey colour with porcellanous bands and are attributed to the Lower Brown Limestone sequence. Except in the extreme north, the main limestone outcrop lies to the east of the River Alyn and forms a chain of hills scarred by the effects of quarrying and lead mining. The White Limestone is well developed and perhaps best seen in high cliffs alongside the Leete footpath at Loggerheads and

below Cilcain. The Upper Limestones are well exposed in the large quarry at Hendre and in quarries in the Alyn Gorge, where the river flows across the outcrop to the village of Rhydymwyn.

The thickness of the White Limestone approaches 500 m and it is generally a thick-bedded crystalline rock. The contact with the Upper Grey Limestone is an erosion surface with a thin shale. The lowest 30 m of the Upper Grey Limestone is a dark, crystalline, well-bedded rock, succeeded in the north by two shale beds, each about 3 m thick, and separated by a further 25 m of dark limestone. Due to their effect on controlling ore deposition in the many E–W lead veins of the district, these shale horizons have been carefully mapped in the Halkyn Mines area (Schnellmann, 1939, 1959). The persistence of the Lower Shale into the south of the area has been deduced from records of former mining (Earp, 1958). The remaining dark grey and coralline limestones of the Upper Grey Limestone sequence are overlain by sandstones and thin limestones of the Sandy Limestone division.

Minerals and mining

Deposits of lead and zinc ores occur widely in fissure veins in the upper Alyn Valley and mines were worked intensively from the mid-eighteenth to the late nineteenth century. The deeper parts of some of the mines in the north of the area, served by the Halkyn Deep and Milwr Sea Level Tunnels, continued producing significant tonnages of ore until 1958. As in the Halkyn area, the limestone is criss-crossed by series of both E–W and N–S faults. Since there is a consistent easterly dip, the E–W faults may be termed dip faults. They have formed channels for ascending mineralizing solutions and form the veins of the area, some having been rich producers of lead and, to a lesser extent, zinc ores. They are considered to be tension fractures, those from Pant y Mwyn northwards downthrow to the south, while from Pant y Buarth southwards they have a northerly downthrow. Their vertical component of movement often exceeds 100 m. The N–S strike faults were known as cross-courses, and while frequently mineralized with vein calcite, sometimes to the extent that mining for 'spar' became viable, they very rarely contained significant quantities of lead or zinc ores. They are considered to be compressional in origin, have a net collective downthrow to the west, and often displace the E–W veins by a few metres. Where unmineralized they may consist of zones of crush breccia or at other times contain sand and clay and communicate with solution channels and open caverns (Smith, 1921). Ten of the veins have been worked on a fairly large scale and, although their most productive period was before 1852, when statutory returns of ore production were instituted, it has been estimated that the richest vein in the Westminster Mine at Llanarmon produced in the region of 100 000 tons of lead concentrates. It is doubtful, however, if production from any other individual vein exceeded half this figure. The siting of the old shafts and information from a number of mine sections confirms that the veins were most productive from the middle of the White Limestone up to the middle of the Upper Grey Limestone under the Lower Shale. In the lower part of the White Limestone discoveries were more sporadic, ore often being

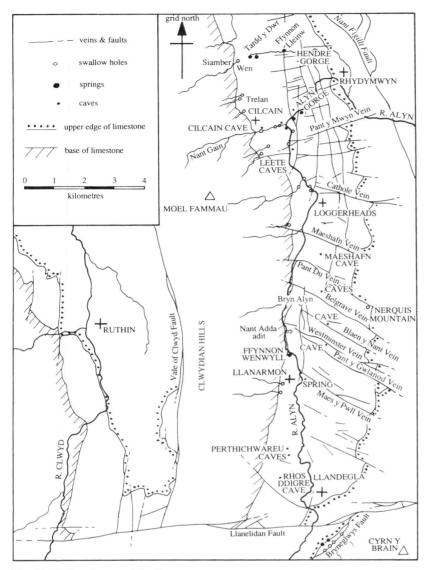

Fig. 20.10. Sketch map of the Upper Alyn Valley and the southern Vale of Clwyd.

found as lumps in a matrix of reddish clay, indicative of past solutional activity. Thin beds of shale at these lower horizons have sometimes been observed to clog narrower vein fissures, producing localized enrichments.

Smaller ore shoots were found in the upper beds of the sandy limestone, particularly where solution had taken place along the contact of a limestone with an overlying sandstone bed. In such situations extensive 'flats' were found at several locations unconnected with the principal veins. Ore was obtained as beds and lumps in a matrix of sand, individual occurrences sometimes amounting to several thousand tons of ore, which might include significant tonnages of cerussite. The predominant gangue mineral is calcite with small quantities of fluorspar occurring at the east end of some of the veins near Llanarmon.

In the north of the area, where the river flows directly over limestone the mines were always troubled with water, while south of Cathole they generally coped by means of adits and a modest pumping capability. The mines worst affected were

those in the Alyn Gorge, the Pen y Fron and Llyn y Pandy Mines (Fig. 20.11). Both had produced substantial quantities of lead ore by the first years of the eighteenth century and still being in rich ore, considerable efforts were made to sink deeper, but even with several steam engines and waterwheels it was only possible to get to the bottom of the workings in dry weather (Williams, 1980). In 1823 these mines, along with those of Pant y Mwyn, Pant y Buarth and Cathole were amalgamated into a single concern, the Mold Mines, under the management of John Taylor. Extraordinary efforts were made to dewater the mines and by 1829 seven steam engines were at work along with four waterwheels, each of 44 ft diameter. Several miles of leats were constructed to supply water to the wheels, the main one carved in places out of sheer cliffs, a remarkable engineering feat in itself. All this power was used to operate 15 pumps discharging into different adits up to a total of 8000 gallons of water per minute from an average depth of '50 fathoms'.

After 1845 the mines were closed and flooded for many

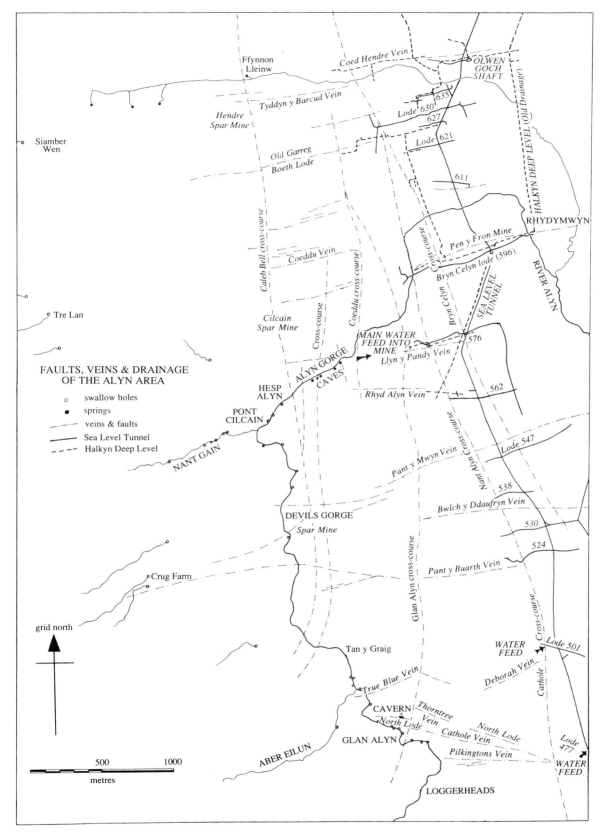

Fig. 20.11. Faults, veins and drainage associated with the Sea Level Tunnel in the River Alyn area.

years, eventually being unwatered in 1901 when the Halkyn Deep Level was extended into the area. The main underground feeders were intercepted by this tunnel at the junction of the Bryn Celyn Cross-course and the Llyn y Pandy Lode in the upper beds of the White Limestone (Fig. 20.11). Although a few thousand tons of ore were produced, mainly from the Bryn Celyn Lode, generally it was found that the mines had been worked out down to Tunnel level during the previous period of working. When the Sea Level Tunnel reached the area in the 1930s, production was resumed by the Halkyn District United Mines. Not only were the known veins drained to about 15 m O.D. but several new and unsuspected veins were discovered. Driving of the Sea Level Tunnel was suspended on reaching the Cathole Lode at Loggerheads some 14 km from the portal, although it had always been planned to continue it south into the Llanarmon district. The Sea Level Tunnel was also used for underground haulage, the ore being raised and dressed at the Pen y Bryn Shaft at Halkyn. So great was the quantity of water flowing through the tunnel in the winter months that tramming was often prevented. The Mill at Pen y Bryn was dismantled soon after 1958, although production of high-grade 'potter's ore' continued on a much smaller scale. Subsequently, a small gravity mill was set up at the Olwen Goch Shaft at Hendre and about 25 tons per month of mixed lead–zinc concentrate was being produced in 1975.

The ore was obtained up until the early 1980s from small-scale workings above tunnel level to the south of the shaft, mainly from an eastward extension of the Pant y Buarth Vein. The Olwen Goch Shaft continued to be used for the maintenance of the Sea Level Tunnel, finally closing in 1986, the end of centuries of lead mining in North Wales.

Spar mining has been carried on for many years, notably along cross-courses east of the village of Cilcain. The two mines most recently active were the Hendre Spar Mine and the Cilcain Spar Mine, both working widths of vein calcite of up to 10 m in what is thought to be a southern extension of the Caleb Bell Cross-course. The Hendre Mine ceased production in the early 1980s.

Karst drainage and caves of the upper Alyn Valley

The Alyn River drainage systems: the River Alyn rises just south of the village of Llandegla flowing north through a wide mature valley between the Clwydian Hills in the west and the main limestone outcrop to the east (Fig. 20.10). Formerly, it continued northwards beyond Cilcain to flow through what is now the Wheeler Valley. Probably in late glacial times, the course of the Alyn was diverted across the limestone outcrop to Rhydymwyn, thereby excavating the Alyn Gorge. It has been suggested that this process was initiated by the action of sub-glacial streams (Embleton, 1964), although it appears that excavation of the rock channel, and consequent rejuvenation upstream to Loggerheads, occurred prior to the last glaciation, since boulder clay occurs in places in the valley floor, and the present river is still downcutting through this.

The River Alyn progressively loses water by leakage into swallow holes from Glan Alyn, near Loggerheads, downstream as fas as Pont Cilcain (Fig. 20.11). During the summer,

the riverbed is often dry below Glan Alyn, but in the winter the river occupies the entire length of its surface course. Previous to lead-mining operations, the water flowing through the underground channels issued at springs in the Alyn Gorge, northeast of a farm called Hesp Alyn (Dry Alyn). The driving of the Halkyn Deep Level had the effect of reducing the flow in the springs to the extent that they sometimes dried up in the summer.

In 1936, the Sea Level Tunnel drained the area to greater depths and the springs became almost totally inactive, flow only occurring in the highest of floods. In such situations, it appeared that the size of the fissures, through which water poured into the mine-workings, was insufficient to pass the high flow, the excess water backing up in the natural system to resume its former underground course.

The main feed of river water into the Sea Level Tunnel comes from workings on Lode 576, the Llyn y Pandy Lode, although other large feeds also enter from Lode 501 and Lode 477. In Lode 501 water issues from a 0.15 m diameter tube in the south wall of a westward drift off the main tunnel (Fig. 20.12). The tube dips northwards and can be seen continuing dry in the opposite wall. Due to the inconvenience caused to mining operations in the 1930s by excessive quantities of underground water, the Halkyn District United Mines under-

Fig. 20.12. A feeder of river water into a branch of the Sea Level Tunnel driven west along Lode 501 in Halkyn Mines, seen under dry weather conditions!

took a great deal of work systematically identifying and sealing swallow holes in the bed of the surface river. Between Logger-heads and Rhydymwyn, where the river ran off the limestone, the 'bad' stretches totalled about 1.5 km of riverbed. Treatment included short diversions of the river, concreting of the riverbed and clogging of individual swallow holes by feeding in mixtures of ashes and mill 'slimes'. Before the latter treatment, swallow holes were cleaned out and were generally found to be immature solution features, often 6 m or more in depth, consisting of networks of vertical fissures often with narrow bedding-plane cavities leading off (Francis & Thompson-Jacob, 1938; Richardson, 1955). During this work a 'small' cavern was found at Glan Alyn at a depth of 7 m. It was followed horizontally by a drift for 20 m before being filled in.

The two main groups of swallow holes were at Glan Alyn and at Tan y Graig. In 1937 dye-tests performed by the mining company proved a connection from each group to the main feed in the Llyn y Pandy Lode at the western end of the workings. In the Glan Alyn test 7 kg of fluorescein were used taking 20 h to travel 2.6 km, and in the Tan y Graig test, 8 kg were used, the colour taking 10 h to travel 2.4 km. Several excavations in swallow holes at Glan Alyn and at Tan y Graig in recent years have failed to reach open cave passage, although one or two sites still appear promising. In 1982, during exploration of the Glan Alyn Mine, which worked the west end of the Cathole Vein, a small cave or 'swallow' was found into which the miners had pumped water from the lower workings. A constricted passage had been formed in the southern or footwall of the vein and while it lay at only 10–15 m below river level, this active passage had been developed entirely by phreatic processes.

The first and still the longest cave system associated with the Alyn drainage was discovered in 1973 by the North Wales Caving Club (Appleton, 1974). The site was found by chance, being merely a patch of gravel above the dry bed of the river in the Alyn Gorge (Fig. 20.13). The presence of a strong draught was considered to be a promising sign, so a shaft was sunk through gravel and stones into fissured limestone, open cave passage being entered at a depth of 7 m. The cave was named Ogof Hesp Alyn (Dry Alyn Cave). The entrance site would have functioned as a rising prior to the lowering of local water levels by mine drainage. With more recent discoveries the length of passage in the cave now totals 2 km, extending over a vertical range of about 90 m. As drainage only took place in the early years of this century, Ogof Hesp Alyn provides a valuable opportunity to study a cave system developed almost entirely by phreatic processes which remains largely unmodified by vadose streams.

The plan of the cave shows a linear arrangement of passages, partly explained in the northern half by a coincidence of strong N–S jointing with a northwards hydraulic gradient (Fig. 20.14). It also appears that the waterflow was confined for considerable distances between N–S faults containing sparry mineralization, as seen both within the cave and on the surface. The longitudinal section illustrates a system of water movement initially down the dip of the limestone beds, where for the northerly moving water to escape via springs in the floor of the Alyn Gorge, passages have been developed at success-ively higher bedding planes. Upward movement of water between bedding-plane conduits has been facilitated by jointing, producing vertical shafts in many parts of the cave. Sometimes entry can be gained to lower passages from the shafts, but often they appear only as steep funnel-shaped cavities or rifts choked with collapse debris.

The initial exploration was halted some 650 m south of the entrance by a perched sump. This obstacle was passed by divers from the Wessex Caving Club in 1982 when a further perched sump was found. The following year this too was passed and the cave was explored to a mineralized E–W fault 400 m further south. A boulder choke in this fault was subsequently dug to reach what seemed to be at least a major part of the Alyn flow, the only place in the entire cave where a permanent stream has been seen, although it could only be followed for a short distance. Access for non-divers was achieved by siphoning off the sumps, when surveying confirmed that the E–W vein was the western end of the Pan y Mwyn Vein, worked extensively for lead further to the east.

The modest length of Ogof Hesp Alyn belies the fact that trips deep into the cave can be both arduous and treacherous. The ever present mud and the awkward nature of some of the climbs and pitches combine to wear down the caver. Most importantly it must not be forgotten that the cave is still linked to the River Alyn, and in wet weather water sinking in the many riverbed swallows backs up in the cave and total flooding can result.

At the north, or resurgence end of the cave the main passage splits, the former flow here diverging to feed two separate springs. The route from the entrance southwards is a muddy crawl at the outset but the roof soon lifts to provide a walking-sized passage. This has been developed from a bedding plane where variable thicknesses of green clay have been etched by solution of the host limestone. A number of solution cavities are passed, some extending to 10 m in height, developed in discontinuous N–S jointing. At a junction of passages the southerly route descends between sediment-covered boulders, the roof likewise descending through different limestone beds in a series of large inverted steps. The cave at the bottom of a short climb has been developed along a joint, although the higher part of the rift so formed widens out into a beautifully sculptured rock tube. Beyond this section the tube form, typically 2.5 m in diameter (Fig. 20.15) continues, now meandering gently uphill until, at one point, so tightly does the passage turn that it cuts back into itself in a solutional analogy to an 'ox-bow' in a surface river. Constrictions occur where the tube loops into the base of joint cavities, heaps of gravelly debris pushed from within confirming the flow direction towards the old resurgence. A dip in the roof produces a low sandy crawl from which the explorer wriggles out into the base of a roomy domed chamber.

From this point the character of the cave changes, the route follows a continuous series of joints southwards through high rift passages (Fig. 20.16). Some fine solution pits occur in the roof high above, stained with rings of black and reddish-brown and creamy-yellow. The hitherto leisurely progress through the cave now takes on a distinctly sporting flavour with scrambles to the highest parts of the rift, the way on gradually

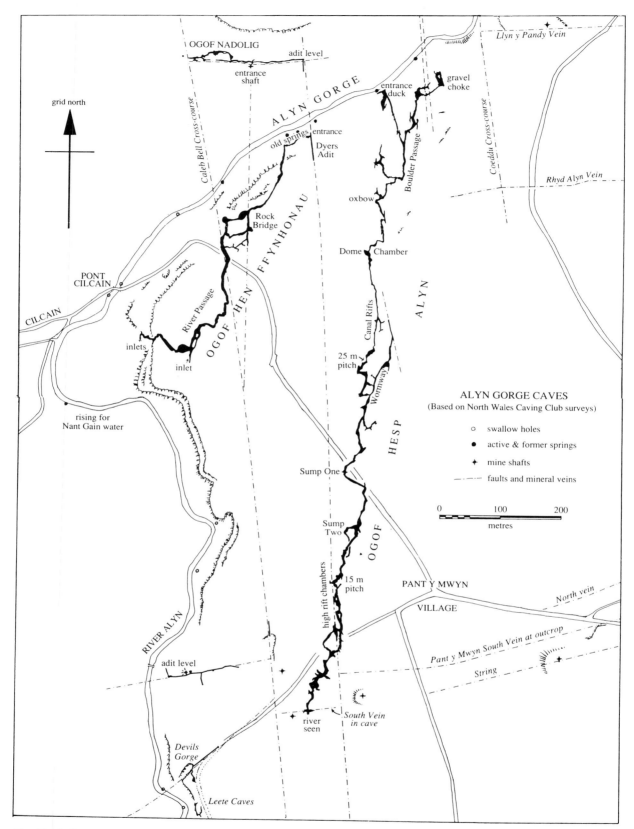

Fig. 20.13. Sketch map of the Alyn Gorge Caves.

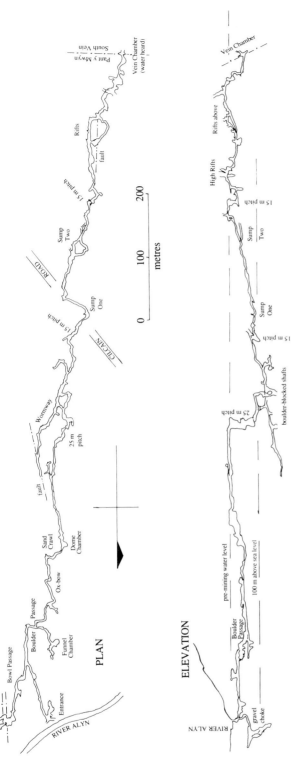

Fig. 20.14. Plan and section of Ogof Hesp Alyn (after North Wales Caving Club surveys).

Fig. 20.15. A phreatic tube passage in Ogof Hesp Alyn.

Fig. 20.16. A rift passage with bedding-plane ledges in Ogof Hesp Alyn.

narrowing until at the start of a 3 m deep canal it is most constricted. The passage walls here lack the smooth contours seen up to this point and with the absence of the usual fine silt, the impression given is that of a short vadose section in the crest of an upward loop in the system. Beyond a low slabby chamber, the cave takes a downward turn, descending a total of 60 m in three slippery pitches to enter a lower main conduit. Passage forms here are more irregular, being typical of slower flow conditions where solution has eaten deeply into joints and bedding planes. The southern route continues with further climbs and a steeply ascending passage to a 15 m pitch. Shortly, after an awkward scramble, the gloom of Sump One is reached (Fig. 20.17).

Beyond Sump One the cave rises steeply, generally through low bedding passages and everywhere dark glutinous mud. Soon Sump Two is reached, being only a slight dip in the upward route. Fins of rock and calcite ribs stand well proud of the walls here, but more mud follows and then a rift climb up into an unusually clean washed and scalloped passage, another high point in the cave, to be followed immediately by a fine circular shaft 15 m deep, formed in N–S jointing (Fig. 20.18). After this pitch the cave becomes more pleasant with roomy passages floored with fine dry silt, although there are several more climbs. As in the central and northerly parts of the system, there are a number of shafts in which water has ascended through joint fissures from lower tributary passages. Most are blocked, but the chambers thus formed can be an

Fig. 20.17. The First Sump in Ogof Hesp Alyn; note calcite veins standing out owing to preferential solution of surrounding limestone.

impressive 20 m or so high. It seems fairly certain that Ogof Hesp Alyn was the former main route for the underground Alyn flow from Glan Alyn and Tan y Graig and, as such, there must still be considerable potential for further discoveries.

The main concentration of springs in the Alyn Gorge was 150 m upstream of Ogof Hesp Alyn, their flow at one time being pumped for water-supply purposes. They occur on the

Fig. 20.18. A circular shaft formed by solution under phreatic lift conditions in Ogof Hesp Alyn.

Fig. 20.19. Recent sediments deposited in a mature water-table passage in Ogof Hen Ffynhonau.

Fig. 20.20. A large stalagmite boss above muddy sediments in Ogof Hen Ffynhonau.

west side of a N–S fault, which appears to downthrow to the east. This fault, or cross-course, contains a 0.5 m width of vein calcite at outcrop, and has been prospected in former times, by the driving of a tunnel 50 m long known as Dyer's Adit. As with the other springs, the flow has been diverted due to mining operations, and it is only during exceptionally wet weather that they again become active.

In 1978 excavation of a sediment-filled tube in the fault plane gave access to Dyer's Adit, deliberately blocked at the surface during the construction of an early nineteenth-century leat. The tube also gave access to a natural rift passage in the fault which appears formerly to have had an exit in the hillside about 10 m above the present river level. Now, from the inside, it is seen to be blocked with clean rounded boulders up to 0.5 m in diameter which can only have tumbled in from a river flowing at a higher level. A further excavation in the floor of the adit by the North Wales Caving Club led to the discovery of the main conduit which fed the old springs. The cave thus found was named Ogof Hen Ffynhonau (Old Springs Cave), and is the only other known cave which is associated with the present Alyn drainage. A total of 800 m of passage was subsequently explored to within a short distance of swallow holes in the Alyn 600 m to the southwest. The main stream first appears from under rubble in the choked western end of the cave. Thick tree roots are present here and surface noises can

be heard. The water flows through a mature meandering passage along a gravelly bed excavated through muddy sediments, deposited at a time when the water level in the cave was higher (Fig. 20.19). Wafer-thin partings of coal can be seen in at least two bedding planes here. Unlike Ogof Hesp Alyn, the main passages in Ogof Hen Ffynhonau lie partly above the phreatic zone as it existed prior to drainage by mining operations. Calcite formations are, therefore, present. The most impressive of these is a large stalagmite boss which provides the focal point in an unexpectedly spacious circular chamber formed at the confluence with a tributary passage (Fig. 20.20). Downstream of the chamber the passage, ankle deep in water, is pleasantly roomy, although the undulating roof dips close to the water in two places.

A change now takes place, the streambed, hitherto almost level, here steepens in a larger passage, gravel giving way to loosely jointed rock and angular blocks of limestone. While once continuing to the springs, the river, now influenced by the lowered water levels due to mine drainage, rushes noisily down through a series of steeply descending tubes to sink in an impenetrable rift in the plane of a N–S fault. The newly erosive power of the water has re-excavated a channel through sedi-

Fig. 20.21. Deeply etched joints and tree roots in the flat-roofed bedding passage not far below the former Alyn resurgence at Ogof Hen Ffynhonau.

ments up to 2 m thick, the lower layers, which had almost filled some of the deeper passages, generally being fine-grained, laminated clays sometimes almost white in colour. The presence of near-vertical tubes in the fault fissure here suggests that at this point a considerable additional flow at one time joined the more locally derived river water in its journey northeastwards.

A large flat tube forming the roof section of the cave at the lower end of the streamway, continues northeast. It is crossed here by a third N–S fault, a southward extension of the Caleb Bell Cross-course, worked for spar 1 km to the north, but here little more than a fissure in the limestone. In high flow conditions the water backs up from the 'sink' to reflood this passage, but being an ascending section the silt layers are undisturbed. Approaching the old springs the cave turns eastwards down the dip, and where flood streams have washed away the sediments, heaps of tiny angular rock fragments have been exposed. These fragments appear to have spalled off the roof probably due to deep frost penetration in former times, at a place where the cave is only a few metres from the surface as shown by tree roots (Fig. 20.21). Nearby deposits of 'moonmilk' occur in roof fissures above the former water level. Ogof Hen Ffynhonau is a straightforward trip in normal weather with only a 5 m pitch at the entrance. In wet weather the explorer needs to be wary, since the cave can flood badly leaving the entrance passages completely submerged.

On the north side of the Alyn, opposite Ogof Hen Ffynhonau, lies the third of the group of Alyn Gorge caves, named Ogof Nadolig (Christmas Cave), due to it being discovered at Christmas in 1973 by the NWCC. It is a fossil system with 200 m of passage associated with the unproductive western extension of the Llyn y Pandy Vein which underlies to the north. Although not connected with the present Alyn drainage, it is convenient to describe it here due to its proximity to the two main cave systems. The cave, although entered via a shallow mine shaft, has been little modified by the 'Old Man', although some sediments have been removed from the lower eastern end, thereby providing access to the main passage. There is a phreatic tube 1–2 m in diameter developed on the north side of the vein, and rising westwards with the dip of the beds. Half-

way through the cave is a 4 m high natural shaft up to a further passage where scalloping on the walls and roof indicates a former westward movement of phreatic water. Digging has revealed more passage and rift chambers decorated with stalactites and ending in a calcreted boulder obstruction nearly on the line of the Caleb Bell Cross-course. Channelling of sediments and the occurrence of cobbles in the upper parts of the cave suggest invasion by vadose streams. Parts of the cave have been scoured clear of sediments and it may be that their redeposition took place in solution cavities found choked with sand at lower levels in the vein fissure.

Caves and drainage east of the Alyn

Leete Caves are a group of small caves 900 m south of Pont Cilcain (Fig. 20.13). They are adjacent to and just south of a strong fault which was stoped from surface for spar and is locally known as the 'Devil's Gorge'. The fault downthrows to the southeast and may be a branch from the west end of the Pant y Mwyn Vein. Although now completely inactive, one passage appears to have been utilized as a swallow hole taking water from the Alyn in the fairly recent past, scalloping inferring an inward flow of water. It is a strategic site for possible access to the Alyn System upstream of Ogof Hesp Alyn and may well respond to a concerted digging effort.

Many solution cavities have been entered during mining operations east of the Alyn, and several of the larger ones are still accessible. One was encountered in the drift west of the Sea Level Tunnel on Lode 621 of the Halkyn Mines (Fig. 20.11). Formed in the vertical vein fissure in the White Limestone, it widens upwards from about 10 m in diameter and extends 25 m or so above the tunnel, here about 15 m above sea level. A downward continuation was inevitably filled with spoil. A series of 'scoop-like' solution pockets in the lower walls indicate a former upward flow of water of considerable volume, the apparent lack of muddy sediment indicating that a percolation-derived source was active prior to mine drainage.

Further south a cavern can be seen in an adit driven east on the Cathole 'North Lode', west of its intersection with the Glan Alyn Cross-course. Several smaller cavities often containing variegated sands and clays occur in workings on the Cathole Vein. South again, solution caverns can be explored in the Maeshafn and Blaen y Nant lead mines (Fig. 20.10). Most of these are ancient phreatic openings in or adjacent to vein fissures. Usually, there is little accessible horizontal development due mainly to natural infilling of the associated conduits by sediments or collapse, or infilling with mine spoil.

Two exceptions are a short stream passage in the Blaen y Nant Mine which seems to have been well decorated at one time, and an interesting stream passage found in 1970 in the Westminster Mine at a depth of 70 m. This latter cave was not entered by the miners due to a constriction, but unlike other 'caves in mines' in this area, it contains a vadose development in the form of a narrow canyon cut in the floor of a 1 m diameter phreatic tube. An intermittently active streambed is littered with large cobbles, which is curious since there are no swallow holes in the near vicinity at the present time. When active, two small streams flow south from the vein fissure through a total of 80 m of passage combining in a small chamber with a stalagmite floor, the passage then closing down to a narrow sump.

In the Alyn Valley upstream of Loggerheads, no water is lost from the river, instead water is fed into the river from springs and old mine drainage adits. No large streams sink on the limestone uplands to the east, although collectively a significant amount of water flows underground through small scattered swallow holes notably in the Sandy Limestone outcrop on Nerquis Mountain. The largest point of discharge is the drainage tunnel of the Westminster Mine, known as the Nant Adda Adit. Only a few tiny caves are known at surface, these being at Maeshafn, on the north slopes of Bryn Alyn, at Llanarmon Bridge and in a quarry 1 km north of Llanarmon.

Karst drainage west of the Alyn

To the west of the river several karst drainage systems are fed by streams flowing off the lower slopes of the Clwydian Hills. The most extensive of these appears to be that which feeds a rising in a small cave at Tardd y Dwr, at the head of the Wheeler Valley and just west of the Hendre Gorge. The catchment of this rising has not been positively established, but due to its close proximity, a stream sinking at Siamber Wen, 600 m to the southwest, probably flows to it. The volume of water sinking here accounts for less than half of the flow from the rising. When studied during a period of drought in 1975, the total hardness level of the water issuing at Tardd y Dwr was 168 ppm, a comparatively low figure in the low flow conditions prevailing, suggesting that the percolation input to the system was small. The only other surface stream of sufficient flow at that time to make up the volume at the rising was the Nant Gain, 2.5 km to the south, which sank completely into its bed near the Cilcain Cave. Other minor inputs to the Tardd y Dwr System are a group of swallow holes northwest of Cilcain around Tre Lan Farm.

Cilcain Cave, 50 m long, seems the best caving prospect in the area, the main feature of which is a northward-trending scalloped rift. It has recently functioned as a swallet cave for the Nant Gain but is now inactive. A two-phase development of the cave is suggested by the presence of remnants of consolidated pebbly sediments occurring high up in the passage. Although it might be expected that the cave would have drained to the near-by Alyn Valley to the east, it may be that waterflow in this direction has been restricted by the presence of mineralized N–S faulting encouraging flow to the north.

In normal weather the Nant Gain Stream flows down the length of a narrow wooded valley to join the Alyn near Pont Cilcain. In Coed Nant Gain are a number of swallow holes and small caves, many of which were excavated during 1966 by local members of the Shropshire Mining Club. Without exception, swallow holes divided into networks of fissures and tiny tubes too small for exploration. Several entrances had a look of greater antiquity, one entering a network of boulder-filled fissures at a depth of 10 m below the stream, others high above the stream being choked completely with sediments. The rising is a mere 200 m distant in the bank of the Alyn and has been explored by divers for about 100 m. It is possible that a group of swallow holes 1.2 km to the southwest also drain to this rising. Several other swallow holes further south have no obvious rising and it may be that their water joins that of the Alyn in its underground course (Fig. 20.10). At Llanarmon swallow holes to the west of the village appear to drain northwards to a strong rising at Ffynnon Wenwyl, 1.2 km distant. In the extreme south of the area, near Llandegla, a few small caves in low-lying outcrops yielded archaeological remains in the nineteenth century. Finally, 2 km southeast of Llandegla, small streams flowing off the northern slopes of Cyrn y Brain sink on reaching a 1 km long lens of limestone let down by the Bryneglwys Fault between Ordovician sediments on the one side and Silurian sediments on the other. Although a dozen or so swallow holes have developed, the narrowness of the outcrop precludes any great speleological interest.

The southern limestone outcrops

Minera to Llangollen

Between the village of Minera, near Wrexham, and the picturesque town of Llangollen on the River Dee the limestone generally dips ESE-ward at an angle of 5–15°, and forms an outcrop a little over a kilometre in width, extending for 10 km in a gentle N–S arc (Fig. 20.22). It is effectively separated from the neighbouring portions of the main eastern limestone outcrop by two major faults. In the north the Llanelidan Fault, with a northwesterly downthrow, brings Cefn y Fedw sandstone of late Brigantian age against the limestone, while in the south the limestone is cut off by the Aqueduct Fault bringing up Ludlow shales on the south side. The entire thickness of the limestone beds is exposed in the imposing Eglwyseg escarpment high above the Vale of Llangollen. Here in the Llangollen Syncline, the limestones apparently attained their most complete development in North Wales, a point which, along with the magnificent scenery, no doubt, inspired George Morton in his stratigraphical studies over a hundred years ago.

At Minera the modest, east-flowing River Clywedog occupies a gorge deeply incised across the limestone outcrop. Described, even at the turn of the century as 'that once beautiful valley', quarrying, especially in recent years, has left a landscape of devastation. The White Limestone is generally a sparry, thick-bedded, light grey rock about 60 m thick. Towards the top of the sequence two palaeokarst surfaces, overlain by green and orange bentonite clays up to 0.5 m in thickness, are conspicuous. The lowest 25 m of the overlying Upper Grey Limestone consist of thinner beds of dark rock separated by partings of grey shale. In the extreme west end of the quarries the lowest beds of the White Limestone are seen to contain several bands of rounded pebbles, mostly of pyritic mudstone or quartz, derived from the older rocks of the neighbourhood. In the bed of the stream here the White Limestone rests directly upon cleaved shales of Ordovician age. These hard shales form the rising ground of Cyrn y Brain to the southwest and lie along the western margin of the limestone outcrop as far as the fault ravine of World's End, some 4 km to the south. Due to the influence of the anticlinal structure of Cyrn y Brain the Lower Brown Limestone, well developed in the Eglwyseg escarpment, is progressively overlapped northwards and thins to nothing at Minera.

The heather-covered moorland between Minera and World's End has a patchy boulder clay cover and peat bogs,

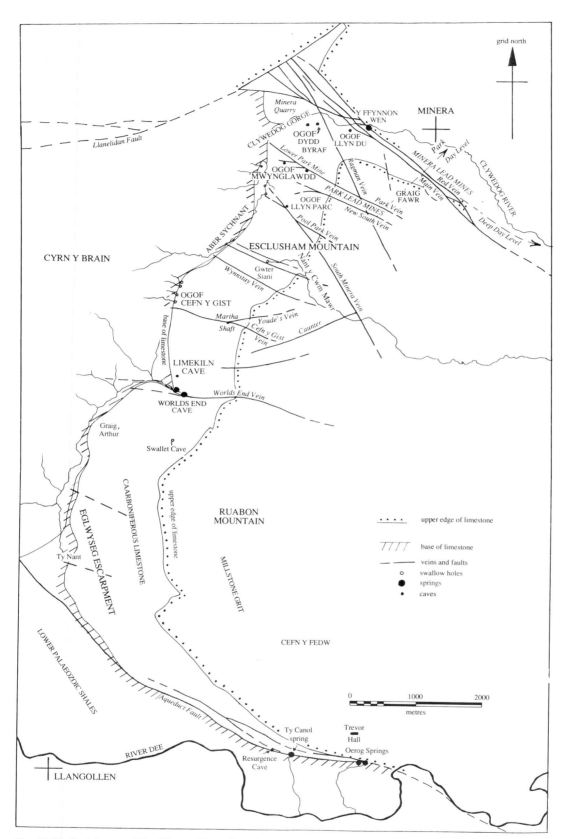

Fig. 20.22. Sketch map of the Minera–Llangollen Limestone area.

and although limestone exposures occur, more frequently southwards, they are restricted in extent and confined largely to the thicker beds of the White and higher limestones. At World's End the limestone is stepped down and westwards by a mineralized fault which can clearly be seen at outcrop and in mine levels at the head of the thickly wooded valley. The thick beds of the White Limestone form the bold cliffs of the Eglwyseg escarpment, at Graig Arthur and Ty Nant. The escarpment can be conveniently explored by the Offa's Dyke Path which follows the foot of the screes as far as Ty Nant, where the Lower Brown Limestone reaches its maximum development. Consisting mainly of thin-bedded muddy limestones with shale partings, they contain porcellanous beds higher in the sequence. Some 25 m below the top of the Lower Brown Limestone is a well-marked palaeokarst surface overlain by a bentonite clay and a thin coal (Somerville, 1979c). To the southeast, high above the River Dee, the Upper Grey Limestones and associated fossiliferous shales are well exposed in old quarries below Trevor Rocks, prior to the limestone being finally faulted out near Trevor Hall.

Minerals and mining The village of Minera was renowned in the nineteenth century as being one of the most prolific producers of lead and zinc ores in the British Isles. Workable orebodies were found in fault fissures occurring as far south as World's End, but the richest mines were developed in the Minera Fault Zone consisting of two main NW–SE fractures with a combined downthrow of up to 200 m to the northeast. The fracture on the north side is known as the North or Red Vein, while that on the south is known as the South or Main Vein (Fig. 20.22). The Main Vein has many branches especially in the northwest, where its identity is lost in a spread of veins which are finally terminated by a branch of the unmineralized Llanelidan Fault.

It seems likely, due to the proximity of sites of Roman occupation, that lead was worked during that period, but the earliest positive evidence of lead mining at Minera relates to the fourteenth century. It was not until around 1720 that mining began in a systematic manner, starting where the ore-bearing strata were close to surface in what became known as the West End Mine. Mining gradually extended eastwards, the ore being followed to deeper levels with the dip of the productive horizons. Several separate companies were involved in the working of the mines and all these ventures seemed to have been successful (Darlington, 1862). Towards the end of the eighteenth century, however, the West End Mines were having increasing water problems, which they attempted to alleviate in 1783 by using a Boulton and Watt pumping engine. A further engine was commissioned a few years later, but in 1799, with the workings only 80 yd deep, the situation was described by Watt (1799) as follows: 'All the workings below (adit) level upon these different veins are inundated with water in wet seasons, and the engines sometimes overpowered; the waters are supposed to accumulate in a natural cavern in the side of the hill of Craig Vawr to the south of the valley, and thence to penetrate through the fissures in the rock into the mine.' With the extension of an adit level to the cavern and the erection of a third engine considerable quantities of ore continued to be

raised until about 1817 when increasing pumping costs and breakdowns rendered the enterprise uneconomical.

The mines at the east end of the Minera veins were worked separately, there being a section of unworked ground between them and the West End Mines about 500 m long. The East End Mines had their own pumping engines and adit system and were worked very profitably until 1815 when 'swallows' were intersected in one of the veins at the '170 yd level', allowing water to pour into the workings from the west. In 1816 seven pumping engines were at work in different parts of the West and East End Mines raising a total of 4000 gallons of water per minute and consuming over 300 tons of coal per week. With little co-operation between the several proprietors over drainage, and mounting financial problems, the last of the Minera Mines closed by 1824.

In 1849 leases were taken up for the whole of the area occupied by the old mines, and the Minera Mining Company was formed. Acting on advice from the mining engineers, John Taylor and Sons, the Deep Day Level was driven to the east end of the property and was extended over the next 10 years, finally intersecting the fissures which poured water into the mines at their western extremity. During the course of this drivage several rich runs of ore were located, and subsequently the mines were deepened, ultimately to a depth of 430 yd at the eastern end of the workings. Between 1852 and 1914, when the mines finally closed, over 200 000 tons of lead and zinc concentrates had been produced (Hughes, 1959; Schnellmann, 1959). Judging from the large scale of the older workings and production figures noted in the incomplete records now available, the quantity of mainly lead ores raised in the hundred years prior to 1824 probably amounted to a further 200 000 tons.

The ore minerals found at Minera were almost exclusively galena and sphalerite (blende), although smithsonite (calamine) occurred in the higher parts of some veins. Chalcopyrite is recorded, but was never apparently found in workable quantities. There was an obvious zoning of the lead–zinc mineralization with depth. Sphalerite, little more than a curiosity in the shallower western parts of the mine, gradually increased in abundance with depth, until at the deepest parts of the mine in the southeast, the ratio of zinc ore to lead ore raised was ten to one. It was this fact, combined with the fall in prices obtained for zinc ore in the early years of this century which forced the final closure of the mine.

Quartz is the commonest gangue mineral, occurring either massive or well crystallized in the many small vughs, or in very finely crystallized form resembling blocks of fine table salt. Calcite also occurs throughout and frequently forms large crystals, the common form being a rather squat six-sided prism surmounted by a flat rhombohedron.

The veins were productive throughout the limestone strata and workable deposits were sometimes found in the overlying Cefn y Fedw sandstone. Yields fell off rapidly, however, as the underlying Ordovician shale or 'bluestone' was approached. There is evidence that shale beds in the Upper Grey Limestone produced local enrichment due to clogging of vein fissure and 'ponding' of mineralizing solutions, as in the mines in the Halkyn area, but in the major fractures at Minera this effect was much less important. In workings of the 'West End' Mines immediately south of the River Clywedog, 'flats' occurred which were reputed to have been very rich. Recent examination has confirmed the great extent of these workings. Some of the flats were found to occur beneath the beds of bentonite clay in the upper part of the White Limestone. Other flats were associated with low angle movement along bedding planes, 'slickensides' being a prominent feature on the roofs of some of these old workings.

Of the veins further to the south, those worked in the Park and Pool Park Mines have been the most productive. They are both fault fissures of small throw lying parallel to and some 900 and 1700 m, respectively, to the southwest of the Minera Fault. Unlike Minera, they were not known until the early nineteenth century. Mine sections and examination of the old workings indicate that there was a marked stratigraphical control of ore deposition, stoping being confined to the upper half of the White Limestone and the lowest 30 m of the Upper Grey Limestone beneath a 7 m thick bed of dark shale (Fig. 20.23). In the west, where the White Limestone outcrops, the vein fissures were often found to be eroded and some lead ore was obtained as detached lumps from among a matrix of boulders and clay, a condition of ground known locally as 'tumblers', due no doubt to its unstable nature. As at Minera, sphalerite became the predominant ore mineral as the ore shoots were followed down-dip to the east.

In the 1820s the Park Mine produced 18 000 tons of mainly lead ores, all raised very cheaply using only horse whims and realizing a profit of £160 000. On attempting to work below the 140 yd level, while still in rich ore, the lowest workings were inundated with water from a river which ran through some large natural chambers at the west end of the mine. The mine was closed only to be taken up again several years later by another company which spent a great deal of money extending one of the old Minera adits to a new shaft sunk at the east end of the mine. This tunnel, now known as the Park Day Level, achieved its object of diverting the underground river in the late 1860s. In this latter period of working, only about 1500 tons of lead and 3000 tons of zinc concentrates were produced and, by 1897 when the mines ceased operations, about £170 000 had been lost.

Although the same river appeared to flow through the Pool Park Mine, drainage was never a problem since the ore shoots worked in this mine were entirely above water level. As at the Park Mine, high profits were made in the earliest period of working, when £100 000 was earned from the production of 10 000 tons of ore, the cost of raising not exceeding eight shillings (40p) per ton.

Quarrying and lime burning At the more accessible extremities of the limestone outcrop, both at Minera and in the Vale of Llangollen, quarrying and lime burning have constituted a long-established local industry. The largest enterprises were in the Clywedog Gorge at Minera, where two companies, the Minera Lime Company and Lester's Lime Works, began operations in the 1850s, the former company on the south side of the river, the latter on the north. Lester seems to have been something of a showman favouring a 'big-blast' technique of breaking the rock, advertised in advance in the local paper. On one such occasion in 1884, two parallel tunnels, 60 yd apart, had each been driven 20 yd into the base of the high quarry face, and one ton of explosives was packed into a small chamber at the end of each heading. Upon detonation, an estimated 20 000 tons of rock was yielded in a most spectacular, if unpredictable, way, some fragments weighing up to 50 tons, and some smaller pieces causing considerable damage to adjoining property!

In the early days, lime burning was performed in open vertical kilns, but in 1868 the Minera Lime Company invested £10 000 in the construction of a 24-chambered kiln working on the Hoffman principle, which was able to produce 100 tons of best quality lime per day with a 50% saving in fuel. A further such kiln was added some years later and both continued in operation until lime burning ceased with the sale of the company in 1954. Today, Minera Quarry is the only active quarry in the area, the product being almost entirely used as aggregate. Output peaked at about 750 000 tons per annum in the early 1970s.

Karst drainage and caves of the Minera catchment
Several small streams rise on the peaty east-sloping moorland of Cyrn y Brain; some sink in swallow holes on meeting the limestone while others collect into a northward-flowing stream known as the Aber Sychnant (Fig. 20.22). At Pool Park this stream flows through a wide flat-bottomed valley, cut into the thick boulder clay which obscures much of the western limits of the limestone outcrop. On Park Farm the stream has cut a short boulder-strewn gorge through a ridge of limestone, and downstream has exposed the unconformity between the lowest beds of limestone and the Ordovician shales, into which a further vertically sided gorge has been incised. The Aber Sychnant Stream now becomes the River Clywedog, turning eastwards, through Minera Quarry. Here remnants of another narrow gorge can still be seen. The origin of this gap in the limestone outcrop is uncertain, but it may have been initiated as a glacial overflow channel, two examples of which occur near by on Esclusham Mountain.

From the normally very slight flow in the Clywedog at Minera, it is difficult to see how, in former times, sufficient power was generated to operate the local corn mill. In those days, however, the main contribution of water to the river must have been from a powerful spring known as 'Y Ffynnon Wen', 1 km west of the village. This spring occurred within the Minera Fault Zone at the lowest point of limestone outcrop in the valley and must have been the main rising for a considerable area. It appears to have become inactive as long ago as the mid-eighteenth century following the driving of the first adit level in an attempt to unwater the lead mines. The flow which once issued from 'Y Ffynnon Wen' now finds its way into two mine adits, the combined average outfall from which is about 3 million gallons per day, but rising quickly to many times that volume after heavy rain. The rapid response of these flows to rainfall and the comparatively low total hardness values of from 70 to 150 ppm under high and low flow conditions,

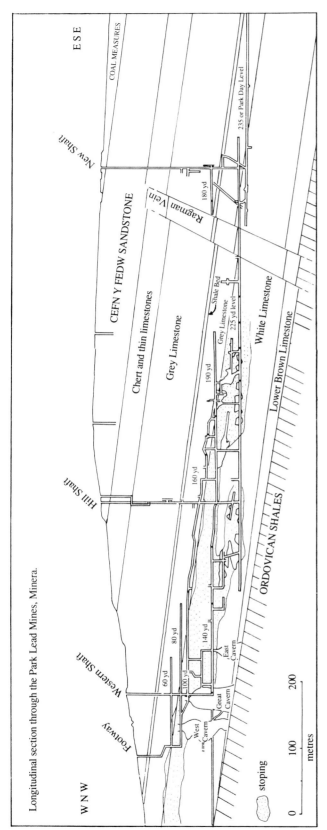

Longitudinal section through the Park Lead Mines, Minera.

Fig. 20.23. Section through the Park Lead Mines and associated caverns, Minera.

respectively, infer the existence of a karst drainage system fed largely by surface streamflow with a limited percolation component.

The streams flowing off the Ordovician rocks provide the most important input to the system. There are two main groups of swallow holes, a northern group occurs on or adjacent to Park Farm, where several small streams sink at the end of short blind valleys, with other swallows and areas of leakage in the near-by Aber Sychnant Stream. The second group of swallow holes is found in a marshy southward-sloping valley near the outcrop of the Cefn y Gist Vein, 1.5 km north of World's End. Water sinks at a number of points along the base of a scarp in the Lower Brown Limestone. Water draining directly from the limestone outcrop constitutes the other main input to the karst system, apart from a single swallow hole at Gwter Siani which drains the marshy land at the head of Nant y Cwm Mawr. The bulk of this water sinks into the innumerable surface solution pits which have formed in extensive lines above the outcrops of the thicker limestone beds, particularly in the Upper Grey and Sandy Limestone formations.

Although caves and 'underground watercourses' appear to have been well known to the lead miners, precise knowledge of their whereabouts and extent seemed to have been lost in the years since the collapse of the local mining industry. When a caving group was started in Wrexham in 1959, emphasis was placed on gaining entry to the Minera Cave System. The most obvious way seemed to be to follow streams, by excavation of swallow holes. Unfortunately, due to the immature nature of the holes selected and the limited resources available at that time, a great effort was put in without any significant discovery being made.

The first digs were in the Aber Sychnant Stream where several swallow holes occur in the lower beds of the White Limestone. They were invariably found to consist of fissures partially choked with razor-edged limestone blocks and pebbly sediments, often consolidated in a peaty matrix. Sometimes, where ground movement had taken place, large blocks had slumped to produce negotiable rifts which pinched out with increasing depth. The main Aber Sychnant dig, which has recently been resumed, has been sunk through 30 m of disjointed limestone to enter a choke of large blocks in what must once have been a mature stream passage developed in the lowest limestone beds. One of the small streams sinking on Park Farm, to the west of the Aber Sychnant, was followed several years ago by excavating for some 30 m through a narrow joint network. The surface depression here contained boulder clay, and tiny side passages in the cave were choked with well-consolidated pebbly sediments, all giving the impression that this was an old swallet cave, formerly choked by glacially derived debris, and now in the process of being re-excavated by the present stream.

The longest swallet cave so far discovered was entered by the NWCC in 1975 after excavation of the most active of the southern group of swallow holes situated 130 m north of a mineralized E–W fracture, the Cefn y Gist Vein. Called Ogof Cefn y Gist, the cave was entered after tunnelling through 25 m of collapse debris. A squeeze between blocks brings the

explorer into the main section of the cave, a 5 m wide and 2 m high flat-roofed passage, with angular undercut walls. The passage can be followed down with the dip of the beds for about 60 m to a boulder choke, the entire floor being littered with large collapsed slabs and smaller blocks, the stream rumbling away underneath. Near the choke a low tributary comes in from the southwest and contains banded stalactite curtains and other calcite formations. The boulder choke was subsequently passed, as was a second one, to give access to a further 60 m of cave followed by a third choke, which is the present limit of exploration.

Ogof Cefn y Gist has been developed in the upper part of the Lower Brown Limestone sequence at its contact with the Ordovician shales, the latter rock being seen at several points in the cave. Water has flowed ESE down the dipping shale surface and lateral solution of the limestone has taken place tending to form a wide low conduit. The thin beds of rock, being incapable of supporting a wide roof, have progressively collapsed and cave development appears to have proceeded as a result of the removal of the debris by a combination of solution and other stream processes. The roof bed near the end of the cave has many conical protuberances about 0.3 m deep and must represent a palaeokarstic erosion surface which was formed between periods of deposition of successive limestone beds. Directly beneath this surface is a coal parting and a layer of fine-textured variegated clay. Among the collapse material in the cave and even at the surface, large fragments of travertine have been found, some pieces having undergone partial re-solution. Much of this material is characteristic of cave pool deposits and includes areas showing random layering formed by the cementing together of fragments of calcite rafts. Remnants of calcite-cemented pebble floors adhere to the cave walls in places, all of which confirms that this is an old swallet cave which has been reactivated following infill and 'fossil' stages.

In 1976 the stream in Ogof Cefn y Gist was dye-traced using sodium fluorescein. Two tests were made, the first using 300 g, the second several months later, using 1 kg of dye. Both gave positive results at the outfall of the Park Day Level 4.5 km away, the dye being extracted from activated charcoal. The second test gave a visible coloration 48 h after the introduction of the dye, persisting for a further 16 h. The southern limit of the catchment of the karst system draining to Minera was thereby proved. The distance between Ogof Cefn y Gist and the old 'pre-mining' resurgence at Y Ffynnon Wen is 4 km and the height difference is 155 m.

No significant caves have been found from the surface solution features, although many were tried in the early 1960s. Some well-developed examples occur in the Sandy Limestone outcrop running north from Martha Shaft of the West Cefn y Gist Mine. Here some of the surface depressions, typically 3–5 m in diameter, open into vertical shafts widening downwards and up to 10 m deep. They are then seen to be formed at joint intersections which have been enlarged by solution with the downward percolation of surface water. More usually, however, they are blocked by peat or drift. Vertical development seems to be arrested at the bottom of the particular limestone bed, although fissures leading off from the base of these shafts can sometimes be followed with difficulty for a few metres.

Exploration of the old lead mines, including the re-excavation and repair of some galleries, has led to the discovery of the more important caves at Minera. Clues do come to light from time to time as more obscure documents relating to mining become available. Such a site was the old Pool Park Lead Mine. An elderly local man claimed that it contained an underground river, and the story was confirmed only comparatively recently when an old Ordnance Survey map, marked with some mining details, was found. The map indicated that a 'subterranean river' ran from the far side of a collapse which terminated the 140 yd level, the deepest in the mine. The spot had been visited some 10 years previously and although draughting strongly its full significance was not appreciated at that time. The route to the collapse involved a 92 m pitch from surface followed by two underground pitches, each 18 m deep. Many visits were made by the North Wales Caving Club over a two-year period, during which a small tunnel was constructed through the blockage. Through this tunnel, completed in 1984, the river was seen for the first time in almost a hundred years (Fig. 20.24). The cave was named Ogof Llyn Parc and is now the longest explored system in North Wales, with over 4 km of passage over a vertical range of 110 m (Fig. 20.25). At the point of entry the passage was about 7 m wide by 3 m high, sweeping round from a massive boulder choke upstream, and continuing low and wide downstream to the north. On the edge of the choke a small tunnel was found, neatly arched in brick and stone, but collapsed after a few metres. It was here that the miners had dug their way through the boulders by the light of candles more than a century before. Nearby were found two names in a fine 'copperplate' hand, traced in the thin layer of mud coating the slabby roof. Soon many more names were seen along with home villages and dates ranging from 1862 to the early 1880s. From these inscriptions and information from the old map, it appears that the miners had explored 400 m of the cave.

Following the river downstream, the roof lifts to a 5 m height, the irregular walls of dark limestone being dissected by minor faults containing sparry mineralization. The streamway steepens with an increase in the dip of the beds, the river rushing noisily down a wide passage strewn with thick slabs of limestone, until after 100 m a sizeable cavity is entered formed along a N–S fault line. From this point abandoned loops of passage occur high on the west side of the river, which now roars between fallen masses of rock on a northeasterly course. Ultimately the river enters a series of immature conduits through which it cannot be followed.

The lower part of the river passage would still have been in the phreatic zone prior to the driving of the Park Day Level, which appears to have lowered water levels by about 50 m. Floodwater still backs up to its former level in wet weather – intending explorers should beware! The river passage has been developed entirely in the thin-bedded Lower Brown Limestone, but a short climb from the lower end of this 200 m long section brings the explorer into a remarkable and contrasting series of quiet, sediment-floored phreatic passages, developed

Fig. 20.24. The subterranean river in Ogof Llyn Parc.

in the lower part of the White Limestone. Much of the river flow would, prior to mining operations, have flowed northeast along the lowest of these passages, combining with the flow of an inlet stream which now alone occupies a steeply dipping passage. This section has been washed clean of its coarse sediments, remnants of which can be seen in wall recesses. Soon even this stream disappears from view leaving the now shingle-floored passage continuing through several small domed chambers, the rock often sculptured into horizontal fins.

Some 700 m from the start of the cave at the Pool Park Shaft, the northeast-trending passage meets another with an ESE trend. A 4 m climb is necessary to reach the floor of this latter gallery which is the most important in the system, most other routes converging on it, and for this reason has been called the Master Passage (Fig. 20.26). It is fairly linear in plan having been developed along the north side of a north-hading fracture, which is probably a westward extension of the new South Vein of the Park Lead Mines. At the junction, the Master Passage measures about 6 m wide by 11 m high and is rectangular in cross-section. Prior to dewatering of the lower parts of the cave, a 'settling-tank' process seems to have operated here. Since this area has now been converted by mine operations into a vadose situation the stream which flows through here in wet weather has carried much of the accumulated material further into the cave leaving a deeply incised streambed between cliffs of sediments, sometimes 5 m in height (Fig. 20.27). These sediments alternate in texture from coarse pebbles to fine sands with a white laminated clay at the bottom of the sequence.

The main conduit soon turns north and then northwest developing into a large ascending tube; scalloping indicates a northward waterflow. Each step up in the roof is accompanied by an obstructing gravel bank below. On approaching the Park Vein the passage, now 6 m in diameter, passes under two 15 m high avens, to be blocked eventually by free-running gravel and sand from above.

There are four main inlet streams to the system, all flowing down with the dip of the beds from the northwest drawing water from the northern group of swallow holes in and to the west of the Aber Sychnant Stream. The northernmost inlet flows for 150 m of its 350 m course over a floor of Ordovician

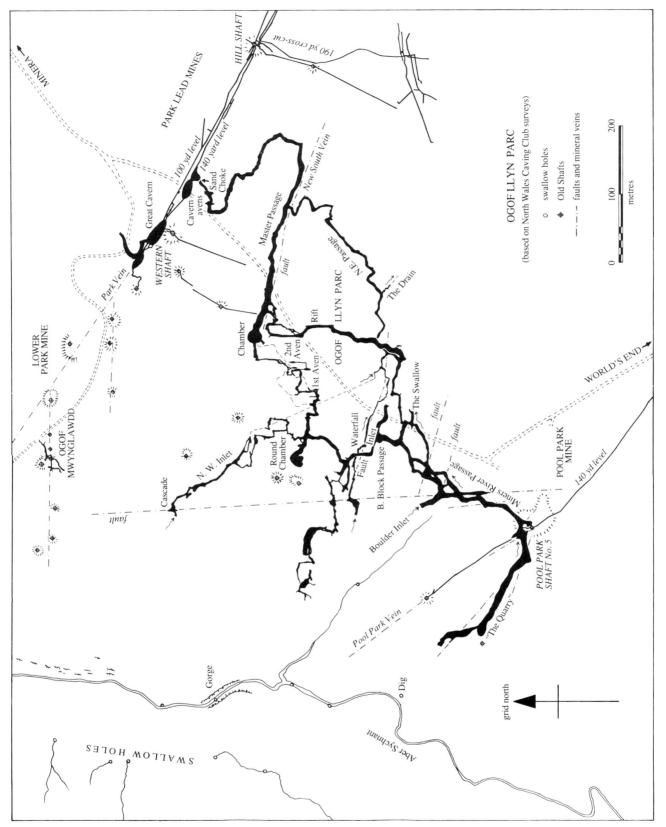

Fig. 20.25. Plan of Ogof Llyn Parc.

OGOF LLYN PARC

(based on North Wales Caving Club surveys)

o swallow holes

⬧ Old Shafts

–··–··– faults and mineral veins

metres

0   100   200

grid north

MINERA

PARK LEAD MINES

HILL SHAFT

190 yd cross-cut

100 yd level

140 yard level

Sand Choke

Great Cavern

Cavern avens

WESTERN SHAFT

Master Passage

New South Vein

NE passage

LOWER PARK MINE

Park Vein

Chamber

2nd Aven

1st Aven

Rift

OGOF LLYN PARC

fault

The Drain

OGOF MWYNGLAWDD

N. W. Inlet

Round Chamber

Waterfall

Fault Inlet

The Swallow

WORLD'S END

Cascade

fault

B. Block Passage

fault

fault

Miners River Passage

POOL PARK MINE

Boulder Inlet

Pool Park Vein

140 yd level

"The Quarry"

POOL PARK SHAFT No. 5

Gorge

SWALLOW HOLES

Dig

Aber Sychnant

Fig. 20.26. The Master Passage of Ogof Llyn Parc.

Fig. 20.27. Sediment cliffs in the Master Passage, Ogof Llyn Parc.

Fig. 20.28. The Northwest Inlet Passage of Ogof Llyn Parc where the stream flows over cleaved Ordovician shales, here stepped up by a fault.

shales, which has suffered little stream erosion. At its north-west end the stream cascades down a 3 m drop where a N–S fault throws up the Ordovician floor on the west side (Fig. 20.28). The lowest limestone beds here contain several pebbly bands as seen in Minera Quarry. The next inlet to the south is the most sporting passage in the cave taking the form of a narrow canyon complete with cascades and a refreshing climb up a 3 m waterfall. This passage follows a poorly mineralized fault fissure trending ENE, the Ordovician rocks appearing in the north or upthrow side. At the northwest end of this passage a 1 m thick bentonite clay, containing tiny pyritic nodules, is well exposed in the passage walls. This clay is variable in thickness, but in the northern inlet the 'feather edge' of the same bed is seen at the contact with the Ordovician shale when it is overlapped westwards by higher beds in the limestone sequence. The third inlet flows directly into the lower end of the river passage; it can be followed for only a few metres through the base of a boulder choke.

The most southerly inlet is the main feed into the cave, the river flowing out from under the boulder choke at Pool Park Shaft. Although the river is not seen again upstream of this point, a large flat-roofed passage can be followed northwest for some 200 m running south of and parallel to the Pool Park Vein. Sometimes measuring 12 m wide by 8 m high, the present passage has formed by progressive roof collapse over its entire

length, aptly named 'The Quarry'; progress is made by picking a route carefully through massive heaps of boulders. It is not clear whether or not the main feed includes water from the southern group of swallets – a dye-test is necessary to prove this point.

In addition to passages connected with the active or recently active system, Ogof Llyn Parc contains an extensive system of well-developed but 'fossil' phreatic conduits (Fig. 20.29). In part, this fossil system has been fed from the upstream sections of the present inlets, before ancient deepening of the vadose zone has diverted the flow to lower levels. Many of the old passages are of flat elliptical shape, but others are high and roomy, and while loops are frequent, all eventually converge on or around a high circular chamber at the head of the Master Passage. Some of the best calcite formations are found in these fossil passages (Fig. 20.30) and include helictites of 'coat-hook' form and one group of a transparent vermiform type. Ancient flowstone also occurs, massive coarsely banded blocks being found in the process of re-solution below a high aven (Fig. 20.31) and partly exhumed from beneath later sediments near by. In another part of the cave light grey, laminated clay has been removed in places revealing a red–brown clay containing sizeable broken and overturned stalagmites. Further exploration and study is needed to piece together a coherent history of the system.

By early 1986 only a preliminary study of cave fauna had been made in the Ogof Llyn Parc System. Main potential food sources are suspended matter, including flood debris brought

Fig. 20.29. A 'fossil' phreatic passage in Ogof Llyn Parc.

Fig. 20.30. Pure white stalactites preserve a ledge composed of old sediments in one of the older passages of Ogof Llyn Parc.

in by the inlet streams. Bat faeces, which sometimes support fungus growth, occur widely in the dry passages. No live bats have yet been observed but a skeleton of unknown species has been found. A pool in the upper end of the main streamway contained a population of the common freshwater shrimp, *Gammarus pulex* (L.), but no troglobitic crustaceans have been recorded. Silted zones subject to periodic flooding are extensively colonized by lumbricid worms, although generally the fauna is poor in both diversity and abundance. Examples of seepage, active flowstone surfaces, and still pools are infrequent in the passages so far explored and suitable habitats for aquatic species are not extensive. Where these features occur at widely separated points in the cave, several species of Collembola have been collected from water surfaces.

Immediately to the north of Ogof Llyn Parc lies the Park Vein, the west end of which has been worked in the Lower Park Mine (Fig. 20.25). During exploration of workings a small fossil cave system containing 150 m of passage was found in 1964 by the former Wrexham Caving Club on a vertical vein branching off from the south side of the main vein. Named Ogof Mwynglawdd it contained two phreatically developed passages, the upper one at 50 m below the surface being well decorated and containing silty sediments which have been partially removed by a tiny stream to reveal a floor of angular rubble. A lower passage had been modified by mining. There

Fig. 20.31. A fallen block of ancient flowstone undergoing re-solution under the First Aven Ogof Llyn Parc.

are reports of other 'swallows' and 'water-courses' being located in the Lower Park Mine, confirmed by the presence of fragments of flowstone floor on several of the old dumps. The main shafts have run in, however, and the workings are now inaccessible.

The central and eastern portions of the Park Vein were worked in the Park Lead Mines, where three natural chambers were found during the early period of working in the first half of the nineteenth century. The largest chamber, known as the

'Great Cavern', was indicated on an old section as being 70 m high, although when visited in 1970 the bottom 30 m were found to have been filled with miner's 'deads' (Fig. 20.23). The chamber, adjacent to Western Shaft, was bell-shaped and about 20 m in diameter. It extended vertically from the Ordovician shales, here known as the 'slate rock', through almost the entire White Limestone sequence, contracting abruptly at its sloping flat roof to a narrow opening in the vein fissure. By following the 140 yd level westwards, a second cavern was entered at a distance of about 50 m. This proved to be a high, narrow chamber, oval in plan, and contained a small lake with the suspicion of an underwater passage.

Since Western Shaft, the only remaining way into the Park Mines, was in a precarious condition, it was felt unwise to risk further exploration at that time. In 1986, with the northern end of Ogof Llyn Parc blocked with sand, it was realized that the only possibility of following the main route was to enter the passage beyond the choke where it passed through the Park Mines. In an ambitious project the NWCC succeeded in casting a concrete lining 22 m deep *in situ* within Western Shaft providing complete stabilization of the dangerous section.

Explorations up to the time of writing have resulted in the discovery of two further chambers to the east of the Great Cavern, both in or adjacent to the vein. One was almost filled with spoil, while the other contained a large conical mound of sand, the other side of the Ogof Llyn Parc sand choke. On re-examining the cavern at the west end of the mine, the lake was not to be seen. Instead the water had receded to reveal a descending 6 m diameter tube passage, which could be followed north for perhaps 80 m (Fig. 20.32). Here, in a downward loop, the passage was obstructed with gravelly sediments needing excavation before further progress could be made.

It appears that prior to being 'captured' by the Park Day Level, the main northward flow of water from Ogof Llyn Parc was diverted westwards on meeting the Park Vein, resulting in the formation of several large chambers before again resuming a northerly course towards the old rising of Y Ffynnon Wen.

Several fossil caves occur in or near the Minera Quarry, the largest of which is Ogof Dydd Byraf. This fine cave, once the largest in North Wales with 700 m of passages, was discovered in 1964 by the Wrexham Caving Club during exploration of lead-workings on a series of E–W mineralized joints (Fig. 20.33). In driving an exploratory tunnel southwards, the miners had hoped to intersect more orebodies, but instead they broke into the top of a 10 m deep solution cavity in a N–S joint, into which they dumped their debris. A brief excavation in the bottom of this cavity gave the cavers access to a series of often beautifully decorated phreatic caverns, which the miners never saw. The cave, although compact, is fairly complex in plan such that it is difficult to unravel its various stages of development. Flow markings indicate that water moved northwards and upwards through the cave to resurge in a former valley which possibly drained to the north. Upper passages in the cave lie approximately 35 m above the rock floor of the present E–W oriented Clywedog Gorge. The ample size of the main conduits in Ogof Dydd Byraf suggests that the system was at the resurgence end of a former main karst flow from Esclusham Mountain. Its period of development was probably

Fig. 20.32. The main phreatic drain extending north from the caverns in the Park Mines, Minera.

contemporary with that of the fossil parts of Ogof Llyn Parc with which there are many similarities.

Accessible passages in the cave have been developed in strata from the middle of the White Limestone to the lower beds of the Upper Grey Limestone, the two seams of bentonite clay at the top of the White Limestone sequence being well exposed. The lower of these is usually a vivid orange colour and is also seen in the adjacent mine-workings where it forms a roof bed to minor runs of lead ore. Passage forms in the White Limestone are sometimes elliptical in cross-section, although more usually are fairly irregular due to solutional enlargement along a host of joints, predominantly of N–S orientation. One passage opens into a three-dimensional joint network developed between three separate bedding-plane horizons. The main passage in the more thinly bedded Upper Grey Limestone has suffered more extensive collapse, but its square cross-section still measures a roomy 8 m by 8 m.

In most passages, floors are deeply covered with a fine dry silt, typical of that deposited from slow-moving water under phreatic conditions. In some places silt has been washed away leaving banks of sand and stream cobbles. That such stream invasion was short-lived can be deduced by the lack of erosion of rock floors thereby exposed. In another part of the cave, in the highest passages, active stalagmite deposits overlie fine silt (Fig. 20.34). In the deeper layers of the same deposit, however, is an old flowstone floor (Fig. 20.35), below which are further sandy layers and finally a reddish clay containing fragments of a yet older flowstone floor. There is clearly scope for a correlation exercise between such sediment sequences and climatic and geomorphological events of the Pleistocene period.

Ogof Dydd Byraf contains many fine calcite formations, with some groups of exceptional beauty. Stalactites are generally of simple form and creamy-white in colour, but many are tinted a delicate green by copper salts. Near shale beds the leaching out of iron salts has produced the deepest of red colorations. Helictites are numerous in one small area, the common form being a short 'finger' or 'coat-hook' shape (Fig. 20.36).

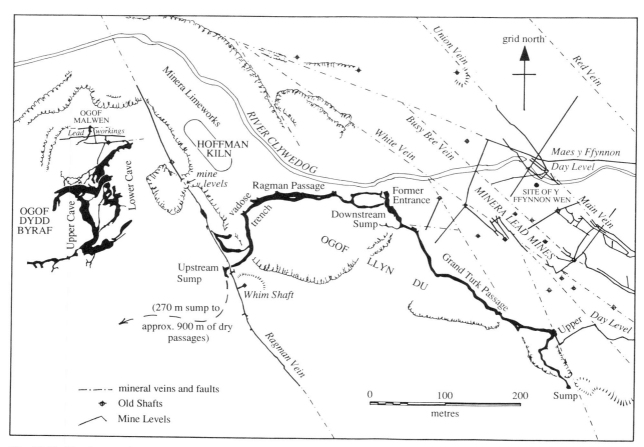

Fig. 20.33. Sketch map of the Minera caves: Ogof Dydd Byraf and Ogof Llyn Du (based on North Wales Caving Club Surveys).

Fig. 20.34. Red stalagmites and flowstone on silts deposited in the phreas, Ogof Dydd Byraf (a Nature Conservancy Council photograph).

Fig. 20.35. Recent stalagmites on old flowstone deposited on earlier silts Ogof Dydd Byraf (a Nature Conservancy Council photograph).

Another fossil system, similar it seems to Ogof Dydd Byraf, was the Minera Quarry Cave, discovered by quarrymen in 1892 and described by Thomas Ruddy of Corwen. It was reputed to cover 2 acres and contained several roomy chambers. Floors were covered with damp clay or 'stalagmitic crusts' and roofs were profusely decorated with stalactites, 'some in the form of cylindrical tubes half a yard in length'.

Souvenir hunters seemed to have been a problem, even in those days, as Ruddy (1892) commented that stalagmites were from '6 inches to 2 feet in height' and 'some were like pillars of alabaster, but most of the best have been carried off'. The entrance, resembling a fox's earth, was apparently blocked up, and its location is now unknown, although the cave probably lies a short distance to the west of Ogof Dydd Byraf.

Fig. 20.36. Small 'coat-hook' helictites and heligmites on flowstone with pendant stalactites, Ogof Dydd Byraf.

The simple phreatic passage of Ogof Malwen opens into the hillside immediately to the north of Ogof Dydd Byraf, with which it may once have connected. A low crawl over a muddy mixture of sediments 30 m into the cave ends at an abrupt downward slope into a walking-sized passage. The sudden end of the sediments which contain sizeable rock fragments is curious, the inference being that the material was introduced by a process of solifluxion.

Sections of cave passage are often found during the working of Minera Quarry. One such cave, an ancient and sizeable phreatic conduit running northwestwards, could be followed for 90 m to a choke. The passage appears to have been truncated by surface erosion in more than one place, allowing infiltration of surface water and coarse sediments. The contents of streambeds in the cave include rounded pebbles of travertine and large quantities of sandstone debris in the form of pebbles and small slightly rounded blocks.

As observed previously, the main rising for the district, Y Ffynnon Wen, is now inactive, and even the site is obscured by an old lime-works tip. The water which formerly fed the spring must have been the same as that which once flooded into the lead mines near by. The miners concluded that the fissures through which the water entered were connected to a natural cavern which lay to the south of the Minera Vein System, and trended down-dip to the southeast. At that time there appeared to have been a surface opening to this cave a short distance up the valley from the spring. Known to the eighteenth-century miners as the Llyn Du Cavern, presumably from a 'black lake' it once contained, it seems to have functioned as the main conduits into which water draining from the cave systems under Esclusham Mountain collected.

Gaining entry to the Cavern proved to be difficult since the original entrance had been buried and its precise location lost under a railway. In 1979 a previously unsuspected western limb of this resurgence system was entered by the NWCC and named Ogof Llyn Du. A wide bedding passage was followed westwards for about 300 m terminating at a near-vertical flooded natural shaft in a mineralized fault. The passage up to

this point follows the same bedding plane for its entire length, but due to its changing direction in relation to the dip of the beds, its long profile is in the form of a gentle hump. The downstream side of the hump is incised into a beautifully decorated vadose canyon, 2.5 m deep and containing remarkable 'gour dams' up to 1.5 m high (Fig. 20.37). Isolated between sections of wholly phreatic development, the canyon soon changes back into smooth bedding passage. Further downstream the passage heads for the Llyn Du Cavern of the old miners, but the presumed connection is obstructed by a sump, contaminated by oil finding its way underground from a near-by quarry. In dry weather the mixture recedes to reveal a boulder choke, curiously containing well-rounded stream boulders which could only have fallen in from the surface.

At the upstream end of the cave the mineralized fault proved to be the Ragman Vein which downthrows about 20 m to the southwest and runs counter to most other veins of the district. The cave is only now active in wet weather when water from the southwest flows up the shaft in the vein. This sump was successfully dived by a member of the Wessex Caving Club in 1985 and proved to be 23 m deep and an awkward 270 m long (Whybro, 1986). An estimated 900 m of dry cave was discovered, trending in the southwesterly direction. Passages ran in two roughly parallel and interconnecting series. The one linking with Ogof Llyn Du consisted of an intermittently active

Fig. 20.37. A gour dam across the vadose canyon of Ragman Passage, Ogof Llyn Du, Minera.

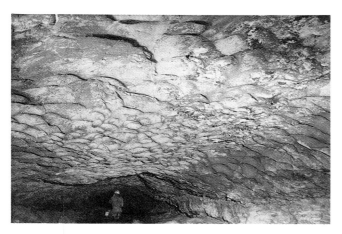

Fig. 20.38. Prominent scalloping demonstrates a former ascending water flow towards the former Minera Risings in the Grand Turk Passage of Ogof Llyn Du.

conduit, generally elliptical in cross-section, fed by several low tributaries which flowed up the dip of the beds. The other series ran at a higher level and contained large fossil passages similar to some of those seen in Ogof Llyn Parc, floored with fine silty sediments and decorated in places with calcite formations. It appears that these passages represent part of the ancient link between the older parts of Ogof Llyn Parc and the now entirely fossil cave of Ogof Dydd Byraf, which was active prior to the incision of the Clywedog Gorge. As a consequence of this event risings were initiated at a lower level to the east, following which the flow was progressively captured by the developing Ogof Llyn Du conduit.

The 'Old' Llyn Du Cavern, which was reputed to run from near the Grand Turk Shaft of the Minera Lead Mines was finally entered, after years of frustrated efforts, in late 1985 by the NWCC. It proved to be a classic bedding-plane conduit, typically 4 m wide by 1.5 m high, with prominent scalloping in many places indicating an upward flow of water from the southeast towards the old rising (Fig. 20.38). After 450 m the lower southeastern end of the passage sumps at a vertical development on meeting a fault fissure downthrowing to the southwest. Mine drainage has allowed vertical penetrations of the former phreatic zone by 50 m but since the cave was flooded until comparatively recently, there are no stalactite formations. Abundant interconnecting calcite veinlets are, however, present in the limestone matrix. They have been etched out and stand proud of the cave walls, sometimes forming intricate latticeworks, detached blocks of which litter the cave floor in places. The significance of the two converging Llyn Du conduits is not entirely clear, although it is fairly certain that the western conduit, the Ragman Passage, took the main flow from Ogof Llyn Parc. The southeasterly conduit, the Grand Turk Passage, is still something of a mystery and whether it represents the latest phase in the grading of the Ogof Llyn Parc drainage to the local base level, or whether it was fed by an entirely separate drainage system under Esclusham Mountain, only time will tell.

The World's End catchment
Water draining to the small resurgences at the World's

End Fault sinks in small swallow holes and surface solution pits on the peat-covered limestone outcrop in a fairly local area to the east and south (Fig. 20.22). The only swallet cave of note consists of a narrow 20 m long series of fissures entered from a shakehole about 800 m south of the ravine and is developed in the higher beds of the Upper Grey Limestone. What seems to be the main route in this hole was found to be blocked with peat-stained calcite pool deposits in a tiny bedding chamber.

The World's End Cave is the immature resurgence of local drainage from the south flowing to the mineralized World's End Fault. The normally small stream flows from a low bedding passage into a fissure in the fault plane to reappear in a mine level below. In 1971 the cave was explored upstream beyond a constricted section of passage occupied by a pool with limited airspace. The short stream passage beyond was found to have been developed along the fault, ribs of vein calcite protruding from the otherwise smooth rock streambed. Only 30 m from the entrance, the passage sumps immediately above a 1.5 m waterfall. Siphoning in dry weather allowed 10 m more cave to be entered, descending to a small elliptical passage still under water, and passed by divers some years later to a narrow choked rift. Limekiln Cave at World's End is an intermittent resurgence cave on the north side of the fault. It is a constricted series of passages in two bedding planes and can be followed with some difficulty for 30 m to a choke.

The Trevor catchment
Although the Eglwyseg escarpment is perhaps the most spectacular landscape feature of North Wales limestone regions, it is one of the poorest areas for known caves. Unlike the Minera catchment to the north, there is no high ground on the up-dip side of the limestone outcrop to the west. Input to the karst drainage system is, therefore, entirely from the outcrop itself being facilitated by the numerous surface solution pits, some of which have become poorly developed swallets for small streams which have collected on the peaty moorland. Most water appears to drain to the southeast, aided by the dip of the beds, accumulating along the Aqueduct Fault and rising in springs at Bryn Oerog, Trevor. The Oerog Springs have a flow of about 0.7 million gallons per day, the water welling up through gravels. In wet weather, water appears to back up along the fault line to increase the flow of a less important spring at Ty Canol, 1 km to the west and 30 m higher. Occasionally a resurgence cave at short distance further to the west becomes active discharging a torrent of peaty water. Here in dry periods, branching passages, developed in a single bedding plane, can be followed for 20 m down with the locally northeasterly dip, before becoming too tight.

Although the Oerog Springs appear to be the rising for what must be a well-integrated karst drainage system extending for 5 km and over a 400 m vertical range, the immature nature of the present surface openings poses severe problems in the search for a route into the presumed cave system.

The Ceiriog Valley
Lower Ceiriog Cave once offered the longest caving trip in North Wales (Baker & Balch, 1907). The entrance is found in a riverside cliff on the south bank of the River Ceiriog,

2.5 km west of Chirk. The cave runs eastwards parallel to the river in beds of the Upper Grey Limestone, and contains 150 m of narrow passages, mostly in a single, steeply inclined fissure. A small stream sometimes flows from tubes and tight rift passages at the end of the cave. The old Upper Ceiriog Cave, 1.7 km to the southwest, was located and reopened in 1962 by the Shropshire Mining Club. It was a 2 m wide oval tube passage with scalloped walls and could be followed for about 50 m before becoming blocked. Only the stumps remained of several large, deep red stalagmites.

Llanymynech

The limestone west and south of Oswestry has disappointingly little cave development, but at Llanymynech, at the extreme southern end of the outcrop, is an interesting Roman copper mine known as the 'Ogof'. In 1965 schoolboy explorers discovered a hoard of silver Roman coins in the mine. Ingots of copper, bearing Roman stamps, have also been found from time to time in the neighbourhood. Scraps of malachite and azurite are the minerals most frequently seen on the old dumps. The pattern of workings seems to be irregular, and

apart from the main gallery in the Ogof, are small. It seems that most of the older workings may now be inaccessible, although the legend of Ned Pugh, the blind fiddler, who was last seen wandering into the mine, and whose music was later heard coming from under the cellar of the Cross Keys Hotel in the village, probably gives an exaggerated impression of their former extent.

Acknowledgements

The writer wishes sincerely to thank members of the North Wales Caving Club and others who, over many years, have contributed their time and considerable efforts towards discovering caves in North Wales, in particular Alan Hawkins, Graham 'Len' Lyon, Peter Law, Phil Hunter, Rod Goslin, Bob Sheppard, Dewi Lloyd, Tony King, Chris Williams, Tony Smith, Humphrey Gibson, Gareth Bryan and Dave Ludlam.

Thanks also to Mr A. W. Boustread and latterly Mr S. Parry of Halkyn Mines, to Mr H. Seddon and the management of the Wrexham and East Denbighshire Water Company, and to Mr C. Dewhurst and the Wynnstay Estate.

References

Anon. (1932). A Lake and Cavern in the Earth. *Daily Sketch*, 5 December.

Appleton, P. (1974). Subterranean courses of the River Alyn, including Ogof Hesp Alyn, North Wales. *Transactions of the British Cave Research Association*, **1**, 29–42.

Baker, E. A. & Balch, H. E. (1907). *The Netherworld of Mendip*. London: Baker & Son.

Baughen, D. J. & Walsh, P. T. (1980). Palaeokarst phenomena in the Carboniferous Limestone of Anglesey, North Wales. *Transactions of the British Cave Research Association*, **7**, 13–30.

Darlington, G. (1862). Mineral and geological sketch of the Minera mining field. *Mining and Smelting Magazine*, **2**, 207–211, 269–77.

Davies, D. C. (1883). The metalliferous deposits of Denbighshire and Flintshire. *Cymmrodor*, **6**, 1–52.

Dawkins, W. B. (1874). *Cave Hunting*. London.

Earp, J. R. (1958). Mineral veins of the Minera–Maeshafn district of North Wales. *Bulletin of the Geological Survey of Great Britain*, **14**.

Embleton, C. (1957). Some stages in the drainage evolution of northeast Wales. *Transactions of the Institute of British Geographers*, **23**, 19–35.

Embleton, C. (1964). Sub-glacial drainage and supposed ice-dammed lakes in northeast Wales. *Proceedings of the Geologists' Association*, **75**, 31–8.

Francis, J. L. & Thompson-Jacob, J. (1938). *An Attempt at Sealing Natural Water Holes or 'Swallows' in the River Alyn*. Report of Halkyn District United Mines Ltd.

George, T. N. (1974). Lower Carboniferous rocks in Wales. In *The Upper Palaeozoic and Post Palaeozoic Rocks of Wales*, ed. T. R. Owen, pp. 85–115. Cardiff: University of Wales Press.

George, T. N., Johnson, G. A. L., Mitchell, M., Prentice, J. E., Ramsbottom, W. H. C., Sevastopulo, G. D. & Wilson, R. B. (1976). *A Correlation of Dinantian Rocks in the British Isles*. London: Geological Society of London Special Report No. 7.

Green, H. S., Stringer, C. B., Collcutt, S. N., Currant, A. P., Huxtable, J., Schwarcz, H. P., Debenham, H., Embleton, C., Bull, P., Molleson, T. I. & Bevins, R. E. (1981). Pontnewydd Cave in Wales – a new Middle Pleistocene hominid site. *Nature, London*, **294**, 707–13.

Hibbert, F. A. & Lacey, W. S. (1969). Miospores from the Lower Carboniferous Basement Beds in the Menai Straits region of Caernarvonshire, North Wales. *Palaeontology*, **12**, 420–40.

Hind, W. & Stobbs, J. T. (1906). The Carboniferous succession below

the Coal Measures in north Staffordshire, Derbyshire and Flintshire. *Geological Magazine*, **Decade 5, 3**, 305–400, 445–59, 496–507.

Hughes, W. J. (1959). The lead–zinc possibilities of the Minera district, Denbighshire. In *The Future of Non-ferrous Mining in Great Britain and Ireland*, pp. 247–57. London: Institution of Mining and Metallurgy.

Morton, G. H. (1876–9). The Carboniferous Limestone and Millstone Grit of North Wales: the country between Llanymynech and Minera. *Proceedings of the Liverpool Geological Society*, **3**, 152–205, 299–352, 371–428.

Morton, G. H. (1883–6). The Carboniferous Limestone and Cefn y Fedw Sandstone of Flintshire. *Proceedings of the Liverpool Geological Society*, **4**, 297–320, 381–403; **5**, 169–97.

Morton, G. H. (1897–8). The Carboniferous Limestone of the Vale of Clwyd. *Proceedings of the Liverpool Geological Society*, **8**, 32–65, 181–204.

Morton, G. H. (1898). The Carboniferous Limestone of the country around Llandudno. *Quarterly Journal of the Geological Society*, **54**, 382–400.

Morton, G. H. (1901). The Carboniferous Limestone of Anglesey. *Proceedings of the Liverpool Geological Society*, **4**, 25–48.

Neaverson, E. (1946). The Carboniferous Limestone Series of North Wales: conditions of deposition and interpretation of its history. *Proceedings of the Liverpool Geological Society*, **19**, 113–44.

Pennant, T. (1796). *The History of the Parishes of Whiteford and Holywell*. London: B. & J. White.

Ramsbottom, W. H. C. (1973). Transgressions and regressions in the Dinantian: a new synthesis of British Dinantian stratigraphy. *Proceedings of the Yorkshire Geological Society*, **39(4)**, 567–607.

Richardson, J. B. (1936). A revival of lead mining at Halkyn, North Wales. *Transactions of the Institution of Mining and Metallurgy*, **46**, 339–461.

Richardson, J. B. (1955). Attempts at sealing off river water from underground workings in North Wales. *Transactions of the Institution of Mining and Metallurgy*, **64**, 211–22.

Ruddy, T. (1892). The exploration of a cave at Minera Limeworks. In *Bye-gones Relating to Wales and the Border Counties*, pp. 369–70. Oswestry: Caxton Press.

Schnellmann, G. A. (1939). Applied geology at Halkyn District United Mines Ltd. *Transactions of the Institution of Mining and Metallurgy*, **58**, 585–650.

Schnellmann, G. A. (1959). Lead–zinc mining in the Carboniferous

Limestone of North Wales. In *The Future of Non-ferrous Mining in Great Britain and Ireland*, pp. 235–46. London: Institution of Mining and Metallurgy.

Smith, B. (1921). *Lead and Zinc Ores in the Carboniferous Rocks of North Wales*. London: Memoirs of the Geological Survey, Special Report on the Mineral Resources of Great Britain, 19.

Somerville, I. D. (1979a). A cyclicity in the early Brigantian (D_2) Limestones east of the Clywedian range, North Wales, and its use in correlation. *Geological Journal*, **14**, 69–86.

Somerville, I. D. (1979b). Minor sedimentary cyclicity in late Asbian (Upper D_1) Limestones in the Llangollen district of North Wales. *Proceedings of the Yorkshire Geological Society*, **42**, 317–41.

Somerville, I. D. (1979c). A sedimentary cyclicity in early Asbian (Lower D_1) Limestones in the Llangollen district of North Wales. *Proceedings of the Yorkshire Geological Society*, **42**, 397–404.

Somerville, I. D. & Strank, A. R. E. (1984a). Discovery of Arundian and Holkerian faunas from a Dinantian platform succession in North Wales. *Geological Journal*, **19**, 85–104.

Somerville, I. D. & Strank, A. R. E. (1984b). The recognition of the Asbian–Brigantian boundary fauna and marker horizons in the Dinantian of North Wales. *Geological Journal*, **19**, 227–37.

Somerville, I. D., Mitchell, M. & Strank, A. R. E. (1986). An Arundian fauna in the Dyserth area, North Wales, and its correla-

tion within the British Isles. *Proceedings of the Yorkshire Geological Society*, **46**, 57–75.

Strahan, A. (1890). *The Geology of the Neighbourhoods of Flint, Mold, and Ruthin*. London: Memoirs of the Geological Survey, GB.

Walkden, G. & Davies, J. (1983). Polyphase erosion of subaerial omission surfaces in the late Dinantian of Anglesey, North Wales. *Sedimentology*, **30**, 861–78.

Walsh, P. T. & Brown, E. H. (1971). Solution subsidence outliers containing probable Tertiary sediment in north east Wales. *Geological Journal*, **7(2)**, 299–320.

Watt, Jnr. (1799). Report upon the state of Maesyffynnon Mine. Birmingham Public Libraries, Boulton and Watt Collection, Box 1 (Minera), Bundle C'.

Wedd, C. B., Smith, B. & Wills, L. J. (1927). *The Geology of the Country Around Wrexham*. London: Memoirs of the Geological Survey, GB.

Whybro, P. (1986). Ogof Llyn Du. *Cave Diving Group Newsletter*, **78**, 44–5; also in *Newsletters*, 80–1.

Wild, P. (1948). Recent cave discoveries in Denbighshire. *British Caver*, **18**, 67.

Williams, C. J. (1980). The lead mines of the Alyn Valley. *Journal of the Flintshire Historical Society*, **29**, 51–87.

Index of caves and related features